Advanced Applications of Micro and Nano Clay Biopolymer-based Composites

Edited by

Amir Al-Ahmed[1] and Inamuddin[2]

[1] Interdisciplinary Research Center for Renewable Energy and Power System (IRC-REPS), King Fahd University of Petroleum & Minerals (KFUPM), Dhahran-31261, Kingdom of Saudi Arabia

[2] Department of Applied Chemistry, Aligarh Muslim University, Aligarh-202002 (UP), India

Published by **Materials Research Forum LLC**
Millersville, PA 17551, USA

Published as part of the book series
Materials Research Foundations
Volume 125 (2022)
ISSN 2471-8890 (Print)
ISSN 2471-8904 (Online)

Print ISBN 978-1-64490-190-8
eBook ISBN 978-1-64490-191-5

Distributed worldwide by

Materials Research Forum LLC
105 Springdale Lane
Millersville, PA 17551
USA
https://www.mrforum.com

Manufactured in the United States of America
10 9 8 7 6 5 4 3 2 1

Table of Contents

Preface

Clays or nano-clays are naturally minerals with layered structure and many of them are silicates (having silicon and oxygen bonds with some other elements). Clay- polymer composites showed novel properties and opened many advanced application opportunities. Depending on the clay, it can be responsible to provide required carrier mobility, or optical band gap, or a range of other properties, such as, electric, magnetic, and dielectric properties and also absorption, and thermal/mechanical stabilities in composite structure. Here, the polymer provides mainly the structural flexibility, convenient processing, and some time required electrical and electronic properties. With the large family of polymers, bio-polymers has a very special position and importance due their characteristic properties, like, biodegradable nature, low cost, non-toxicity etc. Keeping all these important factors in mind, this book has been edited, which contains 13 state-of-the-art articles covering different aspects of the clay-biopolymer based composites and their applications.

Chapter 1: This chapter discusses the processing, properties, characterization, and applications of bio-based nanocomposites with various polysaccharides functionalized by various nanofillers. The major applications of these materials such as packaging, drug delivery, tissue engineering, scaffolds, food additive and orthodontics are also discussed in detail.

Chapter 2: This chapter provides detail the applications of chitosan and HNTs in the real world. The properties and derivatives of chitosan are discussed to provide the insight of chitosan as filler in various composites. The chapter also details about the chitosan-HNT composites used as scaffolds in tissue engineering.

Chapter 3: This chapter provides insight into various applications of Chitosan and montmorillonite-based nanocomposite, including in drug release, antimicrobial activities, wound healing activities, tissue engineering, food packing, water treatment process. The structure and important properties of both chitosan and montmorillonite, are also discussed. Additionally, various processing methods and different forms of chitosan-montmorillonite based nanocomposites are also presented.

Chapter 4: This chapter discusses kaolinite-chitosan nanocomposites and their applications in various fields. The major emphasis is given to applications particularly in medical, pharmaceutical, wastewater treatment and food and packaging industries. The advantages, drawbacks and future feasibility of various kaolinite-chitosan nanocomposites listed in the literature were also discussed.

Chapter 5: This chapter provides detail of various types of scaffolds in tissue engineering particularly nanoscaffolds for fastened wound healing process. Doping of halloysite nanotubes with chitosan has proven to be biocompatible, low in cytotoxicity and mechanically strong. They are low cost with superior performance in tissue engineering, drug delivery, wound healing etc.

Chapter 6: This chapter provides an overview about the ecofriendly vermiculite-starch based nano-composites. It has discussed the properties of vermiculite and processing methods of clay induced nano-composites. It provides the details of some important applications vermiculite-starch induced nano-composites for the pollutant removal, packaging industry and flame retardancy.

Chapter 7: This chapter discusses the properties of halloysite nanotubes and their composites with starch. A brief of different studies reported the development of halloysite-starch nano-composites with a structural illustration are presented. Their application in various fields, including biomedical, food packaging, water treatment, catalyst, and flame retardant, is discussed.

Chapter 8: Biopolymer nanocomposites are the most valuable materials among the existing nanocomposite. These are biodegradable, eco-friendly and low in cost. Compared to pure polymer, clay-polymer nanocomposites exhibit favorable physical, chemical, and mechanical properties since they are dispersed at different sizes and contain improved size dispersion and size distribution.

Chapter 9: This chapter describes the applications of kaolinite-starch based nanocomposites particularly in food packaging, films, paper, etc. have been described. Kaolinite, an important clay material when mixed with starch obtained from different sources, kaolinite-starch nanocomposite is formed. There are three different methods for the synthesis of nanocomposite. In order to have uniform distribution, a plasticizer is mixed.

Chapter 10: This chapter provides detail cellulose materials as an alternative to the toxic and biodegradable polymers that have been used for various applications over the years, paying attention to cellulose at nanometer scale. The various types of cellulose nanomaterials, their modes of synthesis and subsequent use in various applications are reported.

Chapter 11: This chapter provides an extended work of the literature on the different HNT-cellulose based bio-nanocomposites, their structural and functional properties. The chapter also focusses on the applications in wide areas including coatings, fibres, indicators, automobile industry, drug delivery and tissue engineering.

Chapter 12: This chapter discusses the fabrication of kaolinite-cellulose nano-composites. Various characterization techniques to evaluate the desired properties of it are elaborated. The issues and remedies for the fabrication of nanostructured kaolinite, nanostructure cellulose, and their composite as well are presented. The conventional and advanced applications of kaolinite-cellulose nano-composites are also discussed.

Chapter 13: This chapter summarizes widespread applications of montmorillonite-cellulose based nano-composites which include environmental applications for pollutant removal, biomedical applications for wound healing, tissue regeneration, obtaining anti-quorum-sensing and antimicrobial activity and engineering applications for biodegradable, flame retardant bio-based material preparation. Effect of various parameters on the properties of nanocomposites are also discussed.

We thankfully acknowledge all the authors and co-authors for their valued contribution to this book. We would like to take this opportunity to express our gratitude to the publisher, authors, and others for granting us the copyright permission to use their illustrations. At the same time, we also express our sincere apology to any copyright holder if, unknowingly, their right is being infringed, though every effort were made to obtain the copyright permissions from the respective owners to include the citation with the reproduced materials. We would also like to acknowledge the sincere efforts of Mr. Thomas Wohlbier and his team for evolving this book into its final shape.

Dr. Amir Al-Ahmed
Interdisciplinary Research Center for Renewable Energy and Power System (IRC-REPS)
King Fahd University of Petroleum & Minerals (KFUPM)
Dhahran-31261, Kingdom of Saudi Arabia

Dr. Inamuddin
Department of Applied Chemistry
Aligarh Muslim University
Aligarh-202002 (UP), India.

Adv. App. of Micro and Nano Clay – Biopolymer-based Composites   Materials Research Forum LLC
Materials Research Foundations **125** (2022) 1-26   https://doi.org/10.21741/9781644901915-1

Chapter 1

Polysaccharide-Fibrous Clay Bionanocomposites and their Applications

M. Ramesh[1,*], J. Maniraj[1], S. Ganesh Kumar[2]

[1]Department of Mechanical Engineering, KIT-Kalaignarkarunanidhi Institute of Technology, Coimbatore, Tamil Nadu, India

[2]Department of Mechanical Engineering, Sri Eshwar College of Engineering, Coimbatore, Tamil Nadu, India

* mramesh97@gmail.com

Abstract

Bionanocomposites are multifunctional materials, which contain biological origin and particles, have nanometer-scale dimensions (1–100 nm) and can be employed in a vast range of applications in fields like tissue engineering, electronic appliances, biosensors, regenerative medicine, drug delivery systems and food packaging due to their remarkable advantage of exhibiting biocompatibility, antibacterial activity, and biodegradability. To develop naturally biodegradable materials like bionanocomposites, several biopolymers are employed in recent years. Polysaccharides are made up of sugar molecules linked together by glycosidic bonds. These polymeric carbohydrates, which are the most prevalent polymers in nature, are gaining interests as a feasible replacement for synthetic polymers in nanocomposite materials manufacturing. Polysaccharides are promising matrix for the production of green nanocomposites due to their biodegradable nature and biocompatible qualities, hierarchical structure, and high film-forming ability. This chapter discusses the processing, properties, characterisation, and applications of bio-based nanocomposites with various polysaccharides functionalized by various nanofillers.

Keywords

Polysaccharides, Fibrous Clay, Biocompatibility, Biomaterials, Nanocomposites

Contents

1.   Introduction

Recent breakthroughs in material science have stimulate the development of various polymer – nanoclay composites for different applications. Green polymers like polysaccharides, proteins and celluloses are being considered for the production of environmentally feasible plastics as an ecologically forthcoming substitute to fossil materials. This is because of their extensive accessibility, low cost, boundless biocompatibility, biodegradable nature, good film formation and flexible nature. Polysaccharides, a complex carbohydrate made up of monosaccharides linked by glycosidic linkages and a primary structural element of the exoskeleton of plants and animals, or play a key part in energy storage, are the most available macromolecules [1]. In the continuous phase of composite material, matrix is quite important. The polymer matrix is wide range of benefits like chemical stability, thermal stability and flexibility. There are totally four types of polymers based on the extraction sources.

- Agronomic based polymers such as starch, cellulose, proteins, chitin, sea weed derivatives like alginate, carrageenan, and raw seaweed and starch and other agropolymers derived from biomass.

- Polymers produced by microorganisms, such as polyhydroxyalkanoates (PHA).

- Polymers like polylactic acid (PLA) are historically and chemically synthesised, with monomers drawn from agricultural resources.

- Polymers like poly (ethylene oxide), polysulfone, polycarbonate, and polyethylene glycol have monomers that are derived from fossil fuels [2].

For the manufacture of nanophase polymer-silicate hybrids, polymer melt intercalation has recently been more efficient and eco-friendly alternative to standard intercalation processes. The inter layer structure that consists of the aliphatic chains from solid to liquid like state that renders silicate organophilic [3]. Clays are made by incorporating oligomers or surfactants with low molecular weight into their structural sheets. Montmorillonite (MMT) is a layered-silicate consisting of tetrahedral silica and octahedral alumina sheets [4]. The intercalation of polymeric links in silicate galleries reduces their crystallisation susceptibility substantially. The Langmuir-Blodgett transfer of layers onto solid substrates at the air-water interface is a critical step in the production of thin films [5]. Organoclays are the most often utilised commercial nanomaterial in the organophilization method of producing polymer nanocomposites. The inclusion of organoclays into polymeric matrices enhances the mechanical, physical, and chemical characteristics of the matrices while also lowering the cost in some circumstances [6]. Solid surfaces will have the hydrophilic property. The term hydrophilicity refers to the free enthalpy of contact between a solid surface and a liquid medium. The term organophilic has the same meaning as hydrophilic. The use of organoclays makes the nanoclays easily exfoliated for the development of clay–polymer nanocomposites (CPN). This method increases the reinforcing ability of natural clays thereby increase the mechanical strengths of the organoclay – polymeric nanocomposites which leads to the development of new materials [7].

Polymer-layered silicate (PLS) composites can be stiffer, stronger, and light weight than traditionally reinforced polymers. In PLS-based nanocomposites, interactions between the polymeric chains and the silicate result in the organic and inorganic phases being distributed at the nanoscale level. The two main characteristics of PLS is dispersion property with a high aspect ratio and ion exchange property [8]. When compared to pure polymers, polymer-silicate nano composites have higher modulus, lower thermal expansion coefficient, and better solvent resistance. [9] Organically-treated layered silicates (OLS) are composed of organic modifiers in the galleries that convert the initially hydrophilic silicate to an organophilic surface. The production of an exfoliated nano composite by the mixing of polymer and OLS is highly dependent on the properties of the polymer and the OLS. The properties include the polymer's nature as well as the nature, packing volume, and size of organic modifiers on the silicate surface [10].

The first important finding that sparked the resurrection of these polymer/silicate nanocomposites was a study from Toyota research that showed that extremely low inorganic loadings led in contemporaneous and substantial improvements in thermal and mechanical characteristics. Second, it is discovered that polymers and clays may be melt-mixed without the need of organic solvents. However, there is a problem with polymer/MMT interactions that can be addressed in two ways. By "Functionalization" which improves the inter relations between the polymer and the MMT and to reduce the enthalpic interactions of the surfactant with the MMT [11].

1.1 Cellulose

The cellulose are the repeating units of monomer of glucose which has be invented by Anselme Payen, a French chemist in 1838 by segregating from plant cells and determining its chemical composition. The first and foremost thermoplastic polymer, celluloid, was made from cellulose by Hyatt manufacturing industry in 1870. The polymer structure of was determined by Staudinger [12].

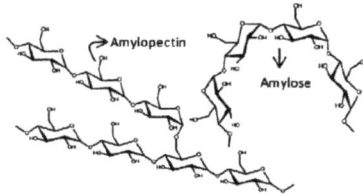

Fig. 1 Structure of starch [2]

1.2   Chitin

Chitin and cellulose are both polysaccharide molecules with a hydroxyl group in cellulose and an acetamide group in chitin. Chitin is abundant in crabs, beetles, worms, and mushrooms. Chitin is a highly stable substance that aids in the protection of insects from harm and pressure. Chitin can be stiff or yielding depending on its thickness. Insect coverings frequently have thick, rigid coatings of chitin. Crustaceans, parasites, fungi, and other pathogens are all protected from the negative impacts of their habitats. The main source of this polymer is shellfish including industry waste (Shrimp or Crab shells), which have chitin concentration ranging from 8% to 33% [13].

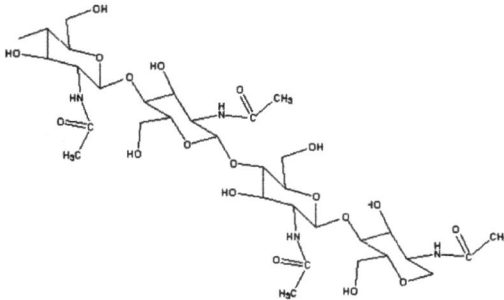

Fig. 2 Molecular structure of chitin [2]

1.3   Chitosan

It is a type of polysaccharide that has protonable amine groups which is opposite to the negatively charged sulphonates of ulvan. Smectites can absorb more chitosan than the CEC allows. As a result, chitosan nanocomposites can convert smectites into anion exchangers.

5

To improve acrylate adsorption, Mt–chitosan was used, and Mt–chitosan–PAA nano-composites were used as the adsorbents [14].

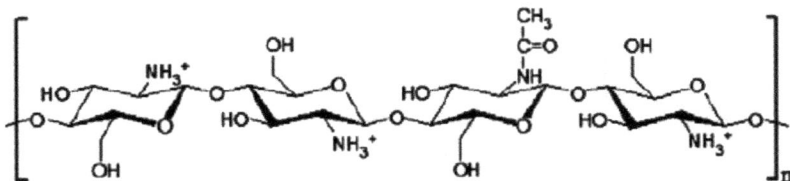

Fig. 3 Structures of chitosan [2]

2.   Modifications of fibrous clays for use as nanofillers

The insertion of nano-sized fillers into composites has been researched. Depending on the nanofiller, nanocomposites can have significant variations in their characteristics, such as enhanced mechanical, thermal, chemical, transparency, and barrier capabilities. The nanofiller geometry and quality play a role in improving characteristics [3]. Although most of the research has aimed on the utilisation of layered-silicates, natural silicates with fibrous morphologies, have recently emerged as promising nano-fillers in the production of bio nanocomposite materials. Sepiolite fibres are normally less than 5 mm long, though this varies depending on the species and the fibres can be significantly longer in some situations. Palygorskite fibres are typically less than 5 mm long, although in hydrothermal palygorskites, this length can reach up to 20 mm. The abundance of these hydroxyl groups in fibrous clays encourages hydrogen bonding interactions with a wide range of polysaccharides and other macromolecules, including neutral or negative charges molecules [12].

3.   Polysaccharides-based bio nanocomposites

Proteins such as gluten and casein and polysaccharides are examples of biopolymers with a direct biological origin (cellulose, starch, galactomannans, dextran, etc.). Polysaccharides are abundant in the natural world. These are complex carbohydrates made up of simple sugars linked together by glycosidic bonds. Polysaccharides, such as cellulose, serve as the plant's primary structural component. Similarly, chitin is a main component of arthropod exoskeletons, fungus cell walls, and plant starch is a source of energy storage [15]. Properties of polysaccharides is given in Table 1 [16].

Table 1 Polysaccharides and their properties [16]

Polysaccharide	Source and composition	Film properties
Plant-based polysaccharides		
Starch based biocomposites	Different plant of the plant Grain – amaranth Rice, wheat, or corn cereals Cashew nut shells Tapioca, potato, or manioc	• Low cost • Easy processibility • Biodegradability • Insoluble in cold water • Oxygen barrier characteristics • Poor mechanical strength • Hydrophilicity nature • Increased crystallinity
Cellulose	β-D-glucopyranose glucose Cotton fibers Wood Sugarcane bagasse	• Mechanical strength • Lower density • Non-toxicity • Cheap • Renewable nature • Durable nature • Biodegradation nature • Good film-forming property
Galactomannans	The availability of mannose and galactose ratio distinguishes different galactomannans. Neutral polysaccharides are found in the endosperm of several dicotyledon plant seeds.	• Edible • Biodegradability • Semi permeable barrier to gases
Animal-based polysaccharides		
Chitin/ Chitosan	N-acetyl-D-glucosamine Extracted from crab and prawn shell	• Transparent material • Antifungal and antibacterial quality • Non-toxic nature • Biocompatible • Biodegradability
Algae-based polysaccharides		
Alginate	Brown seaweeds	• Non-toxicity • Biocompatible • Biodegradable
Carrageenan	Red seaweeds, a-D-1, 3 and b-D-1, 4 galactose	

Microorganism-based polysaccharides		
Xanthan gum	Xanthomonas campestris	• Non-toxic in nature • Water-soluble • High viscosity in aqueous solution causes shear-thinning behaviour. • Stable rheological properties
Gellan gum	Sphingomonas elodea, b-D-glucose, b-D-glucuronic acid, and a-L-rhamnose	• Brittle material • Thermally stable • Non-elastic property
FucoPol	Enterobacter A47	• Water soluble • biodegradable in nature
Pullulan	Aureobasidium pullulans	• Transparent in nature • Homogenic nature • Heat resistant property • Oxygen barrier ability • Antifungal characteristics

4.   Preparation of polysaccharide-fibrous clays bio nano-composites

The CPN are of two types such as intercalated and exfoliated which are of three different types like micro composites, intercalated and exfoliated. Intercalated nanocomposites are prepared by the intercalation of polymeric chains in the intermittent layer space of a clay mineral [17]. The layer stacking is kept, and the layers are arranged along the c-axis. The separation of the planar sides of two adjoining layers is referred to as delamination. The division of the each layer of the mineral results in a nonuniform spread of the mineral layers in the matrix, resulting in exfoliated nanocomposites [18]. There is a weak of crystallographic orientation as the layers become independent of one another. The stripped clay minerals, which are 1-10 % particles randomly spread in the polymeric resin, help to improve the nanocomposite's characteristics and the required properties. The strength and rheological characteristics of the CPN are influenced by the interfacial reaction between the mineral layer and the organic macromolecules [19].

The inorganic component of bio nanocomposites is a silicate from the clay-based minerals. Naturally available and manmade smectites, such as MMT and laponite, are stacked and charged silicates used primarily to make clay-based products. Natural or modified sepiolite and palygorskite have lately been incorporated as inorganic nanofillers in a variety of polymeric matrices, owing to their fibrous nature and nanometric dimensions. Clays with non-lamellar structure arrangements, like tubular halloysite and imogolite, but often

Adv. App. of Micro and Nano Clay – Biopolymer-based Composites   Materials Research Forum LLC
Materials Research Foundations **125** (2022) 1-26   https://doi.org/10.21741/9781644901915-1

sepiolite and palygorskite fibrous materials, have recently gained popularity as nanofillers in the manufacture of new bio nano-composites [17].

4.1   Direct incorporation of polymers

The pristine clay mineral surface has a hydrophilic nature due to its localized layer charges and absorbed moisture. The clay-based nanocomposites are of hydrophilic nature, water soluble bio-based polymers like pectin which has an aqueous dispersion prepared through direct solid state ball milling. The constituents of the CPN, such as clay, surfactant, and polymer are combined in a single step in this method [20].

4.2   In-situ polymerization of monomers

The clay-based polymeric nano-composites are in the form hydrophobic elastomers so there are various difficulties in interaction with the clay surfaces and also there are no stimuli forces for clay dispersion. They are said to be hydrophilic and lipophilic balance which will interact with water and polar solvent [21].

4.3   Polymer-templated clay synthesis

4.3.1   Thin films and coatings

The layer-by-layer (LbL) approach produces thin films and coatings on substrates at a reasonable cost. Its foundation is the adsorption or deposition of polyelectrolyte multilayers (PEMs) on solid substrates. The sequence of layers deposited on the substrate can be controlled by the LbL method creates thin films and coatings with controllable thicknesses typically between 1 and 100 nm [22].

4.3.2   Polyelectrolyte method

To avoid the development of massive precipitates, coacervates are formed by adding polyelectrolyte solutions with low concentration. The electrostatic repulsion between particles prevents solution aggregation and restricts the particle size as they develop [3]. In the formation of polyelectrolyte complexes, the ionic strength of the mixture can play a major role. The limitations of are frequently their poor mechanical characteristics and low durability; however, this may be improved by utilizing polymers with large molecular masses [22].

4.3.3   Ionotropic gelation

Physical materials such as thin layers, hydrogels, and bio-based particles are created by using the ionotropic gelation method. Nanoparticles can be prepared with sodium tripolyphosphate which stabilize the chitosan by the method which is known as electrostatic interactions [20].

4.3.4 Solvent evaporation

It creates films by casting method by pouring polymer solutions into petri dishes, and after solvent evaporation process the polymer thin layers can be obtained by peeling them off from the Petri dishes. This method has advantages like low cost and simple process. But the use of organic and inorganic solvents will results in brittleness of the materials [23].

4.3.5 Freezing–thawing method

This method involves consecutive freezing and thawing process of polymeric solutions. The freezing point must be below 0 °C, thus concentrating the polymeric links in regions between the crystals of solvent and creating self-assembling of polymeric links, and stable assemblies are achieved mainly by H-bonding interactions [15].

4.3.6 Electro-spinning

An electric force is applied or created between the tip of a capillary needle holding a polymeric solution and a grounded metallic collector at a predetermined rate of flow during the electro-spinning process. The electric field between the tip and the collector causes stress that should be greater than the surface force of the polymeric solution, extending the solution drop into a Taylor cone shape [24]. The polymer has come out from the tip to the collector when the electric field spreads a certain magnitude and the repulsive electrical forces outnumber the surface force. This technique is used to create micro and nano fibers [1].

4.3.7 Three-dimensional bio-printing

For tissue repair, 3D bio-printing generates constructs that resemble the organization and structure of cellular matrices of organs. The main goal is to print complicated structures with numerous possible compositions. 3D bio-printed scaffolds have been created using polysaccharide-based bio-inks which are often hydrogels containing live cells in a suitable medium [15].

4.3.8 Preparation of nano-particles

The top-down technique reduces the size of suitable starting materials by using physical modifications such as milling, thermal ablation, or chemical treatments such as etching process. The surface of the MNPs will be affected during preparatory treatment, and higher temperature and pressure employed during size reduction may cause oxidation of the particles, which will influence the physical and surface properties [19]. A metallic precursor is decomposed to a zero-valent state to yield building blocks, which are subsequently nucleated and grown into nano-crystals in a bottom-up manner. Polysaccharides can act as hosts in this process, forming non-covalent bonds with guest metallic ions and MNPs, and then adjusting free energy to offer stability, morphology control, and growth of the nano-materials [25].

Table 2 Method for nanoparticle preparation [24]

S. No	Method	Advantages	Disadvantages
1.	Ionotropic gelation	• Economic and simple • Requires less time and equipment • Organic solvent is not needed	• Poor mechanical strength
2.	Coacervation	• Easily mountable • Simple apparatus	• Toxic cross linkers and organic solvents are used
3.	Electrospinning	• Low-cost equipment • Can influence fiber morphology	• Tough to adjust the size of pores • Obtaining 3D structures is difficult
4.	3D dimensional bio-printing	• Better stability and viability • Improves mechanical and biological properties	• High cost • Difficulty in printing complex structures
5.	The freezing–thawing method	• Easy and fast method	• Time consuming process.
6.	Phase separation	• Morphology is readily manipulated • Consistency level is high	• Less mechanical strength • Used only for few polymers

5.   Properties of polysaccharide-fibrous clay bionanocomposites

5.1   Mechanical properties

The bio nanocomposite films' tensile strength and elongation at break were mostly assessed using an Instron universal testing equipment. Because the machine lacked an extensometer, the elongation was calculated directly from the crosshead speed.

Table 3 Mechanical property of the polysaccharide-based biocomposites [8]

Matrix and reinforcement	Tensile strength	Elongation (%)	Young's modulus
Hydroxypropyl methylcellulose (HPMC) and nano-fibrillated cellulose (NFC)	50	22.5	20
Chitosan and cellulose nanofibers	25.69	31.03	383
Pectin and nano-crystalline cellulose (NCC)	7.12	20	-
Alginate and nano-crystalline cellulose (NCC)	57	10	1800
Carrageenan and cellulose nano-crystal (CNC)	23.4	19	-
Raw seaweed and oil palm shell nanofiller	20.21	19.17	-
Hydroxypropyl methylcellulose (HPMC) and MMT	28.9	8.1	900
Chitosan and graphene oxide (GO)	40.62	13.14	-
Pectin and MMT	24.72	11.15	-
Carrageenan and silver nano-particles	56.5	4	3600
Raw seaweed and calcium carbonate	38.32	23.26	182.81

5.2   Water absorption properties

The water resistance of biopolymer films is an important functional property. Water intake of composites is measured by their water solubility nature, swelling thickness, water quantity, and water vapour permeability in numerous investigations [26, 27]. Water absorption is a major functional attribute of bio nano-composite films based on natural biopolymers, especially for packaging applications and as an alternative for synthetic common polymers [28].

5.3   Intake of heavy metals

Adsorption is widely acknowledged as one of the most efficient strategies for eliminating hazardous metals from the environment. Researchers are looking for bio-based and cost-effective adsorbents to align with environmental concerns caused by heavy metal contamination. Chromium is no exception, as it is well known to be carcinogenic in its +6-oxidation state. Cr(III) is a less dangerous metal that is crucial for glucose metabolism [14].

5.4   Light barrier properties

For direct examination of transmittance, the film samples were sliced into rectangles and inserted in a customized spectrophotometer cell holder. Using a UV spectro-photometer, the light absorbing qualities of the specimens were assessed by the mode of transmittance at wavelengths ranging from 200 to 800 nm [20].

5.5   Gas permeation

Gas permeation property of polymeric-clay bio nanocomposite films were estimated by means of water swollen method. The films were immersed in water for 12 h to make a hydrogel in order to measure utilising these procedures [26]. The BET surface area of the materials is always used for the nitrogen and carbon dioxide adsorption isotherm. The porous characteristic of the cellulose clay material is indicated by the diverse shapes of isotherm [27]. The pore size and maximum pore volume are determined by the Barrett-Joyner-Halenda pore size distribution curve for the adsorbents. The adsorption hysteresis relates to different isotherm like Type I, II, III, IV, etc. The materials microporous and mesoporous nature will be nanopore size, surface area, which will directly reflect the polymeric adsorbent's effective adsorption [2].

6.   Characterisation of the polysaccharide–fibrous clay bionanocomposites

Bio-based polymers like polysaccharides, proteins play strong attention in the manufacturing of green materials. The green material properties can be determined based on the characterization of the polysaccharide and clay. Polysaccharide–fibrous clay materials are the best ecological alternative sources that lead to reduce environmental pollution [29, 30]. The physiochemical characterization plays a vital role in the suitability of scientific and industrial applications. The chemical analysis of polysaccharides reveals ribose, arabinose, xylose, manose, glucose and galactose molecules. Polysaccharides with the structure of polymeric carbohydrates composed of the long chain of monosaccharides. In general, polysaccharides are insoluble in water, do not form crystals when desiccated and not sweet. The chemical structure of polysaccharides leads to the research in alternate fuels due to the availability of hydrocarbons, polysaccharides can be alternative to fossil fuels. Moreover, research has been carried out in polysaccharides as an alternate source of energy storage, adsorption of pollutants from water and many more. From oligosaccharides and disaccharides, these polysaccharides differ in the presence of the number of monosaccharide units. Polysaccharides consist of hydrogen, carbon and oxygen atoms in a 2:1 ratio similar to other carbohydrates. In general, the chemical formula of polysaccharides is exhibited by $(C_6H_{10}O_5)n$. The mixture of polysaccharide and fibrous clay leads to the blending of chemical characteristics, the polymer chain is degradable and hence these nanocomposites can be used as green materials. Hence polysaccharides and fibrous clay material is used to the manufacturing of 3D printing filaments, biomedical applications, ecofriendly food packaging sector, etc.

6.1 Physiochemical characterization of polysaccharides and fibrous clay

Various researches have been explored to investigate the physiochemical characterization of polysaccharides and fibrous clay. From sepiolite and palygorskite polysaccharides are assembled to form the polysaccharides. Inhibits the improved mechanical properties and water resistance. In addition, it possesses a good barrier to UV Light [31]. Sepiolite and palygorskite are assembled to form a polymerization reaction to obtain polysaccharides. The clay–silicate plays the dominant mechanical properties [32]. Alginate potato starch bio nano-composites are blended with micro-fibrous clay mineral sepiolite. The spectroscopic study exhibits the interaction between the polysaccharides and fibrous clay materials, which shows improvement in compression modulus and resistance to fire. Zein bio-hybrid composite material is blended with nanocomposite fibrous clay. Hydrophobic nature protein is separated from corn and others by chemical synthesis. NMR spectroscopy, field emission scanning electron microscopy (FESEM), FTIR spectroscopy and solid-state spectroscopy analysis were carried out in synthesized zein and fibrous clay blend. The interaction between the fibrous clay and polysaccharides were evaluated. Biopolymer matrix composite structure enhances the mechanical properties of fibrous clay and Zein polysaccharides [33]. Bio-adsorbent property of clay nano-composites with sacran mega molecules was investigated to adsorb the neodymium particles from the aqueous media. Sacran to sepiolite bio nanocomposite was assembled with fibrous hydrated magnesium silicate. Sacran – sepiolite films exhibit twice of tensile modulus than that of sacran films. The research concludes that sacran – sepiolite with fibrous clay acts as a very good adsorbent material for neodymium particles in the aqueous solution [34]. Functional characteristics were investigated on nanocomposite materials on fibrous clay. Sepiolite and palygorskite are used as efficient fillers. The inter-relation bonding between polymeric matrices and silicates is more efficient and improves the physio-mechanical strength of the material. The polymer matrix nanocomposite material exhibits good electrical conductivity [35]. Functional characteristics and physio mechanical properties were exhibited for polymer clay nanocomposite films (PCNCHF). The nanocomposites showed PH sensitiveness and research concludes that polymer composite is suitable for drug release in cancer therapy [36]. The chemical properties and applications of bio nanocomposites were reviewed. The improvement of biodegradability, thermal conductivity, tensile properties, and compression modulus is enhanced using the blending of fibrous clay with the polysaccharide materials [37]. Multifunctional applications and chemical characteristics of bio-based nano-composites were investigated. The applications of bio nano-composites extend in the cardiovascular stunt, biomedical engineering, nano-medicines, tissue engineering and many more. The emerging nanostructured bio-composites having excellent mechanical properties enhance in manufacturing of eco-friendly materials due to its excellent biodegradable property [38, 39].

6.2   Spectroscopic analysis

The intermolecular interaction between bio nanocomposites and polysaccharides were analyzed using spectroscopy. Functional carboxymethyl cellulose/Zein bio nano-composite interaction with neomycin supported on sepiolite was analyzed. The strong interaction between both components was exhibited along with inhibiting property of antimicrobial growth [40]. The interaction properties of fibrous clay bio nano-particles. A modified MMT clay with three different aliphatic polyamines experiment and spectroscopic analysis of clay and amines exhibits the number of amines effectively immobilized, moreover X-ray diffraction technique, the thermal analysis describes the enhanced property of novel properties of modified MMT clay compounds [41]. Bacterial inhibition activity was investigated on polysaccharide film-coated nano-fibrous mats. The energy dispersive and X-ray photoelectron spectroscopies exhibit chitosan (CS) and organic rectorite (OREC). The effect of charge on release kinematics was investigated on inorganic halloysite nanotubes (HNT) with chitosan polysaccharides. The spectroscopic results showed the polymeric coating strongly influences the release of probes [42, 43]. Attenuated total reflectance FTIR (ATR-FTIR) was deployed to examine the interaction between the hydroxyl group chitosan and fibrous clay bio nano-composite compounds. The result showed the influence of the nature of clay content controls the characteristics of the nanostructures [44]. Abbasian et al. [45] demonstrated the synthesis of bio nano-composite grafted cellulose and organically modified clay. The experimentation paves the way to alternate to fossil fuels, and synthesis of polymer to use it as bio-renewable resources. Cellulose was synthesized using anhydroglucose added to the solvent mixture. The synthesis of cellulose was done in different batches with varied timings. The spectroscopic analysis exhibited the interaction of synthesized anhyroglucose and biodegradable polymer, which reveals enhancement in bonding properties leads to an increase of mechanical characteristics.

Few scholars researched biopolymer clay nano-composites used as eco-friendly absorbents to reduce water pollution. Investigation exhibits the adsorption phenomena in the water treatment process. Among inorganic materials, layered materials are considered with the capacity to intercalate biopolymers. In adsorption experiments using spectroscopy with biopolymer–clay composites, 96 mg/g of quantity organic or inorganic pollutants are removed from the water when clopyralid is used as the analyte [46, 47]. The interaction of polysaccharide – fibrous clay compounds characteristics is analyzed using spectroscopy and based on the results the polymerization reaction or the synthesizing techniques have been enhanced to improve the chemical properties of fibrous clay – polysaccharide mixture. Solid foams were designed for fire resistant using bio nanocomposite materials. Organic type construction phase is adopted for the polymerization reaction [48, 49]. The ATR-IR spectrum of palygorskite and bio-nanocomposites based on pectin is exhibited in Fig. 4 which shows the possible interactions between polymers and palygorskite [50].

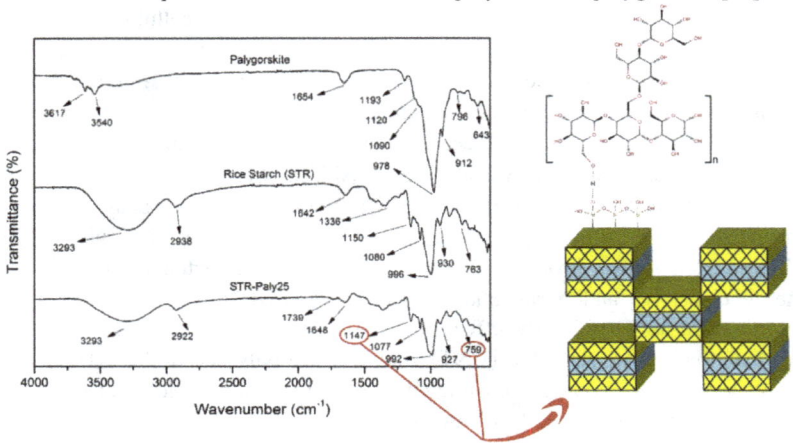

Fig. 4 ATR-IR spectrum bio-based nano-composite and schematic interaction between the species [50]

7.   Applications

7.1   Biomedical applications

Sepiolite and palygorskite based nanocomposites can display the wide range of uses (Fig. 5) [49]. As a result, not only traditional applications, such as reinforcing nanofillers for rubbers and thermoplastics, but also novel applications based on the development of functional composites with magnetic, optical, and conducting properties, as well as active phases for sensor devices, membranes, and bio-plastics for food packaging, drug delivery systems, and vaccine adjuvants, are expected to emerge. Nano-composite antimicrobial system is primarily successful because nano-sized antimicrobial agents have a maximum surface to volume ratio and higher surface reactivity than micro- or macro-scale counterparts, allowing them to inactivate germs more efficiently [50, 51]. These features are critical in regenerative medicine and the processing of eco-friendly bio-based materials, such as green bio-composites [52].

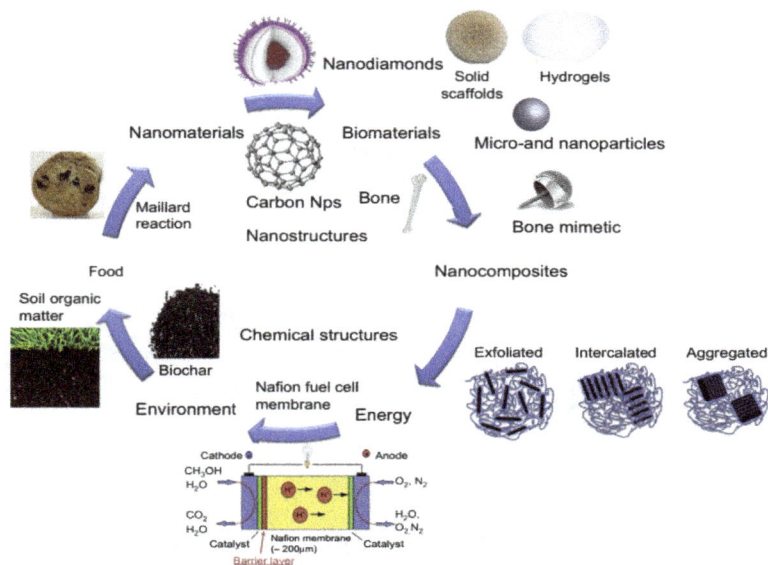

Fig. 5 Applications of bio nano-composites [49]

Drug delivery systems, vaccination process, dressings of wounds, and tissue engineering are all examples of bio-nanocomposites being used in biomedical applications. The reason for this affinity is due to these compounds' well-known biocompatibility and inherent nontoxicity [53]. Bio-based nano-composite thin films can also be used as a low-cost, environmentally friendly, and renewable food packaging material with enhanced antibacterial activity. Starch based cellulose derivatives, polyhydroxybutyrate (PHB), poly-(butylene succinate) (PBS), PLA, and polycaprolactone (PCL) are the most investigated bio-nanocomposites for packaging appliances so far. Layered silicate nanoclays are the most capable nano-scale fillers [54]. Current era shows a tremendous growth in the bio medical field technologies, whereas the clay polymer shows a molecular level property. The benefitable fact about the clay minerals in that they are cheap, non-toxic nature and biocompatibility property. They can absorb greater mass than compare to the own mass which will have a greater water absorption property said to be super adsorbents. This can be also used in the scaffolds in tissue engineering, cell culture and drug delivery system. The biomedical applications of polysaccharide composites is presented in Fig. 6 [55].

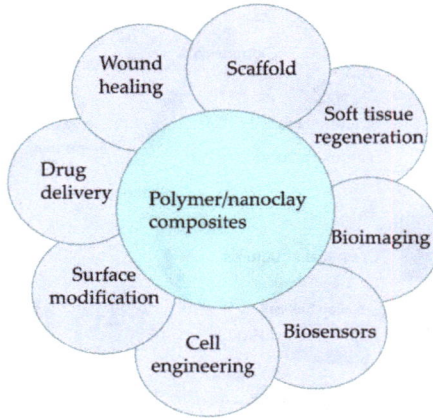

Fig. 6 Biomedical applications of polysaccharide composites [55]

7.2   Biocatalytic applications

Fuel cells are one of the most difficult application sectors for biocatalytic materials. MFC, in which bacteria are utilised as catalysts to oxidise organic and inorganic substances and create current [56], is one of the most impressive examples of bio nano-composite materials and bio-catalytic organisms working together. In a wide range of biological and industrial processes, biocatalysis plays a critical role. Environmental fumigation [57], food industry [58, 59], pharmaceutics synthesis [60], energy generation [61, 62], sensing [63, 64], and, more recently, biocomputing [66–68] are only few of the domains where it is used. In addition to these medicinal applications, the food industry has put a lot of work into developing protein-based systems that can house and transport biocatalytic species [69–71]. Multiple biological macromolecular entities play remarkably varied functions in bio nano-composites employed in biocatalytic applications (Fig. 7) [72].

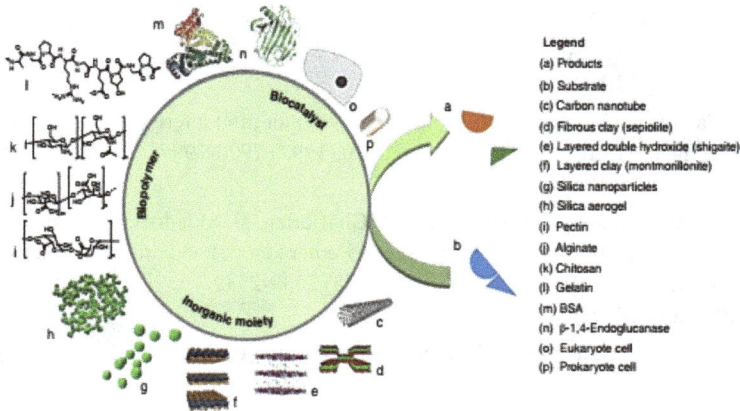

Fig. 7 The elaboration of bio-nanocomposite materials for biocatalytic applications relies on a vast set of possible components, ranging from inorganic solids (c–h) to biopolymers (i–m) and biocatalysts (n–p)

Conclusion

Novel materials with specific qualities have evolved from the fabrication of polymeric nano-composites, which are mostly reliant on the type of modified nano-material used and the synthesis method used. Advantages of bio nano-composites over the other conventional composites will be less weight, flexible nature, higher mouldability, cost effective, acoustical property and plenty of availability. For environmental impact or international demand for greener technology and has a potential replacement for petrochemical based materials. It has a future disposal strategy. The bio nanocomposites have natural fiber reinforcing materials and bio resin is biodegradable. The compounds containing starch and celluloses are the major organic constituents which is cost effective and renewability. The major application such as packaging, biomedical application such as drug delivery, tissue engineering, scaffolds and orthodontics. Polysaccharides are also used as a food additive in order to absorb different toxins in the digestive tract of the human body.

References

[1] E. Ruiz-Hitzky, M. Darder, F. M. Fernandes, B. Wicklein, A. C. S. Alcântara, P. Aranda, Fibrous clays based bionanocomposites, Prog. Polym. Sci. 38 (2013) 1392-1414. https://doi.org/10.1016/j.progpolymsci.2013.05.004

[2] A.C.S. Alcântara, M. Darder, P. Aranda, E. Ruiz-Hitzky, Polysaccharide-fibrous clay bionanocomposites, Appl. Clay Sci. 96 (2014) 2-8. https://doi.org/10.1016/j.clay.2014.02.018

[3] R.A. Vaia, E.P. Giannelis, Lattice model of polymer melt intercalation in organically-modified layered silicates, Macromolecul. 30 (1997) 7990-7999. https://doi.org/10.1021/ma9514333

[4] M.M. Hasani-Sadrabadi, S.H. Emami, R. Ghaffarian, H. Moaddel, Nanocomposite membranes made from sulfonated poly(ether ether ketone) and montmorillonite clay for fuel cell applications, Ener. Fuel. 22 (2008) 2539-2542. https://doi.org/10.1021/ef700660a

[5] R.A. Vaia, S. Vasudevan, W. Krawiec, L.G. Scanlon, E. P. Giannelis, New polymer electrolyte nanocomposites: Melt intercalation of poly(ethylene oxide) in mica-type silicates, Adv. Mater. 7 (1995) 154-156. https://doi.org/10.1002/adma.19950070210

[6] L.B. de Paiva, A.R. Morales, F.R. Valenzuela Díaz, Organoclays: properties, preparation and applications, Appl. Clay Sci. 42 (2008) 8-24. https://doi.org/10.1016/j.clay.2008.02.006

[7] F. Bergaya, M. Jaber, J.F. Lambert, Organophilic clay minerals, in: Maurizio Galimberti (Ed.), Rubber-Clay Nanocomposites: Science, Technology, and Applications, Wiley, 2011, pp. 45-86. https://doi.org/10.1002/9781118092866.ch2

[8] E.P. Giannelis, Polymer layered silicate nanocomposites, Adv. Mater. 8 (1996) 29-35. https://doi.org/10.1002/adma.19960080104

[9] N. Nowak, W. Grzebieniarz, G. Khachatryan, K. Khachatryan, A. Konieczna-Molenda, M. Krzan, J. Grzyb, Synthesis of silver and gold nanoparticles in sodium alginate matrix enriched with graphene oxide and investigation of properties of the obtained thin films, Appl. Sci. 11(9) (2021) 3857. https://doi.org/10.3390/app11093857

[10] R.A. Vaia, E.P. Giannelis, Polymer melt intercalation in organically-modified layered silicates: Model predictions and experiment, Macromol. 30 (1997) 8000-8009. https://doi.org/10.1021/ma9603488

[11] E. Manias, A. Touny, L. Wu, K. Strawhecker, B. Lu, T. C. Chung, Polypropylene/montmorillonite nanocomposites. Review of the synthetic routes and materials properties, Chem. Mater. 13 (2001) 3516-3523. https://doi.org/10.1021/cm0110627

[12] F. Annabi-Bergaya, Layered clay minerals. Basic research and innovative composite applications, Micropor. Mesopor. Mater. 107 (2008) 141-148. https://doi.org/10.1016/j.micromeso.2007.05.064

[13] A. L. Laza, M. Jaber, H. Demais, H. Le Deit, L. Delmotte, L. Vidal, Green Nanocomposites : Synthesis and Characterization. 7 (2007) 3207-3213, doi: 10.1166/jnn.2007.698. https://doi.org/10.1166/jnn.2007.698

[14] A.S.K. Kumar, S. Kalidhasan, V. Rajesh, N. Rajesh, Application of cellulose-clay composite biosorbent toward the effective adsorption and removal of chromium from industrial wastewater, Ind. Eng. Chem. Res. 51 (2012) 58-69. https://doi.org/10.1021/ie201349h

[15] H.P.S.A. Khalil, E.W.N. Chong, F.A.T. Owolabi, M. Asniza, Y.Y. Tye, S. Rizal, M.R. Nurul Fazita, M.K. Mohamad Haafiz, Z. Nurmiati, M.T. Paridah, Enhancement of basic properties of polysaccharide-based composites with organic and inorganic fillers: A review, J. Appl. Polym. Sci. 136 (2019) 47251. https://doi.org/10.1002/app.47251

[16] F. Chivrac, E. Pollet, L. Avérous, Progress in nano-biocomposites based on polysaccharides and nanoclays, Mater. Sci. Eng. R Rep. 67 (2009) 1-17. https://doi.org/10.1016/j.mser.2009.09.002

[17] L.S. Zárate-Ramírez, A. Romero, C. Bengoechea, P. Partal, A. Guerrero, Thermo-mechanical and hydrophilic properties of polysaccharide/gluten-based bioplastics, Carbohyd. Polym. 112 (2014) 16-23. https://doi.org/10.1016/j.carbpol.2014.05.055

[18] L. Vertuccio, G. Gorrasi, A. Sorrentino, V. Vittoria, Nano clay reinforced PCL/starch blends obtained by high energy ball milling, Carbohyd. Polym. 75 (2009) 172-179. https://doi.org/10.1016/j.carbpol.2008.07.020

[19] S. Del Buffa, E. Grifoni, F. Ridi, P. Baglioni, The effect of charge on the release kinetics from polysaccharide-nanoclay composites, J. Nanopart. Res. 17 (2015) 146. https://doi.org/10.1007/s11051-015-2947-z

[20] A. Kocira, K. Kozłowicz, K. Panasiewicz, M. Staniak, E. Szpunar-Krok, P. Hortyńska, Polysaccharides as edible films and coatings: Characteristics and influence on fruit and vegetable quality-a review, Agronomy 11(5) (2021) 813. https://doi.org/10.3390/agronomy11050813

[21] J.R. Capadona, Van Den Berg, O., Capadona, L., Michael Schroeter, Stuart J. Rowan, Dustin J. Tyler, Christoph Weder, A versatile approach for the processing of polymer nanocomposites with self-assembled nanofibre templates, Nat. Nanotechnol. 2 (2007) 765-769. https://doi.org/10.1038/nnano.2007.379

[22] P.R. Souza, A.C. de Oliveira, B.H. Vilsinski, M.J. Kipper, A.F. Martins, Polysaccharide-based materials created by physical processes: From preparation to biomedical applications, Pharmaceu. 13 (2021) 621. https://doi.org/10.3390/pharmaceutics13050621

[23] N.H. Azeman, N. Arsad, A.A. Bakar, Polysaccharides as the sensing material for metal ion detection-based optical sensor applications, Sensors 20 (2020) 1-22. https://doi.org/10.3390/s20143924

[24] H. Park, X. Li, C. Jin, C. Park, W. Cho, C. Ha, Preparation and properties of biodegradable thermoplastic starch/clay hybrids, Macromol. Mater. Eng. 287 (2002) 553-558. https://doi.org/10.1002/1439-2054(20020801)287:8<553::AID-MAME553>3.0.CO;2-3

[25] T.M. Vieira, M. Moldão-Martins, V.D. Alves, Design of chitosan and alginate emulsion-based formulations for the production of monolayer crosslinked edible films and coatings, Foods 10 (2021) 1654. https://doi.org/10.3390/foods10071654

[26] M. Alizadeh-Sani, A. Khezerlou, A. Ehsani, Fabrication and characterization of the bionanocomposites film based on whey protein biopolymer loaded with TiO2 nanoparticles, cellulose nanofibers and rosemary essential oil, Ind. Crop. Prod. 124 (2018) 300-315. https://doi.org/10.1016/j.indcrop.2018.08.001

[27] S.F. Hosseini, M. Rezaei, M. Zandi, F. Farahmandghavi, Development of bioactive fish gelatin/chitosan nanoparticles composite films with antimicrobial properties, Food Chem. 194 (2016) 1266-1274. https://doi.org/10.1016/j.foodchem.2015.09.004

[28] Nahla A. El-Wakil, Enas A. Hassan, Raga E., Abou-Zeid, Alain Dufresne, Development of wheat gluten/nanocellulose/titanium dioxide nanocomposites for active food packaging, Carbohyd. Polym. 124 (2015) 337-346. https://doi.org/10.1016/j.carbpol.2015.01.076

[29] R.N. Tharanathan, Biodegradable films and composite coatings: past, present, and future, Trend. Food Sci. Technol. 14 (2003) 71-78. https://doi.org/10.1016/S0924-2244(02)00280-7

[30] N. Peelman, P. Ragaert, B. De Meulenaer, D. Adons, R. Peeters, L. Cardon, F. Van Impe, F. Devlieghere, Application of bioplastics for food packaging, Trend. Food Sci. Technol. 32 (2013) 128-141. https://doi.org/10.1016/j.tifs.2013.06.003

[31] A.C. Alcântara, M. Darder, P. Aranda, E. Ruiz-Hitzky, Polysaccharide-fibrous clay bionanocomposites, Appl. Clay Sci. 96 (2014) 2-8. https://doi.org/10.1016/j.clay.2014.02.018

[32] E. Ruiz-Hitzky, M. Darder, A.C. Alcântara, B. Wicklein, P. Aranda, Recent advances on fibrous clay-based nanocomposites, in: S. Kalia and Y. Haldorai (Eds.), Organic-inorganic hybrid nanomaterials, Springer, 2014, pp.39-86. https://doi.org/10.1007/12_2014_283

[33] A.C. Alcântara, M. Darder, P. Aranda, E. Ruiz-Hitzky, Zein-fibrous clays biohybrid materials, Euro. J. Inorg. Chem. 32 (2012) 5216-5224. https://doi.org/10.1002/ejic.201200582

[34] A.C. Alcântara, M. Darder, P. Aranda, S. Tateyama, M.K. Okajima, T. Kaneko, M. Ogawa, E. Ruiz-Hitzky, Clay-bionanocomposites with sacran megamolecules for the selective uptake of neodymium. J. Mater. Chem. A 2 (2014) 1391-1399. https://doi.org/10.1039/C3TA14145D

[35] E. Ruiz-Hitzky, M. Darder, A.C. Alcântara, B. Wicklein, P. Aranda, Functional nanocomposites based on fibrous clays, in: Y. Lvov, B. Guo and R.F. Fakhrullin (Eds.), Functional Polymer Composites with Nanoclays, The Royal Society of Chemistry, 2016, pp. 1-53. https://doi.org/10.1039/9781782626725-00001

[36] R.R. Palem, K.M. Rao, G. Shimoga, R.G. Saratale, S.K. Shinde, G.S. Ghodake, S.H. Lee, Physicochemical characterization, drug release, and biocompatibility evaluation of carboxymethyl cellulose-based hydrogels reinforced with sepiolite nanoclay, Int. J. Biol. Macromol. 178 (2021) 464-476. https://doi.org/10.1016/j.ijbiomac.2021.02.195

[37] M.S. Ali, A.A. Al-Shukri, M.R. Maghami, C. Gomes, Nano and bio-composites and their applications: A review, IOP Conf. Ser.: Mater. Sci. Eng. 1067 (2021) 012093. https://doi.org/10.1088/1757-899X/1067/1/012093

[38] S.V. Singh, S. Kumar, A. Sharma, J. Singh, Bionanocomposites and their multifunctional applications, in: R.P. Singh and K.R.P. Singh (Eds.), Bionanomaterials fundamentals and biomedical applications, IOP Publishing Ltd., 2021, pp. 4-1 to 4-46. https://doi.org/10.1088/978-0-7503-3767-0ch4

[39] S. Lee, Y. Hong, B.S. Shim, Biodegradable PEDOT: PSS/clay composites for multifunctional green-electronic materials, Adv. Sustain. Syst. 2021, p. 2100056. https://doi.org/10.1002/adsu.202100056

[40] E.P. Rebitski, A.C. Alcântara, M. Darder, R.L. Cansian, L. Gómez-Hortigüela, S.B. Pergher, Functional carboxymethylcellulose/zein bionanocomposite films based on neomycin supported on sepiolite or montmorillonite clays, ACS Omega 3(10) (2018) 13538-13550. https://doi.org/10.1021/acsomega.8b01026

[41] S. Leporatti, Polymer clay nano-composites, Polym. 11 (2019) 1445. https://doi.org/10.3390/polym11091445

[42] H. Deng, X. Wang, P. Liu, B. Ding, Y. Du, G. Li, X. Hu, J. Yang, Enhanced bacterial inhibition activity of layer-by-layer structured polysaccharide film-coated cellulose nanofibrous mats via addition of layered silicate, Carbohyd. Polym. 83 (2011) 239-245. https://doi.org/10.1016/j.carbpol.2010.07.042

[43] K.C.B.F. Oliveira, A.B. Meneguin, L.C. Bertolino, E.C. da Silva Filho, J.R.D.S. de Almeida, C. Eiras, Immobilization of biomolecules on natural clay minerals for medical applications, Int. J. Adv. Med. Biotechnol. 1 (2018) 31-40. https://doi.org/10.25061/2595-3931/IJAMB/2018.v1i1.8

[44] C. Branca, G. D'Angelo, C. Crupi, K. Khouzami, S. Rifici, G. Ruello, U. Wanderlingh, Role of the OH and NH vibrational groups in polysaccharide-nanocomposite interactions: A FTIR-ATR study on chitosan and chitosan/clay films, Polym. 99 (2016) 614-622. https://doi.org/10.1016/j.polymer.2016.07.086

[45] M. Abbasian, M. Pakzad, K. Nazari, Synthesis of cellulose-graft-polychloromethylstyrene-graft-polyacrylonitrile terpolymer/organoclay bionanocomposite by metal catalyzed living radical polymerization and solvent blending method, Polym. Plast. Technol. Eng. 56(8) (2017) 857-865. https://doi.org/10.1080/03602559.2016.1146905

[46] M. Del Mar Orta, J. Martín, J.L. Santos, I. Aparicio, S. Medina-Carrasco, E. Alonso, Biopolymer-clay nanocomposites as novel and ecofriendly adsorbents for environmental remediation, Appl. Clay Sci. 198 (2020) 105838. https://doi.org/10.1016/j.clay.2020.105838

[47] A.C. Alcântara, M. Darder, Building up functional bionanocomposites from the assembly of clays and biopolymers, The Chem. Record 18(7-8) (2018) 696-712. https://doi.org/10.1002/tcr.201700076

[48] A. Nabgui, T. El Assimi, A. El Meziane, G.A. Luinstra, M. Raihane, G. Gouhier, P. Thébault, K. Draoui, M. Lahcini, Synthesis and antibacterial behavior of bio-composite materials-based on Poly (ε-caprolactone)/Bentonite, Europ. Polym. J. (2021) 110602. https://doi.org/10.1016/j.eurpolymj.2021.110602

[49] R.K. Saini, A.K. Bajpai, E. Jain, Advances in bionanocomposites for biomedical applications, in: N.G. Shimbi (Ed.), Biodegradable and Biocompatible Polymer Composites, Elsevier Ltd., 2018, pp. 379-399. https://doi.org/10.1016/B978-0-08-100970-3.00013-4

[50] J.A. Almeida, , A.S. Oliveira, E. Rigoti, J.C. Neto, A.C. de Alcântara, S.B. Pergher, Design of solid foams for flame retardant based on bionanocomposites systems, Appl. Clay Sci. 180 (2019) 105173. https://doi.org/10.1016/j.clay.2019.105173

[51] E. Ruiz-Hitzky, P. Aranda, M. Darder, Bionanocomposites, in: Kirk-Othmer Encyclopedia of Chemical Technology, Wiley, Hoboken, 2008. https://doi.org/10.1002/0471238961.bionruiz.a01

[52] J.W. Rhim, H.M. Park, C.S. Hac, Bio-nanocomposites for food packaging applications, Prog. Polym. Sci. 38 (2013) 1629-1652. https://doi.org/10.1016/j.progpolymsci.2013.05.008

[53] E. Ruiz-Hitzky, P. Aranda, A. Alvarez, J. Santarén, A. Esteban-Cubillo, Developments in palygorskite-sepiolite research, in: E. Galán and A. Singer (Eds.), A new outlook on these nanomaterials, Elsevier B.V., Oxford, UK, 2011, pp. 393-452. https://doi.org/10.1016/B978-0-444-53607-5.00017-7

[54] A. Sorrentino, G. Gorrasi, V. Vittoria, Potential perspectives of bionanocomposites for food packaging applications, Trend. Food Sci. Technol. 18 (2007) 84-95. https://doi.org/10.1016/j.tifs.2006.09.004

[55] F. Guo, S. Aryana, Y. Han, Y. Jiao, A review of the synthesis and applications of polymer-nanoclay composites, Appl. Sci. 8 (2018) 1696. https://doi.org/10.3390/app8091696

[56] B.E. Logan, B. Hamelers, R. Rozendal, U. Schröder, J. Keller, S. Freguia, P. Aelterman, W. Verstraete, K. Rabaey, Microbial fuel cells: methodology and technology, Environ. Sci. Technol. 40 (2006) 5181-5192. https://doi.org/10.1021/es0605016

[57] X. Wang, Y. Du, J. Luo, B. Lin, J.F. Kennedy, Chitosan/organic rectorite nanocomposite films: structure, characteristic and drug delivery behavior, Carbohyd. Polym. 69 (2007) 41- 49. https://doi.org/10.1016/j.carbpol.2006.08.025

[58] B.K. Singh, Organophosphorus-degrading bacteria: ecology and industrial applications, Nat. Rev. Microbiol. 7 (2009) 156-164. https://doi.org/10.1038/nrmicro2050

[59] M. Puri, D. Sharma, C. J. Barrow, Enzyme-assisted extraction of bioactives from plants, Trends Biotechnol. 30 (2012) 37-44. https://doi.org/10.1016/j.tibtech.2011.06.014

[60] F. Ali, H. Ullah, Z. Ali, F. Rahim, F. Khan, Z.U. Rehman, Polymer-clay nanocomposites, preparations and current applications: a review, Curr. Nanomater. 1 (2016) 83-95. https://doi.org/10.2174/2405461501666160625080118

[61] A. Bruggink, E.C. Roos, E. de Vroom, Penicillin acylase in the industrial production of beta-lactam antibiotics, Organ. Proc. Res. Dev. 2 (1998) 128-133. https://doi.org/10.1021/op9700643

[62] E.T. Johnson, Schmidt-Dannert, Light-energy conversion in engineered microorganisms. Curr. Trend. Biotechnol. 26 (2008) 682-689. https://doi.org/10.1016/j.tibtech.2008.09.002

[63] S.V. Ranganathan, S.L. Narasimhan, K. Muthukumar, An overview of enzymatic production of biodiesel, Bioresour. Technol. 99 (2008) 3975-3981. https://doi.org/10.1016/j.biortech.2007.04.060

[64] E. Casero, M. Darder, F. Pariente, L.E. Anal, Peroxidase enzyme electrodes as nitric oxide biosensors, Analyt. Chim. Act. 403 (2000) 1-9. https://doi.org/10.1016/S0003-2670(99)00555-3

[65] D.R.S. Jeykumari, S.S. Narayanan, Functionalized carbon nanotube-bienzyme biocomposite for amperometric sensing, Carbon 47 (2009) 957-966. https://doi.org/10.1016/j.carbon.2008.11.050

[66] O.B. Ayyub, P. Kofinas, Enzyme induced stiffening of nanoparticle-hydrogel composites with structural color, Amer. Chem. Soc. Nano 9 (2015) 8004-8011. https://doi.org/10.1021/acsnano.5b01514

[67] M. Ikeda, T. Tanida, T. Yoshii, K. Kurotani, S. Onogi, K. Urayama, I. Hamachi, Installing logic-gate responses to a variety of biological substances in supramolecular hydrogel-enzyme hybrids, Nat. Chem. 6 (2014) 511-518. https://doi.org/10.1038/nchem.1937

[68] S. Mailloux, E. Katz, Biocomputing, biosensing and bioactuation based on enzyme biocatalyzed reactions, Biocataly. 1 (2014) 13-32. https://doi.org/10.2478/boca-2014-0002

[69] A. Picot, C. Lacroix, Encapsulation of bifidobacteria in whey protein-based microcapsules and survival in simulated gastrointestinal conditions and in yoghurt, Int. Dai. J. 14 (2004) 505-515. https://doi.org/10.1016/j.idairyj.2003.10.008

[70] S. Gunasekaran, S. Ko, L. Xiao, Use of whey proteins for encapsulation and controlled delivery applications, J. Food Eng. 83 (2007) 31-40. https://doi.org/10.1016/j.jfoodeng.2006.11.001

[71] A. Picot, C. Lacroix, Production of multiphase water-insoluble microcapsules for cell microencapsulation using an emulsification/spray-drying technology, J. Food Sci. 68 (2003) 2693-2700. https://doi.org/10.1111/j.1365-2621.2003.tb05790.x

[72] S. Christoph, F.M. Fernandes, Bionanocomposite Materials for Biocatalytic Applications, in Carole Aimé and Thibaud Coradin (eds.), Bionanocomposites: Integrating Biological Processes for Bioinspired Nanotechnologies, John Wiley & Sons, Inc. 2017, pp. 257-298. https://doi.org/10.1002/9781118942246.ch5.4

Adv. App. of Micro and Nano Clay – Biopolymer-based Composites   Materials Research Forum LLC
Materials Research Foundations 125 (2022) 27-48   https://doi.org/10.21741/9781644901915-2

Chapter 2

Halloysite-Chitosan based Nano-Composites and Applications

G. Santhosh[1*], B. Sowmya [2]

[1]Department of Mechanical Engineering, NMAM Institute of Technology, Nitte-574110

[2]Materials Science Division, CSIR – National Aerospace Laboratories, Old Airport Road, Kodihalli, Bengaluru-560017, India

* drsanthug@gmail.com

Abstract

Chitosan is the most abundant and excellent natural polymer (PMR). The wider usage of chitosan is because of its antimicrobial, non-toxic, biocompatible and biodegradable nature. Chitosan is extracted from crustaceans and squids. Chitosan has been extensively studied in the field of wastewater treatment and biomedical applications. Halloysite nanotube (HNT) is a sort of aluminosilicate nano-clay, famous for their high aspect ratio and hallow configuration, HNT as a nano-filler for polymer matrix can be profitably utilized. HNT with the molecular formula, $H_4Al_2O_9Si_2 \cdot 2H_2O$ have unique tubular structure make them suitable as nano-containers with the intention to store and adsorb with abundant –OH groups. The use of HNT can provide high mechanical strength, high thermal stability and bio-acceptability. With the incorporation of nanosized halloysites nanotubes into chitosan matrix generally leads to desired property enhancement along with the changes in the microstructure. Amongst the most likely available natural materials, the chitosan and halloysite are attractive ones because of their nontoxic and eco-friendly nature. The halloysite was extensively studied as a carrier material in many drug delivery systems, catalytic support, scaffold for tissue engineering and as a nanofiller for food packaging application. In this chapter, the application of chitosan and HNT in the real world are postulated in order to give insights for future studies.

Keyword

Halloysite Nanotubes, Chitosan, Nanocomposite Film

Contents

1. Introduction

The recent research interest is on the combination of green materials with suitable organic/inorganic nanosized additives to achieve desirable properties suitable for particular applications. These natural materials are of prime concern due to their sustainability and cost effectiveness. Such nanocomposites are suitable substitutes for the replacement of composites made out of petrochemical based polymeric materials in various applications (packaging, biotechnology, engineering) [1-10]. Few researchers have demonstrated that there were significant changes noticed in the pure polymeric materials with the addition of clay nanoparticles [11]. The polylactic acid (PLA) based biofilms were fabricated with the inclusion of modified montmorillonite [12] and halloysite nanotubes (HNTs) [13] into PLA matrix by virtue of which the films possessed improved barrier properties and hence, are suitable for packaging applications. Furthermore, in order to improve the thermal stability of the nanocomposites, nanoparticles are incorporated into polymer matrix which are capable of offering higher surface area and thus promotes better interaction between filler and polymer matrix [14]. Additionally, the mesoscopic structure of the composite influences the mechanical, wettability and thermal properties of the biofilms [15,16]. Also, there were noticeable changes observed in the properties and functionalities of multi-layered hybrid materials in comparison with nanocomposites [17]. There are various research works reporting that intermediate clay layers are capable of promoting a flame inhibition on particular nanocomposites. For instance, a multilayer nanocomposite composed of alginate/montmorillonite exhibited extremely good fire as well as flame retardant properties [18]. Additionally, for cotton fabrics a fire-retardant coating was fabricated using polyacrylamide and graphene oxide-based layer-by-layer composite structure [19]. In order to fabricate multi-layered thin films, layer-by-layer (LBL) method serve as an easy and favourable approach. This strategy was advantageous over the Langmuir-Blodgett (LB) method, which is used for amphiphilic molecules. The amphiphilic molecules fabricated using LB method can create monolayers at the liquid-vapor interface [20]. further, LB technique involves the use of special instruments, these instruments have many drawbacks in terms of the size and shape of the substrate along with stability and quality of the films [21, 22]. On the other hand, with the help of the LBL method multilayer structures can be fabricated as a result of electrostatic attractions and hydrophobic interactions between oppositely charged molecules [22-24]. Also, the charge transfer interactions determine the development of multi-layered films [24]. There are various methods like co-extrusion, spin coating, solvent casting and spray coating that can be utilised for preparing multilayer structures via the LBL technique [25]. Herein, a new water casting procedure was used to prepare a multi-layered chitosan/halloysite structure by controlling the pH conditions in order to deposit each layer sequentially to achieve desired morphology in the composite.

The rolling of flat kaolinite sheets generates a hollow tubular morphology namely halloysite [26]. Based on the geological deposit, there are a broad range of sizes of respective halloysite nanotubes (HNTs) [27]. The length of the HNTs may vary from 50 to 1500 nm, whereas 10-15 nm and 20-150 nm, are its internal and external diameters respectively. Over a wide range in pH (2 to 8) conditions, the surface of HNTs exhibits opposite charges due to the difference in their chemical composition. Particularly, the outer surface of HNT is composed of SiO_2 and is negatively charged, whereas its lumen is composed of Al_2O_3 with a positive charge [28]. This combination of charges is capable of influencing the interactions amongst halloysite and ionic biopolymers by virtue of which the resulting properties of the nanocomposites also changes [29]. When the anionic alginate is encapsulated within the cavity of HNTs, there was significant changes in the thermal stability of corresponding nanocomposites in comparison with pure biopolymers. But there were no noticeable changes observed in thermal properties of chitosan/HNTs nanocomposites which were fabricated by the classic casting method in comparison with the adsorption method viz., chitosan was adsorbed on the outer surface of halloysite. The nanocomposite based on chitosan/HNTs are investigated as biofilms for packaging application, as hydrogels and as scaffold for biomedical and tissue engineering applications [30, 31]. The modified chitosan with several reactive groups with biological and synthetic macromolecules for bone regenerative properties is presented in table 1. It was evident that with the inclusion of clay nanoparticles in chitosan matrix, the resultant hybrid nanocomposite (NC) can serve as a potential substrate for drug delivery application, packaging application and also as electrochemical sensor [32a].

Table. 1 Modified chitosan derivatives with their reactive sub-groups used in bone regenerative engineering application [32b].

Derivatives	Sub-Group	Properties
Quaternized Chitosan	N,N,NTrimethyl chitosan (TMC)	• High absorption efficiency
	N-(2-hydroxylsubstituted phenyl)-N,Ndimethyl chitosan	• Strong antifungal activities against many resistant strains
Carboxyalkyl Chitosan	N-Carboxymethyl Chitosan	• High metal chelation and promoted apatite formation
	N,OCarboxymethyl chitosan	• High water retentivity • Negligible immunogenicity
Hydroxyalkyl Chitosan	Hydroxypropyl chitosan	• 99% inhibition against most of the bacteria
	Hydroxybutyl chitosan	• Rapid gelation kinetics with good in vivo stability

1.1   Clay minerals

Clay nanoparticles (NPs) are broadly examined clay minerals to prepare nanocomposites based on polymers. Common clay mineral deposits are used to obtain clay NPs possessing a sheet-like hydrous silicates structure [33]. Silicates are classified into 4 divisions with the crystalline arrangement: such as "kaolinite", "montmorillonite" (MMT), "illite" and "chlorite" groups [34]. MMT are the ones widely explored and utilized form of nano-clay to deliver NCs in light of polymeric matrix. The higher aspect-ratio along with its incomparable 'exfoliation' nature makes MMT an useful material. MMT consist of 2:1 sheet like arrangement with the octahedral sheet of alumina (Al) sharing oxygen atoms between two tetrahedral sheets. The tetrahedral sheets of MMT consist of SiO_4 groups which are connected to a hexagonal frame with repeating unit of Si_4O_{10}. The Al structure composed by closely packed oxygen atoms in two layers; between these layers octahedral aluminum particles are present [35] with halfway from the oxygen atoms. These layers called as tactoids of thickness of around 1 nm with the lateral scale of around 300 Å to a few microns. The layered silicate structure is as shown in Figure 1. As depicted in Figure 1. Vander-Waals of force separate each tactoid called as interlayer. These interlayer are responsible to balance the negative charges created by isomorph addition of crystals. (for instance, in MMT Mg_2^+ instead of Al_3^+) [36]. Unmodified MMTs, the interlayer cations are hydrated Na^+ or K^+ particles. This hydration makes the MMT water soluble and eventually unable to coexist with all polymers that are water insoluble [37]. Accordingly, modification of layered silicates organically by the inorganic cations exchange with alkylammonium, sulfonium, or phosphonium surfactants to improve their compatibility with polymers [38-40]

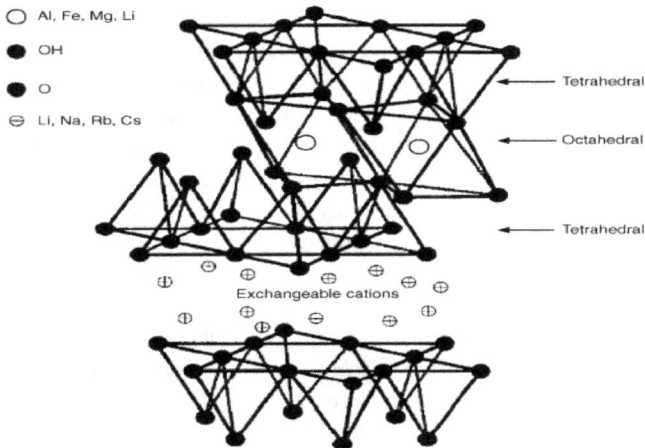

Figure 1. Structure of layered silicate [41]

In general, clay polymer NCs can adopt two strategies based on infiltration of PMR into the interlayer: Figure1.2 represent the morphology of CL/PMR system. In the event that the dispersion of polymeric chains paves the way to a limited extension of the clay sheets under 20-30 Å, the morphology is characterized as exfoliated; if the infiltration of the PMR prompts a delamination of silicate tactoids [42]. In the majority of studies, most of the properties change was seen in exfoliated NCs as they have very good aspect ratio of the tactoids. Further the preparation techniques of the NCs also play an important role in getting total exfoliation of the layers.

1.1.1 Halloysite nanotubes (HNTs)

Halloysite nanotube (HNT) is a sort of aluminosilicate nanoclay, famous for its high aspect ratio and hallow configuration. HNT as a nano-filler for polymer matrix can be profitably utilized. HNT with the molecular formula, $H_4Al_2O_9Si_2 \cdot 2H_2O$ have unique tubular structure make them suitable as nano-containers with the intention to store and adsorb with abundant –OH groups. The use of HNT can provide high mechanical property, high thermal strength and bio-acceptability [43].

1.1.2 Structure of HNT

Halloysite nanotubes as a filler for polymeric materials can be advantageously used, because of its high mechanical strength, thermal stability and biocompatibility [44]. HNT with the molecular formula, $H_4Al_2O_9Si_2 \cdot 2H_2O$, is a class of aluminosilicate nano clay with a hollow tubular structure and high length-to-diameter ratios. The distinctive nano-tubular structure makes these materials to be used as nano-containers for molecular adsorption and storage [45-47]. The unique properties of HNTs make them most convenient in the field of composites to use them as reinforcing nanofillers [48]. However, interfacial decoration remains one of the great challenges in the fabrication of polymer/HNT NCs ascribed lack of effective functional groups on the outer surface of HNT. It has been already proven that HNT can be sufficiently used as flame retardant material [49-51]. However, the big benefit of HNT is their low cost [52] and can be compared with other NPs and easy ecological disposal. Besides all the benefits, HNT is recognized as best dielectric filler which gained ample of attention in the field of micro-circuit components [53-60]. HNT are composed of several siloxane groups and have only a few hydroxyl groups, which provide ability to HNT to form hydrogen bonds and hence, better dispersion potential in polymer matrix. Large luminal diameter of the HNT provides an opportunity to accommodate different polymer molecules to offer polymeric composites[61]. The structure of HNT with different constituents is shown in Figure 2.

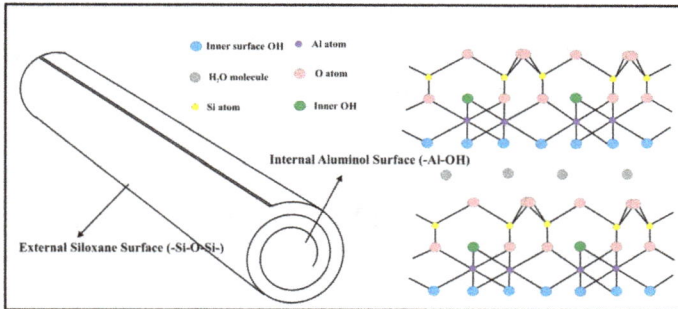

Figure 2. Chemical structure of HNT [62]

2.   Chitosan (CS)

The most abundantly available biopolymer in nature is chitosan. This biopolymer is known for its biodegradability, biocompatibility, lack of toxicity and antibacterial properties. Chitosan is one of the most appropriate materials in biomedical field, drug delivery systems and as edible films for food packaging application.

Chitosan has got good physicochemical and biological properties; due to its unique properties it is used in many industries. The industries such as medical, food, chemical, cosmetics, water treatment, metal extraction and recovery, biochemical, and biomedical engineering are few major fields where chitosan is used. However, in spite of many advantages CS has got some drawbacks as well, chitosan is not soluble in aqueous solutions, hence making it difficult to use in living systems [63]. Though it is not soluble in aqueous solutions thanks to its functional groups that allow the modification that brings the changes in CS with improved properties. the modifications help improve the solubility and consequently broaden the applications window. The modified chitosan creates many derivatives, these derivatives show sustained release properties, non-toxicity, biocompatibility and biodegradability [64].

2.1   Properties of chitosan

Chitin when deacetylated forms chitosan. This is a linear polysaccharide composed of β - (1,4)-linked N -acetyl-glucosamine units [65-67]. Chitin is found in insect exoskeletons, crustaceans as well as in few cell walls [68]. The degree of deacetylation and origin of chitin defines the quality of chitosan [69]. Chitosan is very attractive polymer used in biomedical applications due to its excellent biological, nontoxic, biodegradable and antimicrobial properties. One of the most notable properties of chitosan is its anti-inflammatory response. The toxicity of chitosan can be reduced by deacetylation process, the process is similar to that of succinyl-derived chitosan and chitosan nanoparticles [70, 71].

Deacetylated chitin is a natural polycationic polymer which has bactericidal and bacteriological properties. The positive charge of this chitin adheres on the surface of bacteria which brings about the changes in the membrane wall thereby avoiding growth of microbes [72]. The deacetylation process improves the antibacterial activity of the chitosan especially when chitosan having low pH. Further, the molecular weight reduction also responsible for the increase in antibacterial activities of chitosan [73]. The interaction of bacteria and chitosan depends on hydrophilicity of the cell wall, which makes chitosan a better antitoxic element to mammalian cells [74].

Antioxidants have proven to be most beneficial on health, chitosan and its derivatives as antioxidants has shown their capability in preventing the formation of free active oxygen sites. The antioxidants are primarily used to prevent the membrane destruction, destruction of proteins and DNA [75]. Especially low-weight chitosan molecules have proven to be very effective in eliminating free radicals [76].

3.   Chitosan derivatives

As we all know that chitosan is a natural resource which is a deacetylated product of chitin, however its insolubility in water and other organic solvents makes it difficult to use it in various applications. The better solution to this problem is its derivatives. Chitosan derivatives obtained from chemical modification may increase its solubility and other properties as well. Chitosan with hydroxyl and amino groups undergo many chemical reactions to form pendant groups by destroying crystal structure of chitosan which ultimately increases the solubility of the chitosan. The chitosan derivatives formed by the chemical reactions such as hydroxylation, carboxylation, alkylation, acylation, and esterification will increase the physicochemical and biological behaviours which are essential for the biomedical applications [77].

3.1   Acylated chitosan

This is the most common and widely used to modify chitosan. When chitosan interacts with acids and its derivatives like anhydride and acyl chloride, the process is referred as acylation [78]. The water solubility of the chitosan can be increased by the acylation process, this process destroys inter and intramolecular hydrogen bonding which reduces the crystallinity of chitosan thereby enhancing the solubility.

3.2   Alkylated chitosan

Alkylated chitosan having alkyl group weakens the intermolecular hydrogen bonding leading to better solubility [79]. However, alkyl group is hydrophobic in nature, thereby altering the alkyl chain length solubility of chitosan also gets altered [80, 81]. The alkylation of chitosan takes place by the introduction of alkyl groups to the C_2 –NH_2, C_6 –OH, or C_3 –OH groups of chitosan [82, 83].

3.3   Hydrophilic group

Hydrophilic groups are the most compatible groups with water, which includes phosphoric acid, sulfonic acid, quaternary ammonium, carboxylic acid and many more.

3.3.1   Carboxylated chitosan

Carboxylated chitosan is produced by carboxylation reaction, in the reaction glyoxylic acid or chloroalkanoic acid are reacted with either C_2 –NH_2 or C_6 –OH groups of chitosan, as a result -COOH groups are formed [84]. Carboxylated chitosan has the ability to dissolve in neutral and alkaline solutions with better thickening and heat preventing properties [85]. Carboxylated chitosan fit themselves in agricultural, medical, health, and biochemical applications [86].

3.3.2   Quaternary ammonium chitosan

Quaternary ammonia group in chitosan is a positively charged hydrophilic group. The quaternary ammonia groups in chitosan increases the solubility and chargeability. The process of introducing quaternary groups into chitosan is referred as quaternization, the quaternization occurs with C_2 –NH_2 groups of chitosan. The quaternary ammonia in chitosan can be formed in three different ways [87-91].

1. The direct quaternary ammonium substitution

2. Epoxy derivative open loop method

3. N-alkylation

4.   Esterified chitosan

The esterification process of chitosan take place due to the presence of oxygen containing inorganic acids, the commonly used acids in the esterification process are phosphoric acid, sulphuric acid and chlorosulfonic acid [92]. One of the most widely used chitosan derivative formed through esterification is Sulfated chitosan. Sulfated chitosan is used as substitute for heparin as an anticoagulant and antiviral drug. The sulphated chitosan exhibits the regulatory mechanism same as heparin [93].

5.   Application of chitosan-HNTs (CS-HNT) composites

The suitable scaffold material for biomedical applications should have good porosity, oxygen permeability and wound protecting nature form infection [94, 95]. These qualities can be achieved by using ideal reinforcing nano fillers. The nano fillers are selected not only to meet the biomedical requirements but also to improve the mechanical and thermal stability of the scaffold [96]. HNTs have similar nanostructures to those of carbon nanotubes, but HNTs are biofriendly tubular one dimensional nanotubes exhibit non-toxicity [97]. In addition to these environment friendly nature HNTs have high surface area and large aspect ratio make them demonstrate high mechanical and thermal properties [98, 99]. Furthermore, the HNTs also pose some serious drawbacks for the use in biomedical applications such as brittleness, low fracture strength and density [100-102]. Chitosan-HNTs nanocomposite may solve these problems and provide better solutions to these shortcomings of HNTs. The HNTs reinforced with chitosan enhance blood clotting, platelet activation and also the wound ability can be improved [103]. The functionalised the HNTs has led to the formation of new class of bioactive carriers [104].

One of the major problems in tissues engineering is to establish a scaffold to support 3-D tissue formation [105, 106]. To meet this requirement, the scaffolds must have few specific qualities [107]. The scaffolds must be highly porous and must have enough pore size to accommodate cell seeding and nutrients diffusion. The scaffolds must be biodegradable, with good degradation rate. However, the scaffolds must have good mechanical properties to support the structure, especially used as bone grafts. The chitosan derivatives with bioactive molecules used in bone tissue grafting are listed in table 2.

Chitosan and its derivatives play vital role in tissue engineering applications as these can be easily fabricated in porous 3-D scaffolds. Nevertheless, the use of chitosan alone as scaffolds pose some serious concern as they are mechanically weak [108]. To achieve this goal the scaffolds are used by combining the biodegradable polymer with bioactive materials such as nanoparticles and nanotubes [109]. For instant let us consider the nano-hydroxyapatite with chitosan forming hybrid scaffold shows high mechanical strength, increased porosity and superior physico-chemical and biological properties [110, 111]. Another example is using carbon nanotubes with chitosan to form 3D nano-composite scaffolds with superior mechanical properties [112]. However, the use of such nano particles poses some difficulties to use directly without pre-treatment and also difficulty involved in synthesis procedure limit their use in tissue engineering applications. Hence, naturally available, cost effective, mechanically strong and biocompatible nano materials for fabricating chitosan nanocomposite scaffolds are often emphasised.

Adv. App. of Micro and Nano Clay – Biopolymer-based Composites   Materials Research Forum LLC
Materials Research Foundations **125** (2022) 27-48   https://doi.org/10.21741/9781644901915-2

Table 2. Properties of modified chitosan composites with bioactive molecules for bone tissue engineering [32b].

Sl No.,	Modified chitosan	Molecules	Properties
1	Carboxymethyl-chitosan	With Ulvan biomolecule	Enchances the mechanical Performance of the bone graft
		With genipin carbodiimide biomolecule	Increases Cytocompatibility and proliferation
2	4-phosphorylated-chitosan	With Chitosan-Hydroxyapatite biomolecule	Increases compressive strength
3	Methacryloyloxy-ethyl-carboxyethyl-chitosan	With polyethylene Glycol dimethacrylate And N, Ndimethylacrylamide	Increase mechanical and thermal behaviour of the graft

Halloysite nanotubes (HNTs) are naturally available and biocompatible clay mineral which is suitable for developing chitosan-based nanocomposites. HNTs have positively charged aluminium in the inner layer and negatively charged silicate in the outer layer. The hydrophilicity and small dimensions of HNTs mixed with chitosan forms chitosan-HNTs complex via electrostatic attractions. There is hydrogen bonding between Al-O-H groups of HNTs and the hydroxyl and amino groups of chitosan resulting improved properties of the chitosan-HNTs (CS-HNT) nanocomposites [113]. The HNTs used in chitosan didn't affect the porosity of the chitosan, hence the properties of CS-HNT nanocomposites are appreciated when used in tissue engineering applications.

Conclusion

In this chapter we have discussed brief applications of HNTs and Chitosan based nanocomposites in biomedical and tissue engineering applications. HNTs used in chitosan functions as a reinforcing filler bringing about enormous variations in the resultant properties of chitosan based nanocomposites. The chitosan/HNTs based nanocomposites are simple to prepare and exhibits improvement in thermal and mechanical properties. The good biocompatibility of chitosan-HNTs nanocomposites makes them a potential candidate in the field of biomedicine. When chitosan is reinforced with HNTs, the composite enhances blood clotting, platelet activation and also the wound healing ability. Functionalised HNTs have evolved as a novel bioactive carrier.

Reference

[1] Sapalidis, F.K. Katsaros, G.E. Romanos, N.K. Kakizis, and N.K. Kanellopoulos, "PVA /Montmorillonite Nanocomposites: Development and Properties", Bio-Eng. Compos., 38 (2007) 398-404. https://doi.org/10.1016/j.compositesb.2006.04.005

[2] M. Makaremi, P. Pasbakhsh, G. Cavallaro, G. Lazzara, Y. K. Aw, S. M. Lee and S. Milioto, "Effect of Morphology and Size of Halloysite Nanotubes on Functional Pectin Bionanocomposites for Food Packaging Applications", ACS Appl. Mater. Interfaces, 9 (2017) 17476-17488. https://doi.org/10.1021/acsami.7b04297

[3] Y. Lvov, A. Aerov and R. Fakhrullin, "Clay Nanotube Encapsulation for Functional Biocomposites", Adv Colloid Interface Sci, 207 (2014)189-198. https://doi.org/10.1016/j.cis.2013.10.006

[4] Y. Lvov and E. Abdullayev, "Functional polymer-clay nanotube composites with sustained release of chemical agents" Prog Polym Sci, 38 (2013) 1690-1719. https://doi.org/10.1016/j.progpolymsci.2013.05.009

[5] Y. Stetsyshyn, J. Zemla, O. Zolobko, K. Fornal, A. Budkowski, A. Kostruba, V. Donchak, K. Harhay, K. Awsiuk, J. Rysz, A. Bernasik and S. Voronov, "Temperature and pH dual-responsive coatings of oligoperoxide-graft-poly(N-isopropylacrylamide): Wettability, morphology, and protein adsorption" J. Colloid Interface Sci., 387 (2012) 95-105. https://doi.org/10.1016/j.jcis.2012.08.007

[6] R.F. Fakhrullin and Y. M. Lvov, "Halloysite clay nanotubes for tissue engineering", Nanomed., 11 (2016) 2243-2246. https://doi.org/10.2217/nnm-2016-0250

[7] M. Liu, C. Wu, Y. Jiao, S. Xiong and C. Zhou, "Chitosan-halloysite nanotubes nanocomposite scaffolds for tissue engineering", J. Mater. Chem. B, 1 (2013) 2078-2089. https://doi.org/10.1039/c3tb20084a

[8] H. Wei, K. Rodriguez, S. Renneckar and P. J. Vikesland, "Environmental science and engineering applications of nanocellulose-based nanocomposites", Env. Sci Nano, 1 (2014) 302-316. https://doi.org/10.1039/C4EN00059E

[9] V. K. Thakur and M. R. Kessler, "Self-healing polymer nanocomposite materials: A review" Polymer, 69 (2015) 369-383. https://doi.org/10.1016/j.polymer.2015.04.086

[10] M. Du, B. Guo, Y. Lei, M. Liu and D. Jia, "Carboxylated butadiene-styrene rubber/halloysite nanotube nanocomposites: Interfacial interaction and performance", Polymer, 49 (2008) 4871-4876. https://doi.org/10.1016/j.polymer.2008.08.042

[11] E. Ruiz-Hitzky, P. Aranda, M. Darder and G. Rytwo, "Hybrid materials based on clays for environmental and biomedical applications", J Mater Chem, 20 (2010) 9306-9321. https://doi.org/10.1039/c0jm00432d

[12] G. Ozkoc and S. Kemaloglu, "Morphology, biodegradability, mechanical, and thermal properties of nanocomposite films based on PLA and plasticized PLA", J. Appl. Polym. Sci., 114 (2009) 2481-2487. https://doi.org/10.1002/app.30772

[13] G. Gorrasi, R. Pantani, M. Murariu and P. Dubois, "PLA/Halloysite Nanocomposite Films: Water Vapor Barrier Properties and Specific Key Characteristics", Macromol. Mater. Eng., 299 (2014) 104-115. https://doi.org/10.1002/mame.201200424

[14] I. Blanco and F. A. Bottino, "Thermal study on phenyl, hepta isobutyl-polyhedral oligomeric silsesquioxane/polystyrene nanocomposites", Polym. Compos., 34 (2013) 225-232. https://doi.org/10.1002/pc.22400

[15] V. Bertolino, G. Cavallaro, G. Lazzara, M. Merli, S. Milioto, F. Parisi and L. Sciascia, "Effect of the biopolymer charge and the nanoclay morphology on nanocomposite materials", Ind. Eng. Chem. Res., 55 (2016) 7373-7380. https://doi.org/10.1021/acs.iecr.6b01816

[16] G. Cavallaro, D. I. Donato, G. Lazzara and S. Milioto, "Films of halloysite nanotubes sandwiched between two layers of biopolymer: from the morphology to the dielectric, thermal, transparency, and wettability properties", J Phys Chem C, 115 (2011) 20491-20498. https://doi.org/10.1021/jp207261r

[17] B. Finnigan, D. Martin, P. Halley, R. Truss and K. Campbell, "Morphology and properties of thermoplastic polyurethane composites incorporating hydrophobic layered silicates", J. Appl. Polym. Sci., 97 (2005) 300-309. https://doi.org/10.1002/app.21718

[18] B. Liang, H. Zhao, Q. Zhang, Y. Fan, Y. Yue, P. Yin and L. Guo, "Ca2+ Enhanced Nacre-Inspired Montmorillonite-Alginate Film with Superior Mechanical, Transparent, Fire Retardancy, and Shape Memory Properties", ACS Appl. Mater. Interfaces, 8 (2016) 28816-28823. https://doi.org/10.1021/acsami.6b08203

[19] G. Huang, J. Yang, J. Gao and X. Wang, "Thin Films of Intumescent Flame Retardant-Polyacrylamide and Exfoliated Graphene Oxide Fabricated via Layer-by-Layer Assembly for Improving Flame Retardant Properties of Cotton Fabric", Ind. Eng. Chem. Res., 51 (2012) 12355-12366. https://doi.org/10.1021/ie301911t

[20] J. Zasadzinski, R. Viswanathan, L. Madsen, J. Garnaes and D. Schwartz, "Langmuir-Blodgett films", Science, 263 (1994) 1726-1733. https://doi.org/10.1126/science.8134836

[21] L. Netzer, R. Iscovici and J. Sagiv, "Adsorbed Monolayers versus Langmuir-Blodgett Monolayers-Why and How? I: From Monolayer to Multilayer, by Adsorption",Thin Solid Films, 99 (1983) 235-241. https://doi.org/10.1016/0040-6090(83)90386-3

[22] G. Decher, "Fuzzy nanoassemblies: toward layered polymeric multicomposites", Science, 277 (1997) 1232-1237. https://doi.org/10.1126/science.277.5330.1232

[23] N.A. Kotov, "Layer-by-layer self-assembly: the contribution of hydrophobic interactions", Nanostructured Mater., 12 (1999) 789-796. https://doi.org/10.1016/S0965-9773(99)00237-8

[24] Y. Shimazaki, M. Mitsuishi, S. Ito and M. Yamamoto, "Preparation of the layer-by-layer deposited ultrathin film based on the charge-transfer interaction", Langmuir, 13 (1997) 1385-1387. https://doi.org/10.1021/la9609579

[25] X. Zhang, H. Chen and H. Zhang, "Layer-by-layer assembly: from conventional to unconventional methods", Chem Commun, (2007) 1395-1405. https://doi.org/10.1039/B615590A

[26] E. Joussein, S. Petit, G.J. Churchman, B. Theng, D. Righi and B. Delvaux, "Halloysite clay minerals-a review", Clay Miner., 40 (2005) 383-426. https://doi.org/10.1180/0009855054040180

[27] P. Pasbakhsh, G.J. Churchman and J.L. Keeling, "Characterisation of properties of various halloysites relevant to their use as nanotubes and microfibre fillers", Appl. Clay Sci., 74 (2013) 47-57. https://doi.org/10.1016/j.clay.2012.06.014

[28] V. Bertolino, G. Cavallaro, G. Lazzara, S. Milioto and F. Parisi, "Biopolymer-targeted adsorption onto halloysite nanotubes in aqueous media", Langmuir, 33 (2017) 3317-3323. https://doi.org/10.1021/acs.langmuir.7b00600

[29] R.T. De Silva, P. Pasbakhsh, K.L. Goh and L. Mishnaevsky, "Halloysite nanotubes sandwiched between chitosan layers: Novel bionanocomposites with multilayer structures", Polymer, 55 (2014) 6418-6425. https://doi.org/10.1016/j.polymer.2014.09.057

[30] T. Wu, Y. Li and D.S. Lee, "Chitosan-based composite hydrogels for biomedical applications", Macromol. Res., 25 (2017) 480-488. https://doi.org/10.1007/s13233-017-5066-0

[31] E.A. Naumenko, I.D. Guryanov, R. Yendluri, Y.M. Lvov and R.F. Fakhrullin, "Clay nanotube-biopolymer composite scaffolds for tissue engineering", Nanoscale, 8 (2016) 7257-7271. https://doi.org/10.1039/C6NR00641H

[32] (a) A. Ali and S. Ahmed, "A review on chitosan and its nanocomposites in drug delivery", Int. J. Biol. Macromol., 109 (2018) 273-286. (b) R. LogithKumar A. KeshavNarayan S. Dhivya A. Chawla S. Saravanan N. Selvamurugan, "A Review of Chitosan and its Derivatives in Bone Tissue Engineering", Carbohydrate Polymers, 151 (2016) 172-188. https://doi.org/10.1016/j.carbpol.2016.05.049

[33] J. Zheng, R.W. Siegel, and C.G. Toney, "Polymer crystalline structure and morphology changes in nylon-6/ZnO nanocomposites," J. Polym. Sci. Part B Polym. Phys., 41 (2003) 1033-1050. https://doi.org/10.1002/polb.10452

[34] F. Uddin, "Clays, nanoclays, and montmorillonite minerals," Metall. Mater. Trans. A, 39 (2008) 2804-2814. https://doi.org/10.1007/s11661-008-9603-5

[35] S. Pavlidou and C.D. Papaspyrides, "A review on polymer-layered silicate nanocomposites," Prog. Polym. Sci., 33 (2008) 1119-1198. https://doi.org/10.1016/j.progpolymsci.2008.07.008

[36] L. Torre, M. Chieruzzi, and J.M. Kenny, "Compatibilization and development of layered silicate nanocomposites based of unsaturated polyester resin and customized intercalation agent," J. Appl. Polym. Sci., 115 (2010) 3659-3666. https://doi.org/10.1002/app.31461

[37] M. Alexandre and P. Dubois, "Polymer-layered silicate nanocomposites: preparation, properties and uses of a new class of materials," Mater. Sci. Eng. R Rep., 28 (2000) 1-63. https://doi.org/10.1016/S0927-796X(00)00012-7

[38] K.N. Shilpa, K.S. Nithin, S. Sachhidananda, and B.S. Madhukar, and Siddaramaiah, "Visibly transparent PVA/sodium doped dysprosia (Na2Dy2O4) nanocomposite films, with high refractive index: An optical study," J. Alloys Compd., 694 (2017) 884-891. https://doi.org/10.1016/j.jallcom.2016.10.004

[39] G. Stoclet, "Elaboration of poly (lactic acid)/halloysite nanocomposites by means of water assisted extrusion: structure, mechanical properties and fire performance," RSC Adv., 4 (2014) 57553-57563. https://doi.org/10.1039/C4RA06845A

[40] P. Kiliaris, C.D. Papaspyrides, Polymer/layered silicate (clay) nanocomposites: An overview of flame retardancy, Progress in Polymer Science, 35 (2010) 902-958. https://doi.org/10.1016/j.progpolymsci.2010.03.001

[41] W.O. Yah, A. Takahara, and Y.M. Lvov, "Selective modification of halloysite lumen with octadecylphosphonic acid: new inorganic tubular micelle," J. Am. Chem. Soc., 134 (2012) 1853-1859. https://doi.org/10.1021/ja210258y

[42] P. Yuan, D. Tan, and F.A. Bergaya, "Properties and applications of halloysite nanotubes: recent research advances and future prospects," Appl. Clay Sci., 112 (2015) 75-93. https://doi.org/10.1016/j.clay.2015.05.001

[43] A. Dey and S.K. De, "Large dielectric constant in zirconia polypyrrole hybrid nanocomposites," J. Nanosci. Nanotechnol., 7 (2007) 2010-2015. https://doi.org/10.1166/jnn.2007.759

[44] L.M. Clayton, M. Cinke, M. Meyyappan, and J. P. Harmon, "Dielectric properties of PMMA/soot nanocomposites," J. Nanosci. Nanotechnol., 7 (2007) 2494-2499. https://doi.org/10.1166/jnn.2007.428

[45] N. Raman, S. Sudharsan, K. Pothiraj, Synthesis and structural reactivity of inorganic-organic hybrid nanocomposites - A review, Journal of Saudi Chemical Society. 16 (2012) 339-352. https://doi.org/10.1016/j.jscs.2011.01.012

[46] H. Li, "Polypropylene fibers fabricated via a needleless melt-electrospinning device for marine oil-spill cleanup," J. Appl. Polym. Sci., 131 (2014). https://doi.org/10.1002/app.40080

[47] G. Griffini, M. Levi, and S. Turri, "Thin-film luminescent solar concentrators: A device study towards rational design," Renew. Energy, 78 (2015) 288-294. https://doi.org/10.1016/j.renene.2015.01.009

[48] M. Biswal, S. Mohanty, S.K. Nayak, and P.S. Kumar, "Effect of functionalized nanosilica on the mechanical, dynamic-mechanical and morphological performance of polycarbonate/nanosilica nanocomposites," Polym. Eng. Sci., 53 (2013) 1287-1296. https://doi.org/10.1002/pen.23388

[49] S. Pillai, K.R. Catchpole, T. Trupke, and M.A. Green, "Surface plasmon enhanced silicon solar cells," J. Appl. Phys., 101 (2007) 093105. https://doi.org/10.1063/1.2734885

[50] K. Znajdek, "Zinc oxide nanoparticles for improvement of thin film photovoltaic structures' efficiency through down shifting conversion," Opto-Electron. Rev., 25 (2017) 99-102. https://doi.org/10.1016/j.opelre.2017.05.005

[51] M. Eslamian, "Excitation by acoustic vibration as an effective tool for improving the characteristics of the solution-processed coatings and thin films," Prog. Org. Coat., 113 (2017) 60-73. https://doi.org/10.1016/j.porgcoat.2017.08.008

[52] G. Santhosh, G.P. Nayaka, J. Aranha, and Siddaramaiah, "Investigation on electrical and dielectric behaviour of halloysite nanotube incorporated polycarbonate nanocomposite films," Trans. Indian Inst. Met., 70 (2017) 549-555. https://doi.org/10.1007/s12666-016-1033-2

[53] Y.C. Cao, "Preparation of thermally stable well-dispersed water-soluble CdTe quantum dots in montmorillonite clay host media," J. Colloid Interface Sci., 368 (2012) 139-143. https://doi.org/10.1016/j.jcis.2011.11.044

[54] J. Gaume, C. Taviot-Gueho, S. Cros, A. Rivaton, S. Therias, and J.L. Gardette, "Optimization of PVA clay nanocomposite for ultra-barrier multilayer encapsulation of organic solar cells," Sol. Energy Mater. Sol. Cells, 99 (2012) 240-249. https://doi.org/10.1016/j.solmat.2011.12.005

[55] S. Cros, "Definition of encapsulation barrier requirements: A method applied to organic solar cells," Sol. Energy Mater. Sol. Cells, 95 (2011) S65-S69. https://doi.org/10.1016/j.solmat.2011.01.035

[56] E. Picard, A. Vermogen, J.F. Gérard, and E. Espuche, "Barrier properties of nylon 6-montmorillonite nanocomposite membranes prepared by melt blending: Influence of the clay content and dispersion state: Consequences on modeling," J. Membr. Sci., 292 (2007) 133-144. https://doi.org/10.1016/j.memsci.2007.01.030

[57] H. Kim, Y. Miura, and C.W. Macosko, "Graphene/polyurethane nanocomposites for improved gas barrier and electrical conductivity," Chem. Mater., 22 (2010) 3441-3450. https://doi.org/10.1021/cm100477v

[58] L. Cui, J.T. Yeh, K. Wang, F.C. Tsai, and Q. Fu, "Relation of free volume and barrier properties in the miscible blends of poly(vinyl alcohol) and nylon 6-clay nanocomposites film," J. Membr. Sci., 327 (2009) 226-233. https://doi.org/10.1016/j.memsci.2008.11.027

[59] H. Jing "Preparation and characterization of polycarbonate nanocomposites based on surface-modified halloysite nanotubes," Polym. J., 46 (2014) 307. https://doi.org/10.1038/pj.2013.100

[60] H. Kim and C.W. Macosko, "Processing-property relationships of polycarbonate/ graphene composites," Polymer, 50 (2009) 3797-3809. https://doi.org/10.1016/j.polymer.2009.05.038

[61] C.Y. Lee and J.K. Kil, "Hydrophilic property by contact angle change of ion implanted polycarbonate," Rev. Sci. Instrum., 79 (2008) 02C508. https://doi.org/10.1063/1.2804906

[62] H.S. Kas, Chitosan: Properties, preparation and application to microparticulate systems. J. Microencapsul. 14 (1997) 689-711. https://doi.org/10.3109/02652049709006820

[63] A.K. Singla, M. Chawla, Chitosan: Some pharmaceutical and biological aspects-An update. J. Pharm. Pharmacol. 53 (2001) 1047-1067. https://doi.org/10.1211/0022357011776441

[64] Y. Kato, H. Onishi, Y. Machida, Application of chitin and chitosan derivatives in the pharmaceutical field. Curr. Pharm.Biotechnol. 4 (2003) 303-309. https://doi.org/10.2174/1389201033489748

[65] J. Varshosaz, "The promise of chitosan microspheres in drug delivery systems". Drug Deliv. 4 (2007) 263-273. https://doi.org/10.1517/17425247.4.3.263

[66] G. Galed, B. Miralles, I. Ines Panos, A. Santiago, A. Heras, "N-Deacetylation and depolymerization reactions of chitin/chitosan: Influence of the source of chitin". Carbohydr. Polym. 62 (2005) 316-320. https://doi.org/10.1016/j.carbpol.2005.03.019

[67] M. Huang, E. Khor, L.Y. Lim, "Uptake and cytotoxicity of chitosan molecules and nanoparticles: Effects of molecular weight and degree of deacetylation". Pharm. Res. 21 (2004) 344-353. https://doi.org/10.1023/B:PHAM.0000016249.52831.a5

[68] R. Chien, M. Yen, J. Mau, "Antimicrobial and antitumor activities of chitosan from shiitake stipes, compared to commercial chitosan from crab shells". Carbohydr. Polym. 138 (2016) 259-264. https://doi.org/10.1016/j.carbpol.2015.11.061

[69] Z. Aiping, C. Tian, Y. Lanhua, W. Hao, L. Ping, "Synthesis and characterization of N-succinyl-chitosan and its self-assembly of nanospheres". Carbohydr. Polym. 66 (2006) 274-279. https://doi.org/10.1016/j.carbpol.2006.03.014

[70] C. Yan, J. Gu, D. Hou, H. Jing, J. Wang, Y. Guo, H. Katsumi, T. Sakane, A. Yamamoto, "Synthesis of tat tagged and folate modified N-succinyl-chitosan self-assembly nanoparticles as a novel gene vector". Int. J. Biol. Macromol. 72 (2015) 751-756. https://doi.org/10.1016/j.ijbiomac.2014.09.031

[71] R.C. Goy, D.D. Britto, B.G. Assis Oiii, "A review of the antimicrobial activity of chitosan". Polímeros. 19 (2009) 241-247. https://doi.org/10.1590/S0104-14282009000300013

[72] I. Younes, S. Sellimi, M. Rinaudo, K. Jellouli, M. Nasri, "Influence of acetylation degree and molecular weight of homogeneous chitosan on antibacterial and antifungal activities". Int. J. Food Microbiol. 185 (2014) 57-63. https://doi.org/10.1016/j.ijfoodmicro.2014.04.029

[73] M. Kong, X.G. Chen, K. Xing, H.J. Park, "Antimicrobial properties of chitosan and mode of action: A state of the art review". Int. J. Food Microbiol. 144 (2010) 51-63. https://doi.org/10.1016/j.ijfoodmicro.2010.09.012

[74] D.H. Ngo, S.K. Kim, "Antioxidant effects of chitin, chitosan, and their derivatives". Adv. Food Nutr. Res. 73 (2014) 15-31. https://doi.org/10.1016/B978-0-12-800268-1.00002-0

[75] P.J. Park, J.Y. Je, S.K. Kim, "Free radical scavenging activity of chitooligosaccharides by electron spin resonance spectrometry". J. Agric. Food Chem. 51 (2003) 4624-4627. https://doi.org/10.1021/jf034039+

[76] N. Islam, V. Ferro, "Recent advances in chitosan-based nanoparticulate pulmonary drug delivery". Nanoscale 8 (2016) 14341-14358. https://doi.org/10.1039/C6NR03256G

[77] J. Cai, Q. Dang, C. Liu, B. Fan, J. Yan, Y. Xu, J. Li, "Preparation and characterization of N-benzoyl-O-acetyl-chitosan". Int. J. Biol. Macromol. 77 (2015) 52-58. https://doi.org/10.1016/j.ijbiomac.2015.03.007

[78] Y. Kurita, A. Isogai, "N-Alkylations of chitosan promoted with sodium hydrogen carbonate under aqueous conditions". Int. J. Biol. Macromol. 50 (2012) 741-746. https://doi.org/10.1016/j.ijbiomac.2011.12.004

[79] G. Ma, D. Yang, Y. Zhou, M. Xiao, J.F. Kennedy, J. Nie, "Preparation and characterization of water-soluble N-alkylated chitosan". Carbohydr. Polym. 74 (2008) 121-126. https://doi.org/10.1016/j.carbpol.2008.01.028

[80] T.C. Yang, C.C. Chou, C.F. Li, "Antibacterial activity of N-alkylated disaccharide chitosan derivatives". Int. J. Food Microbiol. 97 (2005) 237-245. https://doi.org/10.1016/S0168-1605(03)00083-7

[81] S.J. Burr, P.A. Williams, I. Ratclie, "Synthesis of cationic alkylated chitosans and an investigation of their rheological properties and interaction with anionic surfactant". Carbohydr. Polym. 201 (2018) 615-623. https://doi.org/10.1016/j.carbpol.2018.08.105

[82] C. Onésippe, S. Lagerge, "Studies of the association of chitosan and alkylated chitosan with oppositely charged sodium dodecyl sulfate". Colloids Surf. A. 330 (2008) 201-206. https://doi.org/10.1016/j.colsurfa.2008.07.054

[83] E. Mohammadi, H. Daraei, R. Ghanbari, S. Dehestani Athar, Y. Zandsalimi, A. Ziaee, A. Maleki, K. Yetilmezsoy, "Synthesis of carboxylated chitosan modified with ferromagnetic nanoparticles for adsorptive removal of fluoride, nitrate, and phosphate anions from aqueous solutions". J. Mol. Liq. 273 (2019) 116-124. https://doi.org/10.1016/j.molliq.2018.10.019

[84] M. Kurniasih, T. Cahyati, R.S. Dewi, "Carboxymethyl chitosan as an antifungal agent on gauze". Int. J. Biol. Macromol. 119 (2018) 166-171. https://doi.org/10.1016/j.ijbiomac.2018.07.038

[85] A. Zhang, Y. Zhang, G. Pan, J. Xu, H. Yan, Y. Liu, "In situ formation of copper nanoparticles in carboxylated chitosan layer: Preparation and characterization of surface modified TFC membrane with protein fouling resistance and long-lasting antibacterial properties". Sep. Purif. Technol. 176 (2017) 164-172. https://doi.org/10.1016/j.seppur.2016.12.006

[86] K. Chen, B. Guo, J. Luo, "Quaternized carboxymethyl chitosan/organic montmorillonite nanocomposite as a novel cosmetic ingredient against skin aging". Carbohydr. Polym. 173 (2017) 100-106. https://doi.org/10.1016/j.carbpol.2017.05.088

[87] X. Huang, C. Xu, Y. Li, H. Cheng, X. Wang, R. Sun, "Quaternized chitosan-stabilized copper sulfide nanoparticles for cancer therapy". Mater. Sci. Eng. C. 96 (2019) 129-137. https://doi.org/10.1016/j.msec.2018.10.062

[88] S.C. Jang, W.C. Tsen, F.S. Chuang, C. Gong, "Simultaneously enhanced hydroxide conductivity and mechanical properties of quaternized chitosan/functionalized carbon nanotubes composite anion exchange membranes". Int. J. Hydrogen Energy, 44 (2019) 18134-18144. https://doi.org/10.1016/j.ijhydene.2019.05.102

[89] M. Rahimi, R. Ahmadi, H. Samadi Kafil, V. Shafiei-Irannejad, "A novel bioactive quaternized chitosan and its silver-containing nanocomposites as a potent antimicrobial wound dressing: Structural and biological properties". Mater. Sci. Eng. C., 101 (2019) 360-369. https://doi.org/10.1016/j.msec.2019.03.092

[90] T.D.A. Senra, S.P. Campana-Filho, J. Desbrières, "Surfactant-polysaccharide complexes based on quaternized chitosan". Characterization and application to emulsion stability. Eur. Polym. J., 104 (2018) 128-135. https://doi.org/10.1016/j.eurpolymj.2018.05.002

[91] P. Ramasamy, N. Subhapradha, T. Thinesh, J. Selvin, K.M. Selvan, V. Shanmugam, A. Shanmugam, "Characterization of bioactive chitosan and sulfated chitosan from Doryteuthis singhalensis". Int. J. Biol. Macromol. 99 (2017) 682-691. https://doi.org/10.1016/j.ijbiomac.2017.03.041

[92] Y. Yang, R. Xing, S. Liu, Y. Qin, K. Li, H. Yu, P. Li, "Immunostimulatory e_ects of sulfated chitosans on RAW 264.7 mouse macrophages via the activation of PI3K/Akt signaling pathway". Int. J. Biol. Macromol. 108 (2018) 1310-1321. https://doi.org/10.1016/j.ijbiomac.2017.11.042

[93] B.S. Atiyeh, M. Costagliola, S.N. Hayek, S.A. Dibo, Effect of silver on burn wound infection control and healing: review of the literature, Burns. 33 (2007) 139-148. https://doi.org/10.1016/j.burns.2006.06.010

[94] N. Akhavan-Kharazian, H. Izadi-Vasafi, Preparation and characterization of chitosan/gelatin/nanocrystalline cellulose/calcium peroxide films for potential wound dressing applications, Int. J. Biol. Macromol. 133 (2019) 881-891. https://doi.org/10.1016/j.ijbiomac.2019.04.159

[95] A. Ashori, R. Bahrami, Modification of physico-mechanical properties of chitosan-tapioca starch blend films using nano graphene, Polym.-Plast. Technol. Eng. 53 (2014) 312-318. https://doi.org/10.1080/03602559.2013.866246

[96] L. Lisuzzo, G. Cavallaro, S. Milioto, G. Lazzara, Layered composite based on halloysite and natural polymers: a carrier for pH controlled release of drugs, New J. Chem. 43 (2019) 10887-10893. https://doi.org/10.1039/C9NJ02565K

[97] F.C. Chiu, Halloysite nanotube-and organoclay-filled biodegradable poly (butylene succinate-co-adipate)/maleated polyethylene blend-based nanocomposites with enhanced rigidity, Compos. Pt. B-Eng. 110 (2017) 193-203. https://doi.org/10.1016/j.compositesb.2016.10.091

[98] V. Bertolino, G. Cavallaro, G. Lazzara, S. Milioto, F. Parisi, Biopolymer targeted adsorption onto halloysite nanotubes in aqueous media, Langmuir. 33 (2017) 3317-3323. https://doi.org/10.1021/acs.langmuir.7b00600

[99] Y. Lvov, R. Price, B. Gaber and I. Ichinose, "Thin film nanofabrication via layer-by-layer adsorption of tubule halloysite, spherical silica, proteins and polycations". Colloids and Surfaces A: Physicochemical and Engineering Aspects, 198 (2002) 375-382. https://doi.org/10.1016/S0927-7757(01)00970-0

[100] S. Levis and P. Deasy, "Use of coated microtubular halloysite for the sustained release of diltiazem hydrochloride and propranolol hydrochloride". International Journal of Pharmaceutics, 253 (2003) 145-157. https://doi.org/10.1016/S0378-5173(02)00702-0

[101] X. Sun, Y. Zhang, H. Shen and N. Jia, "Direct electrochemistry and electrocatalysis of horseradish peroxidase based on halloysite nanotubes/chitosan nanocomposite film". Electrochimica Acta, 56 (2010) 700-705. https://doi.org/10.1016/j.electacta.2010.09.095

[102] M. Liu, Y. Shen, P. Ao, L. Dai, Z. Liu and C. Zhou, "The improvement of hemostatic and wound healing property of chitosan by halloysite nanotubes". RSC Advances, 4 (2014) 23540-23553. https://doi.org/10.1039/C4RA02189D

[103] W.O. Yah, A. Takahara and Y.M. Lvov, "Selective modification of halloysite lumen with octadecylphosphonic acid: New inorganic tubular micelle". Journal of the

American Chemical Society, 134 (2012) 1853-1859.
https://doi.org/10.1021/ja210258y

[104] G. Chen, T. Ushida and T. Tateishi, "Scaffold design for tissue engineering",
Macromol. Biosci., 2 (2002) 67-77. https://doi.org/10.1002/1616-
5195(20020201)2:2<67::AID-MABI67>3.0.CO;2-F

[105] S. J. Hollister, "Porous scaffold design for tissue engineering", Nat. Mater., 4
(2005) 518-524. https://doi.org/10.1038/nmat1421

[106] D. W. Hutmacher, "Scaffolds in tissue engineering bone and cartilage",
Biomaterials, 21 (2000) 2529-2543. https://doi.org/10.1016/S0142-9612(00)00121-6

[107] M. Rinaudo, "Chitin and chitosan: Properties and applications", Prog. Polym. Sci.,
31 (2006) 603-632. https://doi.org/10.1016/j.progpolymsci.2006.06.001

[108] K. Rezwan, Q. Z. Chen, J.J. Blaker and A.R. Boccaccini, "Biodegradable and
bioactive porous polymer/inorganic composite scaffolds for bone tissue engineering",
Biomaterials, 27 (2006) 3413-3431. https://doi.org/10.1016/j.biomaterials.2006.01.039

[109] W.W. Thein-Han and R. D. K. Misra, "Biomimetic chitosan-nanohydroxyapatite
composite scaffolds for bone tissue engineering", Acta Biomater., 5 (2009) 1182-1197.
https://doi.org/10.1016/j.actbio.2008.11.025

[110] D. Depan, P. K. C. Venkata Surya, B. Girase and R. D. K. Misra,
"Organic/inorganic hybrid network structure nanocomposite scaffolds based on grafted
chitosan for tissue engineering", Acta Biomater., 7 (2011) 2163-2175.
https://doi.org/10.1016/j.actbio.2011.01.029

[111] J. Li, H. Sun, D. Sun, Y. Yao, F. Yao and K. Yao, "Biomimetic multicomponent
polysaccharide/nano-hydroxyapatite composites for bone tissue engineering",
Carbohydr. Polym., 85 (2011) 885-894. https://doi.org/10.1016/j.carbpol.2011.04.015

[112] L. J. Sweetman, S. E. Moulton and G. G. Wallace, "Characterisation of porous
freeze dried conducting carbon nanotube-chitosan scaffolds", J. Mater. Chem., 18
(2008) 5417-5422. https://doi.org/10.1039/b809406n

[113] X. Sun, Y. Zhang, H. Shen and N. Jia, "Direct electrochemistry and electrocatalysis
of horseradish peroxidase based on halloysite nanotubes/chitosan nanocomposite
film", Electrochim. Acta, 56 (2010) 700-705.
https://doi.org/10.1016/j.electacta.2010.09.095

Chapter 3

Montmorillonite-Chitosan based Nano-Composites and Applications

Rabinarayan Parhi*[1], Goutam Kumar Jena[2]

[1]Department of Pharmaceutical Sciences, Susruta School of Medical and Paramedical Sciences, Assam University (A Central University), Silchar-788011, Cachar, Assam, India

[2]Department of Pharmaceutics, Roland Institute of Pharmaceutical Sciences, Berhampur, Odisha, India

*bhu_rabi@rediffmail.com, goutam2902@gmail.com

Abstract

Chitosan is a biopolymer gaining widespread attention due to its astounding physicochemical characteristics and properties, including biocompatibility, mucoadhesive, biodegradability, low toxicity, and polycationic nature. Furthermore, chitosan can be processed to obtain different nanostructures such as nanoparticles, nano-vehicles, nanocapsules, scaffolds, etc. Montmorillonite (MMT) is a major constituent of bentonite clay, which has the ability to ameliorate the mechanical strength of polymers including chitosan when combined with it due to its layered structure. Bionanocomposite is a term used to describe composite materials which encompass component(s) with natural inception and the obtained particles with at the minimum one size ranging from 1 nm to 100 nm. Chitosan-based nanocomposite with MMT showed a proven record of wide applications in drug delivery, medical, biomedical, and pharmaceutical fields. This chapter provides an insight into the various characteristics and properties of chitosan and MMT, different methods to develop their composite, and various forms of the resulted composite along a thorough description of applications.

Keywords

Chitosan, Montmorillonite, Bionanocomposite, Biodegradable, Biocompatible, Scaffold.

Abbreviations

MMT: Montmorillonite
DNA: Deoxyribonucleic acid
DD: Degree of deacetylation

FTIR: Fourier-transform infrared spectroscopy
XRD: X-ray diffraction
MNR: Magnetic nuclear resonance
HPLC: High-performance liquid chromatography
CI: Crystallinity index
3D: 3-dimensional
SEM: Scanning electron microscopy
FDA: Food and drug administration
LD50: Lethal Dose 50
PBS: Phosphate Buffer Solutions
PVA: Poly(vinyl alcohol)

Contents

1.   Introduction

The importance of nature-derived polymers or biopolymers grows many-fold over the last two decades owing to their superiority with regard to biodegradability, biocompatibility and toxicity compare to synthetic or petrochemical products. There are four classes of biopolymers including (i) polysaccharides, (ii) proteins, (iii) poly(hydroxyalkanoates, and (iv) deoxyribonucleic acid (DNA) [1]. Naturally occurring polymers such as starch, collagen, gelatin, alginate, cellulose, and chitosan are renewable and sustainable, which can lower environmental risks with their faster degradation [2]. Among all, chitosan is a unique polysaccharide-based polymer with distinct properties such as biodegradability, biocompatibility, hydrophilicity, non-toxicity, abundance presence in nature, exceptionally high-level of affinity for various classes of clays, mucoadhesion, in addition to promising bioactivities including anti-bacterial, anti-fungal, and wound healing [3-5]. Furthermore, the presence of β-(1,4) glycosidic bonds between two units of chitosan such as D-glucosamine, and N-acetyl-D-glucosamine provide a platform for chitosan modification chemically, which led to higher elasticity, flexibility, and reduced anti-inflammatory response [6,7]. All these properties make chitosan a versatile and viable candidate for a broad range of applications ranging from cosmetics to aerospace [8].

Montmorillonite (MMT) is a major constituent of bentonite clay (about 60-95%) that develops when they precipitate as microscopic crystals from water solution [9]. MMT is a smectite and has a distinct layered crystal structure. It consisting of a multi-layered structure with silicon (Si)/aluminium (Al) oxide layers [10]. Bentonite is an already approved excipient, thus the use of MMT, particularly nano-sized ones, has attracted many researchers of the diverse field including biomedical and pharmaceutical field. This is because of its suitable qualities including biodegradability and biocompatibility, and improved mechanical properties when combined with other polymers. Furthermore, abundant availability and ease of nano-size conversion are important factors that made this clay more attractive. In addition, the characteristic multilayer structure of MMT provides a new avenue for sustained and targeted drug delivery in polymeric composite form. Other properties of MMT that suit their use in the synthesis of pharmaceutical products, including good water absorption, swelling, and cation exchange capability [11].

Composite materials are composed of more than one solid phases/materials with contrasting physical and chemical properties. These materials are commonly constituted by a polymer (natural or petroleum-based polymers) and an inorganic substance. When composites are developed with biopolymers such as alginates, chitosan, and starch, etc, called biocomposites. Nano-composites are composite materials where the inorganic component is included in the nanosized state or at the minimum one of the dimensions of the dispersed particles is in the nanometer scale. If the petroleum-based polymer is replaced with biopolymers, it gives rise to bionanocomposites [12]. Composite materials more particularly bionanocomposite exhibit enhanced properties including higher mechanical properties (strength, elastic modulus, and dimensional stability), packaging applications, thermal behaviour, swelling, water uptake, mucoadhesion, and rheology, which provides better biomedical and pharmaceutical applications [13,14]. The nanoparticles involved in nano-composites or bionanocomposites are generally of three types depending on the dimension numbers: (i) nanoparticles (spherical geometry having all three dimensions in the nanometer range), (ii) nanoplatelets (layered geometry with only one dimension i.e., thickness is in the nanometer range) and (iii) fibrous, comprised of nanotubes, nanofibers, nanorods, and whiskers with only two dimensions are in the nanometer range) (Fig. 1) [1]. MMT belongs to the nanoplatelets class with layered structure.

The incorporation of clay-based nanoparticles into biopolymer such as chitosan resulted in bionanocomposite that overcome the conventional drawbacks associated with chitosan and making it efficient to be used in many fields, including medicine, food industry, agriculture, biotechnology, textiles, cosmetics, environmental protection, and so on [15-17]. Nanoclays such as MMT diffuses in the chitosan matrix and develop ideally into an intercalation forming multilayer structure or exfoliation structure inside the chitosan matrix [18-20].

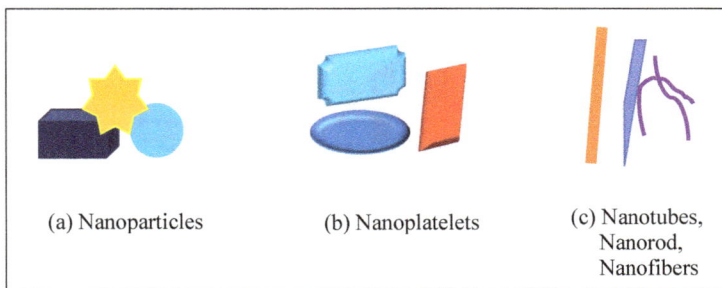

| (a) Nanoparticles | (b) Nanoplatelets | (c) Nanotubes, Nanorod, Nanofibers |

Fig. 1 Different types of nanostructures available to prepare bionananocomposites.

2.   Montmorillonite

2.1   Structure

MMT is a phyllosilicate with having plate-shaped multilayer nanostructure. It is comprised of two tetrahedral sheets with one octahedral sheet sandwiched in between (general structure 2:1 type). Each tetrahedral sheet holds oxide anions at its tip that are facing silicone (Si) atoms with O-Si-O orientation. The Si atoms are often replaced by aluminium and iron. On the other hand, the octahedral sheet has aluminium with oxide ion on either side as O-Al-O orientation. Here, the aluminium atoms are substituted with Si, Mg, iron, and zinc, and additionally surrounded the hydroxy (OH) atoms available at the axial end of tetrahedral planes [21]. The layers are held together primarily by van der Waals force and electrostatic force to form a primary particle of MMT, followed by their aggregation to form secondary MMT particles with size ranging from micrometer scale to millimeter-scale. MMT possesses a net negative charge on its surface due to the isomorphs ionic substitutions in its layered structure with negative oxide ions dominate the cations such as Si, Al, Fe Fe, and Mg present in the interface [22,23].

Like other mineral clays, MMT can combine with polymers that resulted in nano-composite with improved properties. Now it is a major source of material for biomedical and pharmaceutical industries [24]. When combined with polymers, MMT interacts with polymer and may be arranged in three possible manners: tactoid structure, intercalated structure, and exfoliated structure (Fig.2).

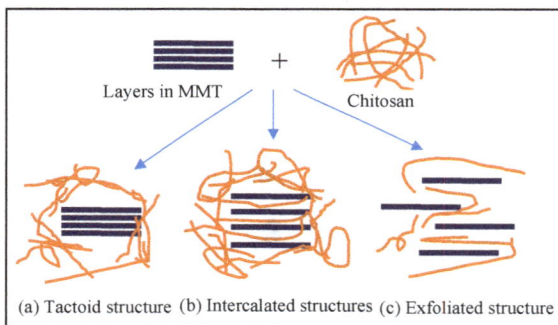

(a) Tactoid structure (b) Intercalated structures (c) Exfoliated structure

Fig. 2 Schematic representation of different nano-composite structures originated from MMT and chitosan.

Tactoid structure: this structure is obtained when the polymer is unable to intercalate in the interlayer space between the unexpanded MMT layers. Usually, this happened due to poor affinity between MMT and polymer. In this, the MMT is dispersed in the polymer matrix and the resulted material is not a true nano-composite.

Intercalated structure: unlike the tactoid structure, this structure expands slightly which resulting in the entry of polymeric chains in to the basal spacing of the clay. However, the layered structure remains intact. There is a moderate affinity between the MMT and polymer.

Exfoliated structures: in this, the layered structure of MMT is lost and a random arrangement of MMT platelets within the continuous polymer phase take place. This is the result of high affinity between MMT and polymer. It is considered as an ideal nano-composite disposition but is difficult to achieve during synthesis or processing [23,25,26].

2.2   Properties of MMT

The accomplished properties of MMT that make it suitable for various applications are sorption, hydration, drug loading, drug release, slightly negative charge on the surface, ion exchange capacity, and interlayer spacing of layers. MMT has overwhelming sorption property with sorption sites residing on the outer surface and edges along with its interlayer space. MMT may contain either Ca or Na based on the place of origin. Both Ca-MMT and Na-MMT have enhanced hydration ability compared to MMT without these ions. However, Na-MMT has more hydration capacity than that of Ca-MMT, which resulted in a higher possibility of delamination of layers to favour exfoliated structure. In addition, the latter exhibited better drug loading as well as drug release. Furthermore, a higher aspect ratio in addition to higher surface area encourage better drug loading. The factors such as pH and ionic concentration of the dispersed medium are found to influence the adsorption and drug release from MMT. The interlayer spacing of nanosheets is in the range of 0.9-1.2 nm which is enough to accommodate organic compounds in its structure [11].

3.   Chitosan

3.1   Structure

Chitosan is not abundantly available on earth. However, it is basically extracted from chitin by partial deacetylation. Chitin is abundantly present polysaccharides in the earth which is only second to cellulose. Chitosan is a $\beta(1\rightarrow4)$ linked semicrystalline, linear co-polymer comprised of 2-acetamido-2-deoxy-D-glucopyraose and 2-amino-2-deoxy-D-glycopyranose. It has two types of functional groups: amine (-NH$_2$) group present on second carbon atom, and hydroxyl (-OH) group present on third carbon (primary -OH groups) as well as on sixth carbon (secondary -OH groups) [27]. Chitin is obtained from marine invertebrates including shrimps, lobsters, oysters, and crabs, and certain fungi and insects. Chitin is insoluble; however, chitosan is soluble in diluted acetic acid, lactic acid, and formic acid [28]. The degree of deacetylation of chitin is a pivotal step that significantly influences various physio-chemical properties of chitosan. The deacetylation of chitin is performed under a severe alkaline condition in the presence of a particular enzyme (chitin deacetylase) [29,30]. In addition, the deacetylation process is conducted under the nitrogen environment or by adding sodium borohydride with NaOH to avoid undesirable side

Adv. App. of Micro and Nano Clay – Biopolymer-based Composites   Materials Research Forum LLC
Materials Research Foundations **125** (2022) 49-86   https://doi.org/10.21741/9781644901915-3

reactions. Chitosan has been used in various fields as a core material because of a number of suitable properties mentioned earlier.

3.2   Characteristics of chitosan

It is very essential to know the characteristics of chitosan as these affect its properties and application. The attributes of chitosan are usually dependent on the source and the involved processing. The important characteristics which establish the quality and application of chitosan are the degree of deacetylation (DD), molecular weight, crystallinity, surface area, and particle size (Fig.3) [31].

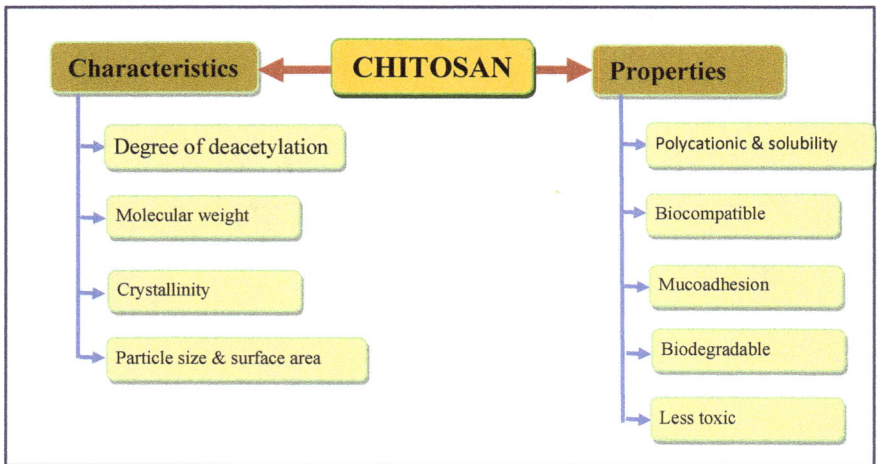

Fig.3 Various characteristics and properties of chitosan

3.2.1 Degree of deacetylation

The DD involves the deletion of an acetyl group (CH_3CO-) is a remarkable chemical attribute of chitosan. In the process of deacetylation, an amino group from within the polymeric chain replaces the acetyl group leading to the generation of D-glucosamine and N-acetyl-glucosamine. In case of chitosan, DD is calculated by dividing combined units of D-glucosamine and N-acetyl D-glucosamine with the number of D-glucosamine units available in the polymeric chain. Therefore, DD of 50 % indicate that the chitosan comprises 50% each of N-acetyl D-glucosamine and D-glucosamine units [32]. When the DD exceeds 50% is typically called chitosan [28] and the commercial chitosan has a DD of 85% [33]. The methods such as Fourier transform infrared spectroscopy (FTIR), X-ray diffraction (XRD) potentiometric titration, and magnetic nuclear resonance (MNR) can be used to measure DD [31,34,35].

3.2.3 Molecular weight

The molecular weight of chitosan is basically depending on the number of monomeric units of the copolymer. The molecular weight of chitosan typically ranges from 300 to 1000 kDa [36]. Based on the molecular weight, chitosan is classified as high, medium, and low molecular weight chitosan. Various properties including viscosity, solubility, and applications are influenced by the variable molecular weight of chitosan [37]. The molecular weight of chitosan can be measured using techniques such as the viscometric method, light scattering, and high-performance liquid chromatography (HPLC) [38].

3.2.4 Crystallinity

The crystallinity of chitosan is quantified as crystallinity index (CI) [39]. Commercial chitosan is present in solid-state and exhibits semi-crystallinity [40]. Chitosan can also exhibit polymorphism and has an orthorhombic unit cell structure. Crystallinity is found to be maximum for fully deacylated chitosan. This crystallinity affects various properties of chitosan such as porosity, hydration, sorption, and swelling [41]. Crystallinity can be obtained by taking XRD of chitosan powder [35].

3.2.5 Particle size and surface area

Chitosan particles of lower than 1 mm are generally used in most applications [42,43]. Chitosan being non-porous, is available generally in powder or flakes form and has a low surface area (< 10 m^2/g). Chitosan can be modified for its non-porous character to enhance its surface area. These characteristics are more relevant when the applications such as adsorption, enzyme immobilization are considered for chitosan [42,44]. Particle size can be determined using scanning electron microscopy (SEM), particle size analyser, etc, whereas N2 adsorption-desorption isotherms are used to measure the surface area of chitosan [41,42].

3.3   Properties of chitosan

3.3.1 Polycationic and solubility

Chitosan has polycationic properties at low pH, the H$^+$ ions present in the solution protonate the NH$_2$ groups of chitosan resulting in the formation of (NH$_3$)$^+$. But at higher pH (>6) deprotonation takes place and chitosan becomes insoluble in water [37]. Due to this polycationic nature, chitosan is soluble in dilute acid solutions including acetic acid (1-3%) and citric acid (3-4%) [45]. Apart from polycationic properties, pKa value and

Adv. App. of Micro and Nano Clay – Biopolymer-based Composites   Materials Research Forum LLC
Materials Research Foundations **125** (2022) 49-86   https://doi.org/10.21741/9781644901915-3

number of NH_2 groups, deacetylation degree, crystalline structure also influences the solubility of chitosan [46,47]. Chitosan is a strong base due to the presence of primary NH_2 groups with a pKa value of 6.3. If the number of NH_2 groups is more, the greater the protonation resulting in higher electrostatic repulsive force between chains. This led to increase in solvation and solubilization of chitosan in an aqueous solvent. Nonetheless, with the increase in pH of the solution to 6, already solubilized chitosan get precipitate [48,49]. As per as the DD is concerned acetylated groups in chitosan are abundant in H-bonding. Therefore, with the increase in acetylated groups in chitosan, the dissolution rate in aqueous medium was found to be decreased. A minimum DD required for its solubility in dilute acid solution is 40-60% [50]. The crystalline structure of chitosan inversely influences its solubility in aqueous solution because of the presence of intra- and inter macromolecular H-bonds in the solid-state [51].

3.3.2 Biocompatibility

Biocompatibility of chitosan is its compatibility with the living tissue or a living system without being toxic or injurious to it and not prompting any immunological reactions both in *in vitro* and *in vivo* [31,52]. Mentioned property made chitosan not only a suitable but also preferred base material for its application in diverse fields including drug delivery, wound dressing, tissue engineering, vaccine and gene delivery, and nasal drug delivery. The biocompatibility of chitosan in its derived materials can generally be measured by using a cytotoxicity test *in vitro* [53] and by genetic toxicity test or intramuscular implantation test *in vivo* [54].

3.3.3 Mucoadhesion

The polycationic property of chitosan provides space for interaction with the anionic mucosa layer in hydrated conditions. The interaction is due to non-covalent bonding such as H-bond and electrostatic interaction [55,56]. Chitosan can adhere not only to mucous but also to hard tissues such as epithelial tissue of skin due to its high positive charge density. As a result, the chitosan-based drug delivery systems will be able to adhere for a prolonged time to the biological tissues and thereby improving the penetration of drug across the membrane [57].

3.3.4 Biodegradability

When a polymer is considered to be used as a drug delivery vehicle, its metabolic fate or biodegradability in the body is given more importance. In this sense, chitosan is a biodegradable polymer because it undergoes hydrolysis by lysozyme present in different body fluids including serum and tears. Lysozyme works by splitting $\beta(1 \rightarrow 4)$ bonds that exist between N-acetyl-D-glucosamine units [58]. Besides lysozyme, enzymes such as chitotriosidase, N-acetyl-β-D-glucosaminidase and acetylchitobiase present in physiological fluids can degrade chitosan [59]. Chitosan can also be degraded by thermal, acidic, and irradiation method [60]. The rate and degree of chitosan degradation *in vivo* is dependent on the DD. Biodegradation of chitosan resulted in the formation of aminosugars, which can be utilized in the metabolic pathways of glycosaminoglycans and glycoproteins in the body [61,62].

3.3.5 Low toxicity

Chitosan has been approved by countries like Japan and Italy to be used as dietary additives. In addition, chitosan is approved by the Food and Drug Administration (FDA) for its use as wound dressing materials [63]. More importantly, chitosan has been approved as a pharmaceutical excipient that led to its wide use in drug delivery [64]. In one study, it was reported that chitosan has a lethal dose of 50 (LD50) comparable to sucrose when orally administered to mice [65]. Many other toxicity and allergic studies were conducted on chitosan, however, the allergic or toxicity was found to be negligible or minimum [64,66,67]. The toxicity of chitosan depends on its origin and process of obtention.

4.　　Montmorillonite-chitosan based nano-composites

Blending MMT with chitosan is a promising way of improving the structural and functional behaviour of chitosan. As mentioned earlier, MMT can diffuse in the chitosan matrix resulting in either of the three structures such as tactoid, intercalation, and exfoliation. The imperfection in the crystal lattice and isomorphous replacement of MMT leads to distribution of negative charges within the plane that was counter-balanced by the positive charge's adsorption on alkaline earth metal ions present in the space between the layers [68]. Mentioned imperfection played the main role in the exchange and activity with organic compounds. In addition, MMT also has a large surface area that demonstrates superior cation exchange capacity along with adhesive capacity, adsorption ability, and drug-carrying properties [69]. In the case of chitosan, the presence of chemically active groups such as NH_2 and OH are vulnerable to modification. Thus, nano-composite of chitosan with MMT may result in ionic exchange and roughness modification of MMT surface [70].

5.　　Processing of MMT-chitosan-based nano-composites

Formation of MMT-chitosan based nano-composite depends on different factors of MMT including sorption, hydration, drug loading, drug release, slightly negative charge on the surface, ion exchange capacity and interlayer spacing, with the particle size and shape, and

59

chitosan including polycationic nature, concentration, molecular weight, etc. Apart from this, the kind of solvent, temperature, and pH of the preparation medium played a crucial rule in the processing of nano-composites based on MMT-chitosan. Followings are the most common methods adopted to develop MMT-chitosan-based nano-composite.

5.1   Intercalation of polymer or pre-polymer from solution

In this method, a suitable solvent is selected in which the polymer is soluble (or pre-polymer if the polymer is insoluble, e.g. polyimide). In the first step, the MMT is dispersed in the selected solvent where the staked layers are converted into individual silicate layers, owing to the weak force of attraction between them. MMT layers are allowed to swell in the solvent and then, the polymer is mixed in the resulted solution which resulted in the intercalation of polymer (in this case chitosan) between the clay layer forming either intercalated or exfoliated nano-composite. Removal of solvent is performed in the last step by using either vaporization under vacuum or by precipitation. The same steps are adopted to generate nano-composite by emulsion polymerization method with the use of aqueous phase. This method has the advantage of forming nano-composites with few polar or non-polar polymers. However, this method has limited industrial applications as it is associated with a large amount of solvents [23].

5.2   In situ intercalative polymerization

In situ intercalative polymerization is otherwise known as interlamellar polymerization. It involves swelling of the layered silicates with the absorption of solution-containing monomer units. This resulted in the intercalation of monomer units between the swollen MMT sheets. Then, the polymerization process is carried out with radiation or heat or organic initiator or catalyst, leading to the formation of nano-composite [23,26,71].

5.3   Melt intercalation

In this method, a molten mass of selected polymer is formed by heating it at a specific temperature and then the MMT is added to the above solution that led to the filling of the interlayer space available between MMT layers. This method has the advantage of forming either an intercalated or an exfoliated nano-composite. In addition, there is no requirement of solvent making the process solvent-free [1,23,26,72].

5.4   Solution intercalation

Solution intercalation method involves preparation of a solution of biopolymer or bio-prepolymer (such as chitosan in a suitable solvent) and swelling of MMT in another solvent such as water, toluene, or chloroform. After mixing the solutions of biopolymer and MMT,

Adv. App. of Micro and Nano Clay – Biopolymer-based Composites   Materials Research Forum LLC
Materials Research Foundations **125** (2022) 49-86   https://doi.org/10.21741/9781644901915-3

the chitosan chains intercalate and displace the solvent used to swell MMT within the interlayer. By removing the solvent from the mixture, the intercalated structure remains, leading to the formation of bionanocomposite [1,26,72].

5.5   Template synthesis

Here, the polymer initiates nucleation followed by the growth of inorganic crystals because of self-assembly forces, when present in the middle of MMT layers. The most frequently used technique to obtain nano-composites is polymer intercalation in layer solid and the sol-gel. The polymer interaction that happened in the former case is due to the formation of H-bond, electrostatic force, dipole coordination, etc, between polymer and MMT [73]. This method is widely used for the preparation of double-layer hydroxide-based nano-composites. Advantages of this method is simple and versatile and suitable for large-scale production [1,74,75].

5.6   Solvent casting method

Solvent casting method is widely employed to develop chitosan-based films. This method is accomplished in the following steps: (1) solubilizing chitosan in acidic solution, (2) mixing with other material such as biopolymer and functional materials/fillers including MMT, (3) removal of any insoluble particles and air bubbles remaining by filtration or centrifugation, (4) casting onto a level surface, (5) removal of the solvent by drying at the fixed condition of temperature, relative humidity and (6) finally, removal from the level surface to get bionanocomposite film. This method has many advantages such as being convenient, simple, and inexpensive. However, the main hurdle of this method is scaling-up of the process of film-formation [76,77].

5.7   Melt-extrusion method

Like the solvent casting method, the melt-extrusion method is also employed to develop nano-composite film and is preferred over the solvent casting method because the melt-extrusion method has faster processing time and the requirement of low energy [78]. Here, the steps are: (1) development of formulations with different compositions of raw materials, (2) mixing the blend with the help of an extruder to form a uniform mixture, (3) the resulted extrudates is converted in to pellets using pelletizer, (4) pellets are dried with using hot air oven, (5) pellets are extruded again to obtain sheets, and (6) blowing of extrudates into a film by using a blown film extruder. This method provided nano-composite films with desired mechanical attributes and better stability towards variable temperature. However, only a limited number of films with chitosan were developed using this technique [79].

5.8   Layer-by-layer assembly (LBL) method

LBL assembly film deposition is a very flexible method which is used most often to develop nano-composite films [32]. This technique involves surface modification that is accomplished with both deposition and mutual attraction existing between each alternating polyelectrolytic layer [80,81]. A biopolymer such as polysaccharides or proteins with net charge is considered as polyelectrolytes. The mutual attraction includes H-bonding, hydrophobic interaction, van der Waals forces along electrostatic interactions between charged polymers. As per as chitosan-clay nano-composite is concerned, chitosan exhibited hydrophilic as well as cationic properties in acidic pH. Thus, it can mix well with MMT (negatively charged) that resulted in the manifestation of electrostatic interactions and H-bonding due to the existence of -OH groups and charged oxygen on its surface [82]. The film deposition can be performed either by using submersion or spraying of solutions onto the substrate. This method has the advantage of forming films with multi-components which do not necessitate any modern instrument. However, there are only a few reports on the preparation of nano-composite films having MMT and chitosan combination [82].

6.   Various forms of Montmorillonite-chitosan based nano-composites

6.1   Film

Nano-composite films are comprised of thin and flexible polymer (e.g., chitosan) layers with nanomaterials (e.g., MMT). These are having various applications including drug delivery, packaging material, etc. The most common method used to prepare nano-composite film is solvent casting [83-85].

6.2   Hydrogel

Hydrogels are comprised of three-dimensional (3D) networks of polymer capable of imbibing water or body fluids in large amounts. It has a number of qualities such as soft consistency, flexibility, biocompatibility that made it widely acceptable in drug delivery applications [83]. In the case of chitosan-MMT hydrogel, the drug release is mainly affected by the cross-linking density between chitosan and MMT [86]. It was also reported that the electrostatic interaction between chitosan and MMT made the hydrogel-based formulation stable and provide sustained drug release [86].

6.3   Scaffold

Scaffolds are basically 3D artificial structures that can perform multiple functions such as cell attachment, providing support for new-tissue formation, and diffusion of nutrients into the cell. The ideal attributes of a scaffold include biodegradability, biocompatibility, and

Adv. App. of Micro and Nano Clay – Biopolymer-based Composites   Materials Research Forum LLC
Materials Research Foundations **125** (2022) 49-86   https://doi.org/10.21741/9781644901915-3

better mechanical properties with porous structure. Scaffolds provide a platform for the regeneration of damaged tissue in our bodies [2]. Scaffolds can be developed with only polymer(s) or polymer with other materials as hybrid. In this context, chitosan is a promising polymer for use as scaffold material in various tissue engineering applications. Furthermore, chitosan with other organic and inorganic materials such as MMT made it more suitable for tissue engineering such as bone, cartilage, liver, brain, and skin [87].

7.   Important applications of Montmorillonite-chitosan based nano-composites

Chitosan is a natural polysaccharide that is available in sufficient quantities in nature. It is biodegradable, biocompatible, and non-toxic and it tends to form films having antimicrobial properties. The films made by chitosan have more permeability to water vapours owing to their hydrophilic nature. The chitosan films have less mechanical strength and also possess barrier properties. Silicate-derived clay (e.g. MMT) has nanoscale particles and also possesses a layered structure. Hence the barrier, as well as mechanical attributes of films developed with chitosan can be enhanced by introducing MMT clay [88]. The hybrid structure of chitosan and MMT has many advantages and hence can be applied in various fields for various purposes which are be portrayed in Fig. 4.

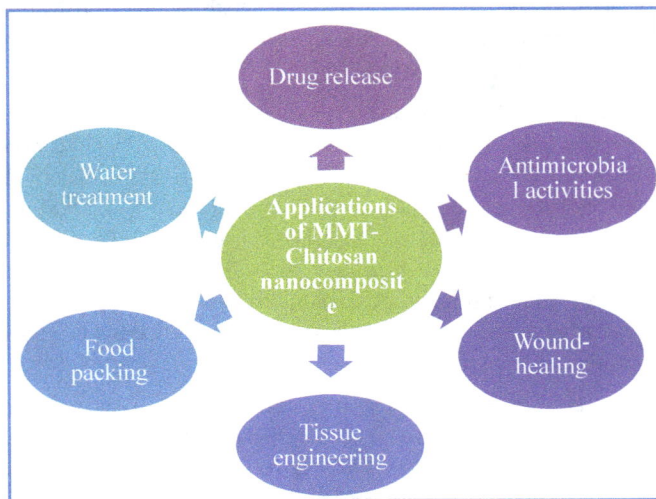

Fig. 4 Different applications of MMT-chitosan-based nano-composite.

Adv. App. of Micro and Nano Clay – Biopolymer-based Composites   Materials Research Forum LLC
Materials Research Foundations **125** (2022) 49-86   https://doi.org/10.21741/9781644901915-3

7.1 Application of MMT-chitosan based nano-composites in drug release

When chitosan and MMT combined leading to the formation of bionanocomposites, which have attracted huge attention in pharmaceutical applications. Due to the biodegradable, biocompatible, and less toxic nature of the chitosan and MMT, they are regarded as safe, renewable, and eco-friendly compounds. These distinguishing characteristics of the hybrid material compel them for various uses including drug delivery, healthcare, and especially in the development of different pharmaceuticals. For effective treatment of any disease, the drug should be released from the formulation in a controlled manner and reach the target site at inadequate concentration. Drug delivery systems are principally designed in such a way that the drug concentration should remain within the therapeutic range over a prolonged period. To achieve this target, chitosan-MMT nano-composites can be employed as controlled drug delivery carriers due to their unique functional and structural characteristics. Chitosan-MMT nano-composites materials are obtained more specifically from organic and inorganic solids interacting at the nanoscale to develop critical formulations for controlled drug delivery matrixes. Particularly, composites based on biopolymer and clay combination showed great promise in the preparation of key formulations for tissue engineering, biomedical applications and controlled drug delivery systems. The nanohybrid materials are derived from both organic and inorganic solids interacting at the nanoscale level. The following two techniques are used to predict drug release from MMT-chitosan nano-composites:

(i) Dialysis bag technique: Dialysis is a very simple and widely used technique in which solute molecules move across the semi-permeable membrane from higher concentration to lower concentration until attainment of equilibrium. The dialysis membrane can selectively pass small molecules but retain only large molecules. Hence, it is the best technique to determine drug release [88]. During the experiment using the dialysis membrane, the chitosan-MMT nano-composites are dispersed in PBS followed by sealing in a dialysis bag that is submerged in a suitable solution. A fixed volume of sampling is done at a specific interval of time followed by the replacement of same volume of freshly prepared diffusion media with appropriate temperature.

(ii) Franz-diffusion cell: The drug diffusion study of nano-composite is also carried out by using Franz-diffusion cell having PBS as drug diffusion media. Dialysis membrane having molecular weight cut-off 10kDa is placed in between donor and receptor compartment cell. A magnetic bead is placed in the receptor compartment and the total assembly is kept on a magnetic stirrer. At regular intervals, sampling is done using a syringe and the exact volume is replaced with freshly prepared diffusion media.

In the last few years, bionanocomposite comprised of chitosan and MMT is being used as a carrier to deliver the drug into the body because they are capable of decreasing side effects of the drug and release the drug in sustained manner. One such study involved the fabrication of sustained-release tablets of aceclofenac. The interaction between -OH groups of MMT and $-NH_2$ groups of chitosan was found to be responsible for the formation of the intercalated structure of the composite. The resulted nanocomposites demonstrated better compression and drug release in sustained manner. The latter was due to the combination of diffusion of the drug across the well-formed matrix and the relaxation of the chain of polymer matrix [89]. The control delivery system in the form of enteric-coated capsules for quinine was successfully fabricated to deliver it in the colon. Here, a nano-composite containing quinine was developed using chitosan and MMT, and then incorporated the resulted nano-composite in the gelatin capsules followed by its coating with eudragit L100. The enteric-coated capsule was found to release the drug in the colon and effective against protozoa [90].

Chitosan and MMT-based composite microsphere were fabricated for tanshinone IIA (isolated from the rhizome of Chinese medicine plant *Danshen*) to avoid its poor bioavailability. The incorporation of MMT into the chitosan matrix enhanced the drug encapsulation and lower the drug release. The microspheres developed with a ratio of chitosan:MMT (10:2) demonstrated the highest encapsulation and slowest cumulative drug release. The microspheres did not show any cytotoxicity dosage below 80µg/ml [91]. Nanoparticles loaded with curcumin were fabricated with chitosan, carboxymethyl starch, and MMT employing the ionic gelation method with the intention to reduce dental bacterial biofilm. MMT was found to enhance drug entrapment efficiency and influence the particle size of nano-composite. In addition, polysaccharide concentration, the ratio of chitosan and starch, and sonication time had affected the particle size significantly. The antibacterial activity of the nanoparticles significantly reduced the formation of biofilm of *Streptococcus mutans* in the teeth [92]. In another study, chitosan and MMT-based nanosystems were developed for sustained release of antiseptic chlorhexidine. The developed nanosystem showed good mucoadhesive properties and was able to adhere to the oral mucosal layer for a considerable period of time, thereby showing sustained drug release. The nanosystem also showed activity against bacteria of either type, molds, yeasts, and viruses, that can be sufficient enough to reduce microbial count in the mouth cavity [93].

A nano-composite film comprised of chitosan and MMT was prepared using cross-linking reaction involving glutaraldehyde for the control release of podophyllotoxin (a cytotoxic drug). The formation of the exfoliated structure of MMT in the matrix and electrostatic interaction between MMT and chitosan improved the swelling behavior, enhanced the drug entrapment efficiency, and decrease the drug release rate [94]. In one study, 5-flurouracil loaded nano-composite film was developed using polymer matrix composed of chitosan and alginate, and MMT as nano-fillers. The films were developed with solution blending method and showed excellent drug loading and release efficiency with 30 wt% loading of MMT. However, there was a significant burst release of the drug with the nano-composite films [95]. In another study, chitosan-MMT-based nano-composite films loaded with diclofenac sodium were developed employing *in situ* solution casting method. Drug incorporated nano-composite films demonstrated controlled drug release up to 72 h in pH-dependent manner. In addition, the films showed excellent antimicrobial activity [96]. 5-amino salicylic acid-loaded nano-composite films were fabricated with chitosan and MMT employing both solution casting and melt mixing methods. The resulted films fabricated by both methods demonstrated prolonged drug release up to 80 h with drug release efficacy of more than 90%. Further, the films showed anticancer efficiency with more than 50% of cell death when human cell lines were used [97].

A nano-composite hydrogel was developed using chitosan as matrix former and MNT as filler for encapsulation of vitamin B12. The release of vitamin B12 is strongly influenced by the cross-linking density between chitosan and MMT, which in turn depends on the presence of cross-linker i.e., the exfoliated structure of MMT. Electrostimulation induces swelling-controlled drug release mechanism, instead of diffusion-controlled drug release in absence of electrostimulation. At 2wt% of MNT in the hydrogel was found to be desirable as per the mechanical property and pulsatile release profile is concerned [98]. In another investigation, ofloxacin-loaded hydrogel containing MNT incorporated chitosan beads were developed. The beads showed decreased drug release which was evident from incomplete disintegration even after 10 h of disintegration study. This result was attributed to electrostatic interaction between chitosan and MNT that enhanced the stability of beads and demonstrated sustained drug release [99].

7.2   Application of MMT-chitosan based nano-composites in antimicrobial activities

Chitosan is a known biopolymer having effective antimicrobial activity. The mechanism of this antibacterial activity is divided into two ways: (i) the positively charged active sites present in chitosan ($-NH_2$) are responsible for antimicrobial activities that interact with bacterial cell wall with negatively charged and thereby killing the bacteria, (ii) The penetration of chitosan (low molecular weight) and its oligomers inside the nuclear core of the cell leads to the inhibition of ribonucleic acid and thereby, disruption of protein synthesis [100]. In addition, the chitosan in the form of nano-composites furthers the levels of enhanced competence. The antimicrobial attributes of chitosan are very vital and therefore it is widely applied in biomedical, cosmetic, food as well as in agriculture. Due to antimicrobial property and most peculiar attributes, chitosan is used for the production of self-preserving material that are under investigation [32]. These principle-based products containing chitosan are designed in the form of films, hydrogels, fibers, and membranes meant for several applications. Followings are some of the important case studies.

In one study, 2D nanomaterial of MMT (delaminated MMT particles) was used to form nano-composite with chitosan to enhance antimicrobial and biocompatibility properties. The resulted nano-composites demonstrated better mechanical strength and good dispersion of MMT particles in the chitosan matrix, which was attributed to the charge interaction between the components. With the addition of MMT into chitosan matrix the antimicrobial activity was found to be augmented and the highest activity was seen when the MMT concentration reached 10^3 ppm. The biocompatibility and biodegradation were also improved with the incorporation of MMT into the chitosan matrix [101].

A novel chitosan and MMT-based nano-in-microparticle platform was fabricated to overcome the problems related to bacterial resistance. Firstly, chitosan was PEGylated to circumvent the problem of poor aqueous solubility and then conjugated with dendritic polymidoamine hyper branches, followed by *in-situ* immobilization of silver nanoparticles. Secondly, MMT was used to encapsulate the anti-inflammatory drug ibuprofen. The former (i.e., silver nanoparticles) demonstrated significant antibacterial properties against multiple bacterial strains including both aerobic and anaerobic bacterial species, whereas the latter showed anti-inflammatory activity against any possible concomitant inflammatory response during bacterial infection [102]. In another study, a novel antibacterial paper was fabricated with silver nanoparticles-loaded quaternized carboxymethyl chitosan and MMT nano-composite using both surface coating and internal additive methods. Here, MMT is used as antimicrobial agent and modified chitosan as a reducing agent for silver nanoparticles. The paper fabricated using the surface coating method exhibited better tensile, tear, and bursting strengths compared to the paper developed using internal additive method. In addition, the paper developed with both methods showed excellent antimicrobial capacity. However, the paper developed with the surface coating method demonstrated higher antimicrobial activities [103]. In another investigation, the impact of poly(vinyl alcohol) (PVA) on the chitosan and MMT-based nano-composite in terms of mechanical and antimicrobial properties was studied. In this, nano-composite films were fabricated using PVA, MMT (with both hydrophilic and organo-modified types), and chitosan employing the method called reflux-solution-heat pressing. The concentration of PVA at 20 and 30% demonstrated augmented elongation at break and declined stiffness and strength. The antimicrobial activity of the film was not affected by the incorporation of PVA into the film, rather the water and oxygen barrier properties were found to be increased. The presence of both types of MMT in films has exerted better stiffness, enhanced barrier as well as antimicrobial properties [104].

7.3   Application of MMT-chitosan based nano-composites in wound-healing activities

The complexity of wound-healing appears a great challenge to generate skin tissues by integrating anatomic as well as functional activities after tissue injury. In last few decades, work has been extensively done for constructing nano-composites by combining organic polymers with inorganic materials to improve the process of healing. Wound healing is an intricate process which starts automatically following injury and stops with curing wounds. This event consists of the following steps. Step I: Hemostasis and inflammations, Step II: Cellular proliferation and new tissue formation, and Step III: Tissue remodelling [105]. The process of wound healing depends on many factors such as chemokines and cytokines that are involved in the complex integration of signals coordinating multiple cellular behaviours. The nano-composite biomaterials are employed effectively for the generation of skin tissue due to their most desirable properties including antibacterial activity, enhanced angiogenesis, and mechanical attributes.

To prevent cytotoxic effects towards fibroblasts and keratinocytes and to maintain antimicrobial efficacy, silver sulfadiazine was incorporated into chitosan and MMT-based nano-composite. The resulted nano-composites were intended to be used in the form of powder or dressing for the treatment of skin ulcers. The nano-composites demonstrated better biocompatibility, enhanced wound healing (formation of fibroblast) with maintained antimicrobial properties of silver sulfadiazine, particularly against complex bacteria *Pseudomonas aeruginosa* [106]. To improve the mechanical strength of polymeric hydrogels that are used on the body under-tension, a nano-composite hydrogel based on PVA and chitosan as polymers and MMT as nano-fillers was fabricated for wound dressing applications. Nano-composite hydrogel with 3wt% of incorporated MMT exhibited the highest improved mechanical properties, which is ideal for use as dressing material on wounds under high stress. In addition, combined properties of both PVA and chitosan such as biocompatibility, antimicrobial activity, blood coagulation, swelling, and stimuli response behavior were also with the nano-composite hydrogel [107].

7.4   Application of MMT-chitosan based nano-composites in tissue engineering

For promoting regrowth, regenerating tissue in *in-situ*, and filling irregular defects, chitosan-MMT-based hydrogels can be injected. There is a great challenge for controlling stem cells in a 3D microenvironment for bone generation. It was reported that a microporous networked photo-cross-linkable chitosan *in situ* forming hydrogel was developed by adding 2D nanoclay particles [108]. The nanosilicates present in hydrogels increased Young's modulus and the rate of degradation. These nanosilicates also demonstrated that the reinforced hydrogels promoted proliferation and attachment. In addition, they prompted the differentiation of mesenchymal stem cells *in vitro*. Furthermore, they investigated the influence of the resulted hydrogels *in vivo* using mouse model. The results revealed that the hydrogel induces better healing without delivering therapeutic agents or stem cells.

Nano-composites based on chitosan and MMT in scaffolds forms have been used widely in bone tissue engineering. Few important investigations are mentioned here. A nano-composite scaffold was fabricated with MMT, hydroxyapatite, and chitosan using microwave irradiation and gas foaming methods. The resulted scaffold was found to exhibit the synergistic effect of hydroxyapatite and MMT on the mechanical and other biological properties such as degradation, protein adsorption, bioactivity, and swelling. Moreover, the scaffolds were found to be non-toxic to MG63 osteoblast cell lines. The above results indicated that the scaffold has promising potential in non-load-bearing bone tissue engineering [109]. In another study, xylan, chitosan, and MMT-based nano-composite scaffold with highly porous structures were fabricated by the freeze-drying process to further their properties for bone tissue engineering application. A study on the resulted scaffold with 5% MMT showed needle-like morphology of deposited apatite that reflected the synergistic response in increasing the mechanical properties when in contact with body fluid. This resembled the natural bone apatite. In addition, the scaffolds were non-cytotoxic in nature and showed good cell viability [110].

7.5   Application of MMT-chitosan based nano-composites in food packaging

The more demands, as well as preference for convenience by the consumers for food throughout the year have increased interest in developing the latest technology for the packaging of foods in order to safeguard accessibility of food with healthy and safety. The key objectives of food packaging involve the segregation of food items from the environment prevailing around, avoiding exposure to elements that spoil food such as the influence of microorganisms, temperature, oxygen as well as humidity to prevent the wastage of nutritional value and quality and to extend their self-life. Petroleum-based synthetic plastics have been used traditionally in food packaging industries and also takes lion's share of the total market because of their inexpensive, low weight, better mechanical strength, and highest grade of barrier attributes [32]. Although plastic in the packaging industry has played a significant role, it has harmful effects on the environment. The key challenges for food manufactures are to maintain the quality of foods with free from any types of synthetic chemicals. Hence, alternative for plastics has been tried as a safe packaging material. As chitosan is a biopolymer, it has been regarded as favourable substance as antimicrobial agent in food industries. Hence, chitosan-MMT nano-composite has been chosen as one of the best packing materials in food industries. There is a vast number of investigations have been conducted on the performance of chitosan and MMT-based nano-composite or by incorporating additional material into nano-composite as food packaging material. The followings are the few important studies discussed below.

A film based on chitosan and MMT loaded with tocopherol was fabricated to investigate its potential for active food packaging materials. The resulted nano-composite film with loading up to 10wt% of tocopherol influenced the better intercalation of MMT layers in the chitosan matrix. This resulted in the increased hydrophobicity, reduced water vapour permeability, and lower equilibrium moisture content of the prepared film. These are the ideal properties for an active food packaging material. However, above 10wt% of tocopherol, the films became more irregular, less transparent with yellow colour [88]. In one study, a promising active food packaging film was fabricated using MMT and copper oxide as nanofillers in the chitosan matrix. The resulted nano-composite films loaded with 3% of MMT-copper oxide showed higher mechanical strength, enhanced water solubility, and water barrier properties, in addition to increased antibacterial activity against both Gram-positive and Gram-negative bacteria. Thus, the fabricated films have intense antibacterial activity against food-borne pathogens, thus could be a promising novel active food packaging material [111]. In another study, eco-friendly and non-toxic food packaging films were fabricated with date palm fiber as core material, which is coated with nano-composite material composed of chitosan, MMT, and thiabendazolium. The latter was intercalated in between layers of MMT with the intention to improve the antibacterial activity of the films. The resulted films showed enhanced mechanical properties, increased surface hydrophobicity, and exhibits inhibitory effects against diverse bacteria compared to uncoated films [112].

Wang et al., 2018 developed composite food packaging films comprised of MMT, chitosan and poly(ethylene oxide) nanofibrous membrane embedded in the polymeric matrix of PVA-co-ethylene. The resulted films showed significant improvement in mechanical and thermal stability and also exhibited better oxygen barrier and moisture barrier properties by lengthening the tortuous paths for molecular penetration in the films [113]. Xu et al., 2018 successfully preserved fresh tangerine fruits with chitosan and MMT-based coating. The resulted coating containing 1wt% of MMT found to exhibit lower weight loss and lower decay rate, higher titratable acidity, and total soluble solids compared to coating with chitosan film during the total storage period. Therefore, coating with chitosan and MMT film, the stability, and self-life of tangerine fruits were enhanced [114].

In order to improve the dispersion property of MMT in a polymer matrix containing starch and chitosan, MMT was modified with polydopamine. The polydopamine provides exfoliated structure to MMT with the degree of exfoliation increased with the concentration of polydopamine from 0.5 to 3 mg/mL. The obtained nano-composite films had enhanced mechanical properties [115]. In another study, the effect of MMT content on mechanical, thermal, and barrier properties of nano-composite films composed of chitosan and MMT. The results showed that the thermal and mechanical properties of the film improved significantly, whereas the barrier property against water vapour and gas permeability decreased significantly compared to films made of only chitosan [116].

7.6   Application of MMT-chitosan based nano-composites in water treatment process

Chitosan-MMT nano-composite materials are eco-friendly which can replace metallic salts in water treatment for the production of drinking water [117]. There are several steps in the production of drinking water from surface fresh water, including coagulation, flocculation, sedimentation, filtration, and disinfection. Coagulation, as well as the flocculation process, necessitates minimization of polymer release. Chitosan-MMT nano-composites can also be used for the removal of heavy metals, fluoride, dye, other chemicals, nanofiltration and antibacterial process.

The paucity of water resources and the pollution of wastewater attracted many researchers worldwide to focus on research pertaining to a sustainable and environmentally friendly adsorbent that can remove heavy metals from water. One such study described the fabrication of novel aerogel adsorbent with chitosan, cellulose nanofiber and MMT using liquid nitrogen directional freezing method. The developed aerogel displayed 3D pore structure and demonstrated better mechanical properties with maintaining stable structure in a strong acid environment, excellent adsorption performance towards heavy metals such as Pb^{2+}, Cu^{2+} and Cd^{2+} and good reusability with retention of 80% of their initial adsorption capacity even after 5-cycles [118]. In another investigation, the promising use of chitosan and MMT-based films as iron adsorption from groundwater was investigated. Among various weight percentage ratios, 40:60 of chitosan and MMT was found to be best for the iron adsorption that could manage to attain 89.2% at the optimal conditions. Moreover, the process of adsorption was found to be endothermic and spontaneous [119].

Like heavy metals, the removal of a dye such as methylene blue from an aqueous medium can be eliminated by using nano-composite involving chitosan and MMT. A super adsorbent nano-composite was fabricated with modified chitosan (i.e. chitosan-g-poly(acrylic acid) and MMT for the elimination of methylene blue from the aqueous medium. The addition of MMT, the weight ratio of the acrylic acid component to chitosan, and increased pH value had a great impact on the adsorption capacity of the nano-composite against the dye. In addition, the nano-composite adsorbent displayed the potential for regeneration and reuse after the adsorption [120]. Table 1 presented different applications of chitosan-MMT nano-composites with formulation types, composite materials, incorporated drugs with results.

Table 1. Various applications of chitosan-MMT nano-composites with formulation types, composite materials, incorporated drugs with results.

Formulation types	Composite material (excluding chitosan and MMT)	Drug	Results	References
Application of MMT-chitosan based nano-composites in drug release				
Tablets	----	Aceclofenac	Sustained release	[89]
Enteric-coated capsules	Eudragit L100	Quinine	Controlled delivery of quinine in the colon	[90]
Microsphere	----	Tanshinone IIA	Sustained release of the drug, and no cytotoxicity	[91]
Nanoparticles	Carboxymethyl starch	Curcumin	Significantly reduced the formation of biofilm of *Streptococcus mutans* in the teeth	[92]
Nanoparticles	----	Chlorhexidine	Sustained release and Reduce microbial count in the mouth cavity	[93]
Film	----	Podophyllotoxin	Decreased drug release rate	[94]
Film	Alginate	5-flurouracil	Slow release	[95]
Film	----	Diclofenac sodium	Controlled drug release up to 72 h in a pH-dependent manner, and excellent antimicrobial activity	[96]
Film	----	5-amino salicylic acid	Prolonged drug release up to 80h, and anticancer efficiency with more than 50% of cell death of human cell lines	[97]
Hydrogel	----	Vitamin B12	Improved mechanical property and pulsatile release profile	[98]
Hydrogel		Ofloxacin	Sustained drug release	[99]

Application of MMT-chitosan based nano-composites in antimicrobial activities				
----	----	----	Improved biocompatibility, biodegradation, and antimicrobial activity	[101]
Nano-in-microparticle	Silver nanoparticles, Polymidoamine, and PEG	Ibuprofen	Significant antibacterial properties and inflammatory response during bacterial infection	[102]
Paper	Silver nanoparticles and quaternized carboxymethyl chitosan	----	Better mechanical strength and excellent antimicrobial activity	[103]
Film	Poly(vinylalcohol)	----	Improved water and oxygen barrier properties and mechanical and antimicrobial properties	[104]
Application of MMT-chitosan based nano-composites in wound-healing activities				
Powder or wound dressing material	----	Silver sulfadiazine	Enhanced biocompatibility and wound healing with maintaining antimicrobial properties	[106]
Hydrogel	PVA	----	Increased mechanical properties and wound healing under high stress.	[107]
Application of MMT-chitosan based nano-composites in tissue engineering				
Scaffold	Hydroxyapatite	----	Repairing of non-load bearing bone tissue and non-toxic to MG63 osteoblast cell lines	[109]
Scaffolds	Xylan	----	Improved mechanical properties, non-cytotoxic and good cell viability	[110]
Application of MMT-chitosan based nano-composites in food packaging				
Film	----	Tocopherol	Increased hydrophobicity, reduced water vapour permeability, and lower equilibrium moisture content.	[88]
Films	Copper oxide	----	Higher mechanical strength, enhanced water solubility, water barrier properties, and increased antibacterial activity	[111]

Film	Palm fiber	Thiabendazolium	Enhanced mechanical properties, increased surface hydrophobicity, and exhibits inhibitory effects against various bacteria.	[112]
Film	Poly(ethylene oxide)	----	Significant improvement in mechanical and thermal stability, and better oxygen and moisture barrier properties	[113]
Film	----	----	Improved stability and self-life of tangerine fruits	[114]
Film	Starch and polydopamine	----	Enhanced mechanical properties.	[115]
Film	----	----	Mechanical strength and thermal properties of the increased significantly, and barrier property against water vapour and gas permeability decreased	[116]
Application of MMT-chitosan based nano-composites in the water treatment process				
Aerosel	Cellulose nanofiber	----	Excellent adsorption performance towards heavy metals such as Pb2+, Cu2+ and Cd2+ and good reusability.	[118]
Films	----	----	Iron adsorption from groundwater	[119]
Super-adsorbent	Chitosan-g-poly(acrylic acid)	----	Removal of a dye such as methylene blue from aqueous solution, and reusability	[120]

Conclusions

The field of bionanocomposite is an exciting and emerging potential research area for a decade or so. From the environmental point of view, bionanocomposite containing biopolymers seem to be the only and best alternative material to petroleum-derived products as they provide sustainable development. In addition, they are being a big contributor to decreasing greenhouse gas emissions. In this regard, bionanocomposite developed with chitosan as biopolymer demonstrates all the important requirements such as non-toxic, biodegradable, biocompatible, etc. Combining chitosan with layered clay such as MMT further the qualities to tune the bionanocomposites for various biomedical and pharmaceutical applications. The use of chitosan and MMT-based nano-composite is also extended to fabricate food packaging material and water treatment processes. It is evident that there is significant growth in the bionanocomposite field, however, constant innovation and development are essential to make them more tunable to precisely fit various industries and also be cost-effective.

References

[1] Y. Shchipunov, Bionanocomposites: Green sustainable materials for the near future, Pure Appl. Chem. 84 (2012) 2579-2607. https://doi.org/10.1351/PAC-CON-12-05-04

[2] R. Parhi, Fabrication and characterization of PVA-based green materials, in: S. Ahmed (Ed.), Advanced Green Materials, Woodhead Publishing, Duxford, United Kingdom, 2021, pp. 133-168. https://doi.org/10.1016/B978-0-12-819988-6.00009-4

[3] V. Sencadas, D.M. Correia, C. Ribeiro, S. Moreira, G. Botelho, J.G. Ribelles, S. Lanceros-Méndez, Physical-chemical properties of cross-linked chitosan electrospun fiber mats, Polym. Test. 31 (2012) 1062-1069. https://doi.org/10.1016/j.polymertesting.2012.07.010

[4] C.K.S. Pillai, W. Paul, C.P. Sharma, Chitin and chitosan polymers: Chemistry, solubility and fiber formation, Prog. Polym. Sci. 34 (2009) 641-678. https://doi.org/10.1016/j.progpolymsci.2009.04.001

[5] P. Monvisade, P. Siriphannon, Chitosan intercalated montmorillonite: Preparation, characterization and cationic dye adsorption, Appl. Clay. Sci. 42 (2009) 427-431. https://doi.org/10.1016/j.clay.2008.04.013

[6] R. Parhi, Drug delivery applications of chitin and chitosan: a review, Env. Chem. Let. 18 (2020) 577-594. https://doi.org/10.1007/s10311-020-00963-5

[7] Z. Gu, H. Xie, C. Huang, L. Li, X. Yu, Preparation of chitosan/silk fibroin blending membrane fixed with alginate dialdehyde for wound dressing, Int. J. Biol. Macromol. 58 (2013) 121-126. https://doi.org/10.1016/j.ijbiomac.2013.03.059

[8] S. Pradhan, A.K. Brooks, V.K. Yadavalli, Nature-derived materials for the fabrication of functional biodevices, Mater. Today Bio. 7 (2020) 100065. https://doi.org/10.1016/j.mtbio.2020.100065

[9] C. Rodrigues, J.M. Muneron de Mello, F. Dalcanton, D.L.P. Macuvele, N. Padoin, M.A. Fiori, C. Soares, H.G. Riella, Mechanical, Thermal and Antimicrobial Properties of Chitosan Based Nanocomposite with Potential Applications for Food Packaging, J. Polym. Env. 28 (2020) 1216-1236. https://doi.org/10.1007/s10924-020-01678-y

[10] C. Aguzzi, P. Cerezo, C. Viseras, C. Caramella, Use of clays as drug delivery systems: Possibilities and limitations, Appl. Clay Sci. 36 (2007) 22-36. https://doi.org/10.1016/j.clay.2006.06.015

[11] S. Jayrajsinh, G. Shankar, Y.K. Agrawal, L. Bakre, Montmorillonite nanoclay as a multifaceted drug-delivery carrier: A review, J. Drug Deliv. Sci. Technol. 39 (2017) 200-209. https://doi.org/10.1016/j.jddst.2017.03.023

[12] A.M. Youssefa, Samah. M. El-Sayed, Bionanocomposites materials for food packaging applications: Concepts and future outlook, Carbohydr. Polym. 193 (2018) 19-27. https://doi.org/10.1016/j.carbpol.2018.03.088

[13] E. Günister, D. Pestreli, C.H. Ünlü, O. Atici, N. Güngör, Synthesis and characterization of chitosanMMT biocomposite systems, Carbohydr. Polym. 67 (2007) 358-365. https://doi.org/10.1016/j.carbpol.2006.06.004

[14] K. Kudumula, Scope of polymer nano-composite in bio-medical applications, J. Mechanic. Civil Eng. 13 (2016) 18-21. https://doi.org/10.9790/1684-1305021821

[15] H.M.C. De Azeredo, Nanocomposites for food packaging applications, Food Res. Int. 42 (2009) 1240-1253. https://doi.org/10.1016/j.foodres.2009.03.019

[16] J.-W. Rhim, H.-M. Park, C.-S. Ha, Bio-nanocomposites for food packaging applications, Prog. Polym. Sci. 38 (2013) 1629-1652. https://doi.org/10.1016/j.progpolymsci.2013.05.008

[17] X. Liu, Q. Hu, Z. Fang, X. Zhang, B. Zhang, Magnetic chitosan nanocomposites: a useful recyclable tool for heavy metal ion removal, Langmuir 25 (2008) 3-8. https://doi.org/10.1021/la802754t

[18] S.B. Ghelejlu, M. Esmaiili, H. Almasi, Characterization of chitosan-nanoclay bionanocomposite active films containing milk thistle extract, Int. J. Biol. Macromol. 86 (2016) 613-621. https://doi.org/10.1016/j.ijbiomac.2016.02.012

[19] M.H. Lee, S.Y. Kim, H.J. Park, Effect of halloysite nanoclay on the physical,

mechanical, and antioxidant properties of chitosan films incorporated with clove

essential oil. Food Hydrocoll. 84 (2018) 58-67.
https://doi.org/10.1016/j.foodhyd.2018.05.048

[20] Y. Wang, S. Yi, R. Lu, D.E. Sameen, S. Ahmed, J. Dai, Y. Liu, Preparation, characterization, and 3D printing verification of chitosan/halloysite nanotubes/tea polyphenol nanocomposite films, Int. J. Biol. Macromol.166 (2021) 32-44. https://doi.org/10.1016/j.ijbiomac.2020.09.253

[21] C.T. Edelman, J.C. Favejee, On the Crystal Structure of Montmorillonite and halloysite, Zeitschriftfür Kristallographie-crystalline Mater. 102, (1940) 417-431. https://doi.org/10.1524/zkri.1940.102.1.417

[22] M. Segad, B. Jonsson, T. Åkesson, B. Cabane, Ca/Na montmorillonite: structure, forces and swelling properties, Langmuir 26 (2010) 5782-5790. https://doi.org/10.1021/la9036293

[23] R. Onnainty, G. Granero, Chitosan-clays based nanocomposites: promising materials for drug delivery applications, Nanomed. Nanotechnol. J. 1 (2017) 114.

[24] D.F. Xie, V.P. Martino, P. Sangwan, C. Way, G.A. Cash, A. Gregory, E. Pollet, K.M. Dean, P.J. Halley, L. Averous, Elaboration and properties of plasticised chitosan-based exfoliated nanobiocomposites, Polym. (United Kingdom) 54 (2013) 3654-3662. https://doi.org/10.1016/j.polymer.2013.05.017

[25] S. Maisanaba, S. Pichardo, M. Puerto, D. Gutiérrez-Praena, A.M. Cameán, A. Jos, Toxicological evaluation of clay minerals and derived nanocomposites: A review, Environ. Res.138 (2015) 233-254. https://doi.org/10.1016/j.envres.2014.12.024

[26] T.C. Yadav, P. Saxena, A.K. Srivastava, A.K. Singh, R.K. Yadav, Harish, R. Prasad, V. Pruthi. Potential Applications of Chitosan Nanocomposites: Recent Trends and Challenges, in; Shahid-ul-Islam and B.S. Butola (Eds.), Advanced Functional Textiles and Polymers, Scrivener Publishing LLC, 2020, pp. 365-403. https://doi.org/10.1002/9781119605843.ch13

[27] P.K. Dutta, J. Dutta, V.S. Tripathi, Chitin and chitosan: chemistry: properties and applications, J. Sci. Ind. Res. 63 (2004) 20-31

[28] M. Rinaudo, Chitin and chitosan: properties and applications, Prog. Polym. Sci. 31 (2006) 603-632. https://doi.org/10.1016/j.progpolymsci.2006.06.001

[29] R. Jayakumar, D. Menon, K. Manzoor, S.V. Nair, H. Tamura. Biomedical applications of chitin and chitosan-based nanomaterials. A short review, Carbohydr. Polym. 82 (2010) 227-32. https://doi.org/10.1016/j.carbpol.2010.04.074

[30] J. Venkatesan, S.-K. Kim, Chitosan composites for bone tissue engineering-an overview, Mar. Drugs. 8 (2010) 2252-66. https://doi.org/10.3390/md8082252

[31] G.L. Dotto, L.L.A. Pinto, General consideration about chitosan, in: G.L. Dotto, S.P. Campana-Filho, L.A.A. Pinto (Eds.), Chitosan Based Materials and its Applications. Frontier in biomaterials, Vol 3, Bentham Science Publishers, 2017, pp. 3-33. https://doi.org/10.2174/9781681084855117030004

[32] S. Kumar, A. Mukherjee, J. Dutta, Chitosan based nanocomposite films and coatings: Emerging antimicrobial food packaging alternatives, Trends in Food Sci. Technol. 97 (2020) 196-209. https://doi.org/10.1016/j.tifs.2020.01.002

[33] V.S. Yeul, S.S. Rayalu, Unprecedented chitin and chitosan: A chemical overview, J. Polym. Environ. 21 (2013) 606-614. https://doi.org/10.1007/s10924-012-0458-x

[34] M.R. Kasaai, Determination of the degree of N-acetylation for chitin and chitosan by various NMR spectroscopy techniques: A review, Carbohydr. Polym. 79 (2010) 801-10. https://doi.org/10.1016/j.carbpol.2009.10.051

[35] J. Kumirska, M. Czerwicka, Z. Kaczynski, A. Bychowska, K. Brzozowski, J. Thoming, P. Stepnowski, Application of spectroscopic methods for structural analysis of chitin and chitosan, Mar. Drugs. 8 (2010) 1567-636. https://doi.org/10.3390/md8051567

[36] X. Liu, L. Ma, Z. Mao, C. Gao, Chitosan-based biomaterials for tissue repair and regeneration. Chitosan for biomaterials II, in: R. Jayakumar, M. Prabaharan, R.A.A. Muzzarelli (Eds.), Advances in polymer science, Heidelberg, Springer Berlin, 2011, pp. 81-127. https://doi.org/10.1007/12_2011_118

[37] K.V.H. Prashanth, R.N. Tharanathan, Chitin/chitosan: modifications and their unlimited application potential: an overview, Trends Food Sci. Technol. 18 (2007) 117-31. https://doi.org/10.1016/j.tifs.2006.10.022

[38] K. Kurita, T Sannan, Y. Iwakura, Studies on chitin 4: Evidence for formation of block and random copolymers of N-Acetyl-D-glucosamine and D-glucosamine by hetero and homogenous hydrolyses, Makromol. Chem. 178 (1977) 3197-202. https://doi.org/10.1002/macp.1977.021781203

[39] V.V.C. Azevedo, S.A. Chaves, D.C. Bezerra, M.V.L. Fook, A. Costa, Chitin and chitosan: Applications as biomaterials, Rev. Elet. Mater. Proc. 2 (2007) 27-34.

[40] S.K. Shukla, A.K. Mishra, O.A. Arotiba, B.B. Mamba, Chitosan-based nanomaterials: a state-of-the-art review, Int. J. Biol. Macromol. 59 (20113) 46-58. https://doi.org/10.1016/j.ijbiomac.2013.04.043

[41] G. Crini, P.M. Badot, Application of chitosan, a natural aminopolysaccharide, for dye removal from aqueous solutions by adsorption processes using batch studies: a review of recent literature, Prog. Polym. Sci. 33 (2008) 399-447. https://doi.org/10.1016/j.progpolymsci.2007.11.001

[42] J.S. Piccin, M.L. Vieira, J.O. Goncalves, G.L. Dotto, L.A. Pinto, Adsorption of FD & C Red No. 40 by chitosan: Isotherms analysis, J. Food Eng. 95 (2009) 16-20. https://doi.org/10.1016/j.jfoodeng.2009.03.017

[43] G.L. Dotto, M.L. Vieira, L.A. Pinto, Kinetics and mechanism of tetrazine adsorption onto chitin and chitosan, Ind. Eng. Chem. Eng. 51 (2012) 6862-8. https://doi.org/10.1021/ie2030757

[44] A.A. Mendes, P.C. Oliveira, H.F. Castro, R.L. Giordano, Application of chitosan as support for immobilization of enzymes of industrial intrest, Quim Nova. 34 (2012) 831-40.

[45] D.L. Hawary, M.A. Motaleb, H. Farag, O.W. Guirguis, M.Z. Elsabee, Water-soluble derivatives of chitosan as a target delivery system of Tc-99m to some organs in vivo for nuclear imaging and biodistribution, J. Radioanal. Nucl. Chem. 290 (2011) 557-567. https://doi.org/10.1007/s10967-011-1310-9

[46] R. Hejazi, M. Amiji, Chitosan-based gastrointestinal delivery systems, J. Control. Release. 89: (2003) 151-165. https://doi.org/10.1016/S0168-3659(03)00126-3

[47] K.K. Byung, H.J. Shim, M.H. Sang, E.S. Park, Chitin-based embolic materials in the renal artery of rabbits: pathologic evaluation of an absorbable particulate agent, Radiol. 236 (2005) 151-158. https://doi.org/10.1148/radiol.2361040669

[48] P.V. Kumar, A.A. Bricey, V.V. Selvi, C.S. Kumar, N. Ramesh, Antioxidant effect of green tea extract in cadmium chloride intoxicated rats, Adv. Appl. Sci. Res. 1 (2010) 9-13.

[49] J.A. Jennings, J.D. Bumgardner, Chitosan based biomaterials (Vol 2), in: J.A. Jennings and J.D. Bumgardner (Eds.), Tissue engineering and therapeutics, Woodhead Pub Ltd, Elsevier Science & Technology, Duxford, UK, 2017.

[50] E.F. Franca, L.C.G. Freiras, R.D. Lins, Chitosan molecular structure as a function of n-acetylation, Biopolym. 95 (2011) 448-460 https://doi.org/10.1002/bip.21602

[51] A.V. Il'ina, V.P. Varmalov, Chitosan-based polyelectrolyte complexes: A review, Appl. Biochem. Microbiol. 41 (2005) 5-11. https://doi.org/10.1007/s10438-005-0002-z

[52] A.S. Halim, L.C. Keong, I. Zainol, A. Hazri, A.H.A. Rashid, Biocompatibility and biodegradation of chitosan and derivatives, in: B. Sarmento, J.D. Neves, (Eds.) Chitosan-based systems for biopharmaceuticals. Delivery, targeting and polymers therapeutics, Wiley, Chihester, 2012, pp. 57-74. https://doi.org/10.1002/9781119962977.ch4

[53] W.A. Sarhan, H.M. Azzazy, High concentration honey chitosan electrospun nanofibers: biocompatibility and antibacterial effects. Carbohydr. Polym. 122 (2015) 135-43. https://doi.org/10.1016/j.carbpol.2014.12.051

[54] Y. Chen, Y. Zhou, S. Yang, et al., Novel bone substitute composed of chitosan and strontium-doped α-calcium sulfate hemihydrate: Fabrication, characterization and evaluation of biocompatibility, Mater. Sci. Eng. C Mater. Biol. Appl. 66 (2016) 84-91. https://doi.org/10.1016/j.msec.2016.04.070

[55] I. Bravo-Osuna, C. Vauthier, A. Farabollini, G.F. Palmieri, G. Ponchel, Mucoadhesion mechanism of chitosan and thiolated chitosan-poly(isobutyl cyanoacrylate) core-shell nanoparticles, Biomater. 28 (2007) 2233-2243. https://doi.org/10.1016/j.biomaterials.2007.01.005

[56] E. Meng-Lund, C. Muff-Westergaard, C. Sander, P. Madelung, J. Jacobsen, A mechanistic based approach for enhancing buccal mucoadhesion of chitosan, Int. J. Pharm. 461 (2014) 280-285. https://doi.org/10.1016/j.ijpharm.2013.10.047

[57] C. Saikia, P. Gogoi, T.K. Maji, Chitosan: a promising biopolymer in drug delivery applications, J. Mol. Genet. Med. S4:6 (2015) 1-10. https://doi.org/10.4172/1747-0862.S4-006

[58] F. Laffleur, F. Hintzen, D. Rahmat, G. Shahnaz, G. Millotti, A. Bernkop-Schnürch, Enzymatic degradation of thiolated chitosan, Drug Develop. Indu. Pharm. 39 (2013) 1531-1539. https://doi.org/10.3109/03639045.2012.719901

[59] E. Szymańska, K. Winnicka, A. Amelian, U. Cwalina, Vaginal chitosan tablets with clotrimazole-design and evaluation of mucoadhesive properties using porcine vaginal mucosa, mucin and gelatine, Chem. Pharm. Bull. 62 (2014) 160-167. https://doi.org/10.1248/cpb.c13-00689

[60] C.L. Domínguez-Delgado, I.M. Rodríguez-Cruz, E. Fuentes-Prado, J.J. Escobar-Chávez, G. Vidal-Romero, L. García-González, R.I. Puente-Lee, Drug Carrier Systems Using Chitosan for Non Parenteral Routes, in: Pharmacology and Therapeutic. Infotech, 2014, pp. 273-325. https://doi.org/10.5772/57235

[61] D.J. Ormrod, C.C. Holmes, T.E. Miller, Dietary chitosan inhibits hypercholesterolaemia and atherogenesis in the apolipoprotein E-deficient mouse model of atherosclerosis, Atherosclerosis. 138 (1998) 329-334. https://doi.org/10.1016/S0021-9150(98)00045-8

[62] S.B. Jing, L. Li, D. Ji, Y. Takiguchi, T. Yamaguchi, Effect of chitosan on renal function in patients with chronic renal failure, J. Pharm. Pharmacol. 49 (1997) 721. https://doi.org/10.1111/j.2042-7158.1997.tb06099.x

[63] Wedmore, J.G. McManus, A.E. Pusateri, J.B. Holcomb, A special report on the chitosan-based hemostatic dressing: Experience in current combat operations. J. Trauma. 60 (2006) 655-658. https://doi.org/10.1097/01.ta.0000199392.91772.44

[64] K. Arai, T. Kinumaki, T. Fujita, Toxicity of chitosan, Bull. Tokai Region Fish Res. Lab. 56 (1968) 88-94.

[65] K. Sonaje, Y.H. Lin, J.H. Juang, S.P. Wey, C.T. Chen, H.W. Sung, In vivo evaluation of safety and efficacy of self-assembled nanoparticles for oral insulin delivery, Biomater. 30 (2009) 2329-39. https://doi.org/10.1016/j.biomaterials.2008.12.066

[66] L. Illum, N.F. Farraj, S.S. Davis, Chitosan as a novel nasal delivery system for peptide drugs, Pharm. Res. 11 (1994) 1186-9. https://doi.org/10.1023/A:1018901302450

[67] K.H. Waibel, B. Haney, M. Moore, B. Whisman, R. Gomez, Safety of chitosan bandages in shellfish allergic patients, Military Medicine. 176 (2001) 1153-1156. https://doi.org/10.7205/MILMED-D-11-00150

[68] R. Abdeen, N. Salahuddin, Modified chitosan-clay nanocomposite as a drug delivery system intercalation and in vitro release of ibuprofen, J. Chem. (2013) 2013. https://doi.org/10.1155/2013/576370

[69] S. Olivera, H.B. Muralidhara, K. Venkatesh, V.K. Guna, K. Gopalakrishna, K.Y. Kumar, Potential applications of cellulose and chitosan nanoparticles/composites in wastewater treatment: A review, Carbohyd. Polym. 153 (2016) 600-618. https://doi.org/10.1016/j.carbpol.2016.08.017

[70] Y. Xu, X. Ren, M.A. Hanna, Chitosan/clay nanocomposite film preparation and characterization, J. Appl. Polym. Sci. 99 (2006) 1684-1691. https://doi.org/10.1002/app.22664

[71] V.V. Pande, V.M. Sanklecha, Bionanocomposite: A Review, Austin J. Nanomed. Nanotechnol. 5 (2017) 1045.

[72] R. Zhao, T. Peter, P. Halley, Emerging biodegradable materials: starch- and protein-based bio-nanocomposites, J. Mater. Sci. 43 (2008) 3058-3071. https://doi.org/10.1007/s10853-007-2434-8

[73] M. Darder, E. Ruiz-hitzky, Investigación Química Bio-nanocomposites: nuevos materiales ecológicos, biocompatibles y funcionales. 103 (2007) 21-29.

[74] P. Camargo, K. Satyanarayana, F. Wypych, Nanocomposites: Synthesis, Structure, Properties and New Application Opportunities, Mater. Res. 12 (2009) 1-39. https://doi.org/10.1590/S1516-14392009000100002

[75] G. Siqueira, J. Bras, A. Dufresne, Cellulosic Bionanocomposites: A Review of Preparation, Properties and Applications, Polym. 2 (2010) 728-765. https://doi.org/10.3390/polym2040728

[76] K. Basumatary, P. Daimary, S.K. Das, M. Thapa, M. Singh, A. Mukherjee, S. Kumar, Lagerstroemia speciosa fruit-mediated synthesis of silver nanoparticles and its application as filler in agar-based nanocomposite films for antimicrobial food packaging, Food Packaging and Shelf Life. 17 (2018) 99-106. https://doi.org/10.1016/j.fpsl.2018.06.003

[77] A. Naskar, H. Khan, R. Sarkar, S. Kumar, D. Halder, S. Jana, Anti-biofilm activity and food packaging application of room temperature solution process-based polyethylene glycol capped Ag-ZnO-graphene nanocomposite, Mater. Sci. Eng. C. 91 (2018) 743-753 https://doi.org/10.1016/j.msec.2018.06.009

[78] S. Estevez-Areco, L. Guz, L. Famá, R. Candal, S. Goyanes, Bioactive starch nanocomposite films with antioxidant activity and enhanced mechanical properties obtained by extrusion followed by thermo-compression, Food Hydrocoll. 96 (2019) 518-528. https://doi.org/10.1016/j.foodhyd.2019.05.054

[79] M. Matet, M.-C. Heuzey, A. Ajji, P. Sarazin, Plasticized chitosan/polyolefin films produced by extrusion, Carbohydr. Polym. 117 (2015) 177-184. https://doi.org/10.1016/j.carbpol.2014.09.058

[80] G.B. Khomutov, Interfacially formed organized planar inorganic, polymeric and composite nanostructures, Adv. in Colloid and Interface Sci. 111 (2004) 79-116. https://doi.org/10.1016/j.cis.2004.07.005

[81] D. Zhang, C. Jiang, Y.E. Sun, Q. Zhou, Layer-by-layer self-assembly of tricobalt tetroxide-polymer nanocomposite toward high-performance humidity-sensing, J. Alloys Comp. 711 (2017) 652-658. https://doi.org/10.1016/j.jallcom.2017.03.365

[82] F.B. Dhieb, E.J. Dil, S.H. Tabatabaei, F. Mighri, A. Ajji, Effect of nanoclay orientation on oxygen barrier properties of LbL nanocomposite coated films, RSC Adv. 9 (2019) 1632-1641. https://doi.org/10.1039/C8RA09522A

[83] R. Parhi, Chitin and Chitosan in Drug Delivery, in; G. Crini, E. Lichtfouse (Eds.), Sustainable Agriculture Reviews 36-Chitin and Chitosan: Applications in Food, Agriculture, Pharmacy, Medicine and Wastewater Treatment, Springer Nature Switzerland AG, 2019, pp. 175-240. https://doi.org/10.1007/978-3-030-16581-9_6

[84] L.N. Mengatto, I.M. Helbling, J.A. Luna, Recent advances in chitosan films for controlled release of drugs, Recent Pat. Drug Deliv. Formulation 6 (2012) 156-170. https://doi.org/10.2174/187221112800672967

[85] S.P. Noel, H. Courtney, J.D. Bumgardner, W.O. Haggard, Chitosan films a potential local drug delivery system for antibiotics, Clin. Orthop. Relat. Res. 466 (2008) 1377-1382. https://doi.org/10.1007/s11999-008-0228-1

[86] A. Ali, S. Ahmed, A review on chitosan and its nanocomposites in drug delivery, Int. J. Biol. Macromol. 109 (2018) 273-286. https://doi.org/10.1016/j.ijbiomac.2017.12.078

[87] N. Sultana, M. Mokhtar, M.I. Hassan, R.M. Jin, F. Roozbahani, T.H. Khan, Chitosan-Based Nanocomposite Scaffolds for Tissue Engineering Applications, Mater. Manufact. Process. 30 (2015) 273-278. https://doi.org/10.1080/10426914.2014.892610

[88] M.V. Dias, V.M. Azevedo, S.V. Borges, N. de Fátima Ferreira Soares, R.V. de Barros Fernandes, J.J. Marques, É.A.A. Medeiros, Development of chitosan/montmorillonite nanocomposites with encapsulated a-tocopherol, Food Chem. 165 (2014) 323-329. https://doi.org/10.1016/j.foodchem.2014.05.120

[89] G. Thakur, A. Singh, I. Singh, Chitosan-Montmorillonite polymer composites: formulation and evaluation of sustained release tablets of aceclofenac, Sci. Pharm. 84 (2016) 603-617. https://doi.org/10.3390/scipharm84040603

[90] G.V. Joshi, B.D. Kevadiya, H.M. Mody, H.C. Bajaj, Confinement and controlled release of quinine on chitosan-montmorillonite bionanocomposites, J. Polym. Sci. Part A Polym. Chem. 50 (2012) 423-430. https://doi.org/10.1002/pola.25046

[91] C. Luo, Q. Yang, X. Lin, C. Qi, G. Li, Preparation and drug release property of tanshinone IIA loaded chitosan montmorillonite microspheres, Int. J. Biol. Macromol. 125 (2019) 721-729. https://doi.org/10.1016/j.ijbiomac.2018.12.072

[92] S. Jahanizadeh, F. Yazdian, A. Marjani, M. Omidi, H. Rashedi, Curcumin-loaded chitosan/carboxymethyl starch/montmorillonite bio-nanocomposite for reduction of dental bacterial biofilm formation, Int. J. Biol. Macromol. 105 (2017) 757-763. https://doi.org/10.1016/j.ijbiomac.2017.07.101

[93] J. Kolahi, A. Soolari, Rinsing with chlorhexidinegluconate solution after brushing and flossing teeth: a systematic review of effectiveness, Quintessence Int. 37 (2006) 605-612.

[94] S. Sedaghat, Preparation of chitosan/ Montmorillonite (MMt) nanocomposite as a drug delivery carrier of podophyllotoxin, Asian J. Appl. Sci. 6 (2018) 86-96.

[95] F.F. Azhar, A. Olad, A study on sustained release formulations for oral delivery of 5-fluorouracil based on alginate-chitosan/montmorillonite nanocomposite systems, Appl. Clay Sci., 101 (2014) 288-296. https://doi.org/10.1016/j.clay.2014.09.004

[96] D. Cheikh, F. García-Villén, H. Majdoub, C. Viseras, M.B. Zayani, Chitosan/beidellite nanocomposite as diclofenac carrier, Int. J. Biol. Macromol. 126 (2019) 44-53. https://doi.org/10.1016/j.ijbiomac.2018.12.205

[97] C. Aguzzi, P. Capra, C. Bonferoni, P. Cerezo, I. Salcedo, R. Sanchez, C. Caramella, C. Viseras, Chitosan-silicate biocomposites to be used in modified drug release of 5-aminosalicylic acid (5-ASA), Appl. Clay Sci. 50 (2010) 106-111. https://doi.org/10.1016/j.clay.2010.07.011

[98] K.-H. Liu, T.-Y. Liu, S.-Y. Chen, D.-M. Liu, Drug release behavior of chitosan-montmorillonite nanocomposite hydrogels following electrostimulation, Acta Biomater. 4 (2008) 1038-1045. https://doi.org/10.1016/j.actbio.2008.01.012

[99] S. Hua, H. Yang, W. Wang, A. Wang, Controlled release of ofloxacin from chitosan-montmorillonite hydrogel, Appl. Clay Sci. 50 (2010) 112-117. https://doi.org/10.1016/j.clay.2010.07.012

[100] Z. Sun, C. Shi, X. Wang, Q. Fang, J. Huang, Synthesis, characterization, and antimicrobial activities of sulfonated chitosan, Carbohydr. Polym. 155 (2017) 321. https://doi.org/10.1016/j.carbpol.2016.08.069

[101] S.-h. Hsu, M.-C. Wang, J.-J. Lin, Biocompatibility and antimicrobial evaluation of montmorillonite/chitosan nanocomposites, Appl. Clay Sci. 56 (2012) 53-62. https://doi.org/10.1016/j.clay.2011.09.016

[102] E.A. Abd Elsalam, H.F. Shabaiek, M.M. Abdelaziz, I.A. Khalil, I.M. El-Sherbiny, Fortified hyperbranched PEGylated chitosan-based nano-in-micro composites for treatment of multiple bacterial infections, Int. J. Biol. Macromol. 148 (2020) 1201-1210. https://doi.org/10.1016/j.ijbiomac.2019.10.164

[103] Y. Ling, Y. Luo, J. Luo, X. Wang, R. Sun, Novel antibacterial paper based on quaternized carboxymethyl chitosan/organic montmorillonite/Ag NP nanocomposites, Industr. Crops Prod. 51 (2013) 470-479. https://doi.org/10.1016/j.indcrop.2013.09.040

[104] A. Giannakas, M. Vlacha, C. Salmas, A. Leontiou, P. Katapodis, H. Stamatis, N.-M. Barkoula, A. Ladavos, Preparation, characterization, mechanical, barrier and antimicrobial properties of chitosan/PVOH/clay nanocomposites, Carbohydr. Polym. 140 (2016) 408-415. https://doi.org/10.1016/j.carbpol.2015.12.072

[105] L. Zhang, J. Chen, W. Yu, Q. Zhao, J. Liu, Antimicrobial Nanocomposites Prepared from Montmorillonite/Ag+/Quaternary Ammonium Nitrate, J. Nanomater. 2018;2018. https://doi.org/10.1155/2018/6190251

[106] G. Sandri, M.C. Bonferoni, F. Ferrari, S. Rossi, C. Aguzzi, M. Mori, P. Grisoli, P. Cerezo, M. Tenci, C. Viseras, C. Caramella, Montmorillonite-chitosan-silver sulfadiazine nanocomposites for topical treatment of chronic skin lesions: In vitro biocompatibility, antibacterial efficacy and gap closure cell motility properties, Carbohydr. Polym. 102 (2014) 970-977. https://doi.org/10.1016/j.carbpol.2013.10.029

[107] S. Noori, M. Kokabi, Z. M. Hassan, Nanoclay enhanced the mechanical properties of Poly(Vinyl Alcohol)/Chitosan/Montmorillonite nanocomposite hydrogel as wound dressing, Procedia Mater. Sci. 11 (2015) 152-156. https://doi.org/10.1016/j.mspro.2015.11.023

[108] Z.K. Cui, S. Kim, J.J. Baljon, B.M. Wu, T. Aghaloo, M. Lee, Microporous methacrylated glycol chitosan-montmorillonite nanocomposite hydrogel for bone tissue engineering, Nat. Commun. 10 (2019) 1-10. https://doi.org/10.1038/s41467-019-11511-3

[109] S. Kar, T. Kaur, A. Thirugnanam, Microwave-assisted synthesis of porous chitosan-modified montmorillonite-hydroxyapatite composite scaffolds, Int. J. Biol. Macromol. 82 (2016) 628-636. https://doi.org/10.1016/j.ijbiomac.2015.10.060

[110] A.A.S. Bano, S.S. Poojary, D. Kumar, Y.S. Neg, Effect of incorporation of montmorillonite on Xylan/Chitosan conjugate scaffold, Colloids and Surfaces B: Biointerf. 180 (2019) 75-82. https://doi.org/10.1016/j.colsurfb.2019.04.032

[111] A. Nouri, M.T. Yaraki, M. Ghorbanpour, S. Agarwal, V.K. Gupta, Enhanced Antibacterial effect of chitosan film using Montmorillonite/CuO nanocomposite, Int. J. Biol. Macromol. 109 (2018) 1219-1231. https://doi.org/10.1016/j.ijbiomac.2017.11.119

[112] F.-Z. Semlali, A. Hassani, K. El Bourakadi, N. Merghoub, A. el kacem Qaiss, R. Bouhfid, Effect of chitosan/modified montmorillonite coating on the antibacterial and mechanical properties of date palm fiber trays, Int. J. Biol. Macromol. 148 (2020) 316-323. https://doi.org/10.1016/j.ijbiomac.2020.01.092

[113] P. Wang, H. Wang, J. Liu, P. Wang, S. Jiang, X. Li, S. Jiang, Montmorillonite@chitosan-poly (ethylene oxide) nanofibrous membrane enhancing poly (vinyl alcohol-co-ethylene) composite film, Carbohydr. Polym.181 (2018) 885-892. https://doi.org/10.1016/j.carbpol.2017.11.063

[114] D. Xu, H. Qin, D. Ren, Prolonged preservation of tangerine fruits using chitosan/montmorillonitecomposite coating, Postharvest Biol. Technol. 143 (2018) 50-57. https://doi.org/10.1016/j.postharvbio.2018.04.013

[115] M. Zhou, Q. Liu, S. Wu, Z. Gou, X. Wu, D. Xu, Starch/chitosan films reinforced with polydopamine modified MMT: Effects of dopamine concentration, Food Hydrocoll. 61 (2016) 678-684. https://doi.org/10.1016/j.foodhyd.2016.06.030

[116] Y. Kasirga, A. Oral, C. Caner, Preparation and characterization of chitosan/montmorillonite-K10 nanocomposites films for food packaging applications, Polym. Compos. 33 (2012) 1874-1882. https://doi.org/10.1002/pc.22310

[117] J. Salvé, B. Grégoire, L. Imbert, F. Hubert, V. Karpel, N. Leitner, M. Leloup, Design of hybrid Chitosan-Montmorillonite materials for water treatment: Study of the performance and stability, Chem. Eng. J. Adv. 6 (2021) (September 2020) 100087. https://doi.org/10.1016/j.ceja.2021.100087

[118] N. Rong, C. Chen, K. Ouyang, K. Zhang, X. Wang, Z. Xu, Adsorption characteristics of directional cellulose nanofiber/chitosan/montmorillonite aerogel as adsorbent for wastewater treatment, Separat. Purific. Technol. 274 (2021) 119120. https://doi.org/10.1016/j.seppur.2021.119120

[119] A.M. Shehap, R.A. Nasr, M.A. Mahfouz, A.M. Ismail, Preparation and characterizations of high doping chitosan/MMT nanocomposites films for removing iron from ground water, J. Environ. Chem. Eng. 9 (2021) 1104700. https://doi.org/10.1016/j.jece.2020.104700

[120] L. Wang, J. Zhang, A. Wang, Removal of methylene blue from aqueous solution using chitosan-g-poly(acrylic acid)/montmorillonite superadsorbent nanocomposite, Colloids Surfaces A. Physicochem. Eng. Asp. 322 (2008) 47-53. https://doi.org/10.1016/j.colsurfa.2008.02.019

Adv. App. of Micro and Nano Clay – Biopolymer-based Composites   Materials Research Forum LLC
Materials Research Foundations **125** (2022) 87-102   https://doi.org/10.21741/9781644901915-4

Chapter 4

Kaolinite-Chitosan based Nano-Composites and Applications

Narayana Saibaba KV

Department of Biotechnology, GITAM Institute of Technology, GITAM University, Visakhapatnam-530045, Andhra Pradesh, India

skvn@gitam.edu

Abstract

Chitosan is a naturally available biopolymer, having numerous applications due to its unique physico-chemical properties. Kaolinite is one of the abundantly available clay, combining these two materials imparts excellent properties suitable for various applications. Although various chitosan-based composites are developed greater attention is focused on kaolinite- chitosan composites in recent years. This chapter discusses developments in various kaolinite- chitosan nanocomposites and their applications particularly in medical, pharmaceutical, wastewater treatment and food and packaging industries. First, discussion on chitosan and kaolinite is given, then kaolinite- chitosan nanocomposite fabrication, advances, development and followed by various applications of chitosan-kaolinite nanocomposites. Finally concluding remarks are presented to summarize latest developments in this area. Challenges and future perspectives are also given in conclusions to guide the research directions.

Keywords

Chitosan, Kaolinite, Biopolymer, Nanocomposites, Clay

Contents

1. Introduction

Petrochemical polymer-based composites are toxic, non-renewable and pose environmental pollution problems. Moreover, large amounts of waste are generated after their use and create disposal problems due to their non-biodegradable characteristics. Recently, interest has been growing towards the usage of naturally available, eco-friendly green composite materials. Composites fabricated from biopolymers are gaining importance due to their non-toxic nature and abundant availability. Biopolymer composites are biodegradable, available at a cheaper cost, flexible towards chemical modification suitable for various applications. They have been used in numerous industrial applications due to their adhesive, ion exchange and adsorption properties [1].

2. Kaolinite-chitosan composites

Chitosan is derived from the deacetylation of chitin. Chitosan is a naturally available polymer; its excellent sorption characteristics find applications in wastewater treatment, industrial effluent treatment, photography, packaging etc. Despite having numerous applications, its usage is still below the expected range due to drawbacks such as weak mechanical properties, gel formation tendency, high solubility in acidic media and low surface area. Their hydrophilic nature makes them soft, swells and floats in aquatic solutions and affects their performance in applications [1,2,3,4]. Modification of physicochemical properties is thus required to increase the scope of chitosan industrial applications.

Immobilization of chitosan on clay materials overcomes the above limitations and hence gains importance. Clay-chitosan based composites are easy to prepare, inexpensive and abundantly available. Nanomaterials prepared from the biopolymers and clay shows significant improvement in mechanical and thermal properties, antimicrobial activity, bioadhesive property and lowered solubility and swelling properties [5].

Adv. App. of Micro and Nano Clay – Biopolymer-based Composites　　　　Materials Research Forum LLC
Materials Research Foundations **125** (2022) 87-102　　　　https://doi.org/10.21741/9781644901915-4

Kaolinite is one the most abundant clay material on earth, and its eco-friendliness, economic availability, small particle size and good surface interaction characteristics, high mechanical strength makes it a potential material for various industrial, food, cosmetic, medical and pharmaceutical applications. Besides, Kaolinite is biocompatible, non-toxic and has good cation removal properties [6,7,8,9].

Usage of composites prepared from the chitosan polymers and kaolinite clay has been growing over the past few years due to their distinctive properties and wide applications. The negative charge on the kaolinite surface and positive charge on the chitosan surface in acidic solutions provides versatile charge distribution on Kaolinite-chitosan biocomposites. Abundant availability, cost-effective nature, providing large surface area and a large number of active sites on the surface of Kaolinite-chitosan nanocomposite makes this nanocomposite attractive for various industrial applications.

Combining chitosan and kaolinite clay may result in nanocomposites through three processes. They are intercalation, exfoliation, and conventional distribution. Exfoliated nanocomposites are known to be true nanomaterials because of delamination and expanded size. Exfoliated nanocomposites layers are spatially separated from each other and provide dispersion in the polymer matrix [10]. Fig 1 shows the nanocomposite preparation method through the exfoliation method.

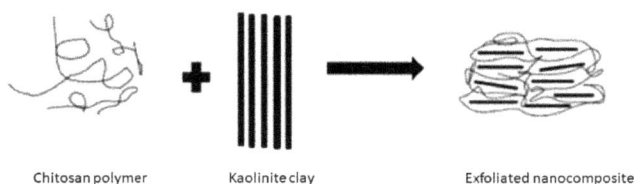

Chitosan polymer　　　　Kaolinite clay　　　　Exfoliated nanocomposite

Figure 1. Exfoliated kaolinite-chitosan nanocomposite

Adv. App. of Micro and Nano Clay – Biopolymer-based Composites   Materials Research Forum LLC
Materials Research Foundations **125** (2022) 87-102   https://doi.org/10.21741/9781644901915-4

Physicochemical properties of the kaolinite-chitosan bio composite depend mainly on their fabrication techniques.   Method of fabrication and physicochemical modification of biocomposites is thus required for widening the applications of biocomposites.   Changes in proportions of Kaolinite-chitosan and physical conditions such as pH, temperature etc., has a significant effect on composite efficiency. Physico-chemical modification improves the thermal stability of the composite. Studies showed that kaolinite-chitosan composite is stable even at 3180C temperature. Crystallinity and functional groups attached to the biocomposite surface has a vital role in the efficiency of composite.   Incorporation of kaolinite to chitosan improves the thermal stability and favours the disappearance of crystallinity of chitosan and thus widens the environmental, textile, food, and pharmaceutical applications of composite [1,5].

Kaolinite-chitosan nanocomposites are widely used in various biomedical applications, drug delivery systems, hemostatic agents, microbial fuel cells, tissue engineering, agriculture, and cosmetic and food industries.   It is also used in various membrane separation techniques such as reverse osmosis, pervaporation etc., for the purification of wastewater and saline water.

3.   Applications of kaolinite-chitosan nanocomposites

Several studies on nanoparticles proved that nanocomposites are advantageous to conventional composites due to their versatile interfacial and morphological characteristics.   Nanocomposites have a high surface area and smaller interparticle distances, a small fraction of polymer matrix, and provide high strength, higher thermal stability, high fire resistance, lower gas permeability, etc. In addition to these, their cost-effectiveness and eco-friendly nature, nanocomposites application has been increasing in various fields such as medical, pharmaceutical industries, wastewater treatment, food and beverage industries, automotive industries, sports industries, plastic and fibre industries etc. [11,12,13]. Table-1 lists various applications of kaolinite-chitosan composite.

Table 1. Applications of Kaolinite-chitosan composites

S.No.	composite	Application	Reference
1.	chitosan kaolinite nanocomposite	Adsorption analysis of Pb (II)	[14]
2.	chitosan/kaolin/γ-Fe2O3	Adsorption of an anionic azo dye	[15]
3.	chitosan-coated kaolinite beads	Removal of copper (II)	[16]
4.	kaolin based chitosan-g-PHEMA nanocomposite hydrogel	Biodegradability and swelling capacity	[17]
5.	Silver/Kaolinite Nanocomposite and Its Chitosan-Capped Derivative	Healing of Infected Wound	[18]
6.	kaolin/chitosan/titanium dioxide	Removal of CV dye.	[19]
7.	chitosan/kaolin/γ-Fe2O	Removal of MO	[20]
8.	chitosan and kaolinite biocomposite	Removal of Heavy Metal and Dye from Industrial Effluent	[21]
9.	chitosan and kaolinite nanocomposite	Hemostatic dressings	[22]
10.	Crosslinked Chitosan-epichlorohydrin/kaolin composite	Reactive blue removal and COD reduction	[23]
11.	Chitosan/kaolin composite	Cu(II) removal	[24]

3.1   Medical applications

Novel, non-toxic, and effective therapeutic drug delivery systems are primary requirements for treating many fatal diseases like cancer, dengue, influenza, SARS, etc. Inefficient drug carriers, biodistribution and effective delivery systems are significant challenges for drug effectiveness. Nanocomposites prepared from kaolinite-chitosan derivatives have an excellent surface area, biocompatibility, thermal stability. They have potential applications in medical applications such as gene therapy and tissue engineering. These nanocomposites have been used as drug carriers and delivery systems for antibiotics, antihypertensive drugs, antipsychotic and anticancer drugs. These nanocomposites are used to control drug delay time and drug release, and drug distribution [25].

Blood loss due to injury is considered one of the significant causes of death. Efficient hemostatic dressings should be applied immediately after the injury to stop bleeding and reduce deaths. Kaolinite and chitosan have been used in various medical applications due to their excellent ability to stop bleeding quickly from injuries. These nanocomposite materials are gaining importance due to their excellent biocompatible properties and not introducing any clinical or biochemical changes after their applications. Studies proved that combining more than one hemostatic agent such as kaolin-chitosan composite have improved the efficiency of bleeding control [22]. This could be due to the increased surface area and multiple components responsible for more than one mechanism in treating wounds. Studies conducted on rats and rabbits showed Kaolin-chitosan-based nanocomposite's ability as an efficient hemostatic agent. These nanocomposites effectively stopped the bleeding from the injuries in less time compared to the conventional dressing methods.

Clay materials have been used as antimicrobial agents since ancient times, and they have been used for wound healing. Kaolinite is one of the most widely used clay minerals in pharmaceutical and medical applications. Kaolinite has been used as an antimicrobial agent and blood clotting agent. When applied to the wound site as dressing, Kaolinite initiates the clotting process through transforming and activating the coagulation and prevents bleeding [26, 27].

Nanocomposites prepared from kaolinite-chitosan could be utilized in medical applications for would healing. Kaolin-chitosan nanocomposites show efficient wound healing characteristics; moreover, they are non-toxic. Mahmoudabi et al. [18] have successfully synthesized silver/kaolinite/chitosan nanocomposites of sizes 7-11nm and investigated various physico-chemical properties. Cytotoxicity and antibacterial activities of synthesized nanocomposites were also investigated. Clinical studies such as would area, bacterial count, histological parameters, protein expression etc., conducted on the usage of these nanocomposites exhibited excellent wound healing properties. Silver/kaolinite/chitosan nanocomposites showed good antibacterial activity, increased proliferative phase and were found to be very effective in lowering inflammation and wound healing.

Studies conducted on the hemostatic ability of chitosan-kaolinite composite demonstrated improved hemostatic ability compared to chitosan alone. Combined inverse emulsion and thermally induced phase separation methods were used for the synthesis of composite. The addition of Kaolinite to chitosan significantly improved the surface area. Blood clotting time was significantly reduced pm rat tail amputation and liver laceration models. These studies suggested the potential use of chitosan-kaolinite composite as a pro-coagulant agent for traumatic haemorrhaging control [28].

Adv. App. of Micro and Nano Clay – Biopolymer-based Composites   Materials Research Forum LLC
Materials Research Foundations **125** (2022) 87-102   https://doi.org/10.21741/9781644901915-4

3.2   Pharmaceutical applications

Chitosan is a natural biopolymer produced through the deacetylation of chitin, which is the second most abundantly available biopolymer in nature. Chitin is the most important component of the defensive epidermises of many shellfish. It is an integral part of many insects, crabs, shrimp, lobsters etc. It exhibits excellent physico-chemical and biological properties such as non-toxic, mucoadhesive, biocompatible and biodegradable. Because of its excellent immunological and antibacterial activities are widely used in pharmaceutical industries [17]. However, chitosan has poor mechanical strength and poor water resistance. Hence, it is necessary to combine chitosan with other materials to improve its mechanical strength and decrease its water resistance.

In acidic to neutral conditions, protonation of the amino group present in chitin polymer imparts cationic character, which enables strong electrostatic interactions with negatively charged pollutants present in the wastewater. This biopolymer is available naturally, biocompatible, non-toxic, and shows excellent antimicrobial and fungistatic effects. Thus, it is considered as one of the potential green polymers used for many industrial and municipal applications [29, 30].

Chemical modification of clay minerals such as intercalation of ions into the interlayer space of clay minerals or surface modification of clay minerals is a strategy to tune the properties of nanoclays for the loading and release of a drug. The modified nanoclay can take up drugs by encapsulation, immobilization, ion exchange reaction, or electrostatic interactions. Controlled drug release from the drug–clay originates from the incorporation and interactions between the drug and inorganic layers, including electrostatic interactions and hydrogen bonding.

Chitosan-Kaolinite nanocomposite exhibits excellent antioxidant and antimicrobial activities. Studies conducted on chitosan-kaolinite nanocomposite films showed excellent inhibitory activity against various gram-positive and gram-negative bacteria. The addition of kaolinite clay to the chitosan biopolymer resulted in a decrease in particle size and water solubility. It improved the colour and light transmission characteristics. The addition of clay material to chitosan significantly improved the mechanical strength of the biocomposite [9]. The results show that the chitosan-kaolinite biocomposite has potential for use in therapeutic applications. Chitosan biopolymer incorporated with kaolinite nanofilm improves the biopolymer Physico-chemical, mechanical and rheological properties. It makes it a potential material for use in wastewater treatment, drug delivery systems, tissue scaffolds, wound healing applications, biosensors, etc.

Kaolinite-chitosan nanocomposites have superior physical and chemical properties over unmodified chitosan or Kaolinite. These composites have combined advantages of both

Kaolinite and chitosan and find various applications in the pharmaceutical and medical industries. Due to their improved thermal stability and biocompatibility, they are widely used in drug delivery systems. These nanocomposites are also widely used in vaccine and antimicrobial applications, callus, and tissue regeneration applications [31]. Kaolinite-chitosan nanocomposites are widely used for the diagnosis and treatment of various diseases. These composites are excellent drug carriers and can be used to control the release of drugs effectively. They are very efficient in carrying the single drug and multiple drugs and can target molecules with specific receptors [31].

Chitosan/kaolin/magnetite composite prepared by emulsion crosslinking was studied for the removal of ciprofloxacin from an aqueous solution [32]. The results indicated good adsorption characteristics of chitosan/kaolin/magnetite composite. Experiments conducted on reuse of this adsorption showed very little deterioration in efficiency of removal of ciprofloxacin even after four repeated cycles. The biocomposite prepared from this method was cheap, efficient and economical. It is biocompatible and has good adsorption capacity and regeneration characteristics. It can be safely used in wastewater treatment, biological molecule separation and drug delivery.

3.3   Wastewater and industrial effluent treatment

Exponential growth in industrialization and urbanization triggered environmental pollution problems worldwide. Particularly water adulteration poses a severe threat to the fauna, flora, and other living animals. Water pollution spreads very quickly and creates serious health issues such as liver failure, kidney diseases, infertility, vomiting, cholera, cancer, eye problems etc. and may lead to death also. As a result, numerous studies have been conducted to develop efficient and economical materials and technologies for the removal of contaminants from wastewater. Among such methods, adsorption was found to be a simple, efficient, and economical method.

Various conventional and non-conventional adsorbents have been investigated for the removal of contaminants from the wastewater [33,34, 35]. Among such adsorbents, natural clay materials and biopolymers were very efficient in the wastewater treatment process due to their biocompatibility, surface characteristics, non-toxic, abundant availability, and cost-effectiveness. The use of biopolymers as adsorbents is one of the emerging trends in the adsorption of contaminants from wastewater. Chitosan is found to be very efficient in the removal of dyes and heavy metals. The presence of many hydroxyl and amino groups on its surface makes chitosan a very efficient adsorbent. Amino groups on the chitosan surface become cationic in acidic conditions and strongly attract anionic dyes through electrostatic forces [36]. However, its use is restricted due to poor mechanical strength, low specific gravity, soluble in acidic solutions, sensitivity to pH, easy agglomeration, or gel formation characteristics [32].

Kaolinite is the principal mineral constituent in kaolin, which is one of the abundantly available clay. Kaolinite has excellent adsorption properties and is widely used in wastewater and industrial effluent treatment [37,38,39]. Kaolinite has a little affinity towards the anionic pollutants' removal; hence, it must be modified to improve its ability to remove pollutants with anionic behavior [40]. Hence, chitosan modified with Kaolinite overcomes this deficiency and efficiently removes anionic dyes and metals.

Kaolinite is a little affinity towards the removal of anionic pollutants. Hence modification with chitosan improves the affinity towards anionic pollutants. Chitosan surface contains higher amounts of hydroxyl and amino groups than many conventional adsorbents. In acidic conditions, chitosan surface protonates and effectively removes anionic pollutants. Thus, kaolinite-chitosan nanocomposites are widely used in various dyes and metal removal.

Nanocomposites prepared from natural clay (Kaolinite) and biopolymer (chitosan) based materials are drawing attention due to their large surface area, excellent thermal, mechanical, and chemical stability and wide range of functional groups on their surface. Kaolinite-chitosan bio composite polymer membranes play a significant role in wastewater treatment and environmental pollution control. Kaolinite-chitosan bio composite shows excellent surface chemistry, morphological properties, chemical stability, efficient water resistance, and water permeability characteristics.

In recent years, kaolinite-chitosan based magnetic beads and nanocomposite materials have been gaining importance in wastewater treatment. Many studies were conducted to evaluate the ability of kaolinite-chitosan based immobilized beads for the removal of targeted contaminants from wastewater. Chitosan enhances the stability, and Kaolinite enhances active sites on the surface, which enable the chitosan-kaolinite composites strong interact with various contaminants present in the wastewater [41].

The advanced engineering of chitosan-kaolinite biocomposites has opened new doors in the fields of wastewater treatment. Studies conducted on removing Remazol Red using chitosan-kaolinite nanocomposites showed more than five times of higher biosorption capacity than conventional activated carbon adsorbents. Regeneration studies were also carried out, and results showed that 99% of adsorbents could be regenerated. The results proved that chitosan-kaolinite nanocomposites could be used as potential adsorbents to remove azo dyes [42].

Crosslinked chitosan epichlorohydrin/kaolin biocomposite was synthesized and studied for its reactive blue colour removal and COD reduction ability by Ali et al. [23]. Response surface methodology using Box-Behnken design was used to optimize the percentage of colour removal efficiency and COD reduction efficiency of biocomposite material. The thermodynamic studies proved the endothermic nature and spontaneity of the composite material. Studies proved that bio-composite material provided more electrostatic interactions, hydrogen bonding and facilitated good adsorption characteristics.

Kaolinite-chitosan nanocomposites are non-toxic, environment friendly and exhibit extraordinary physicochemical characteristics such as mechanical and thermal stability, high surface area, high porosity, chemical reactivity, biodegradability etc. [43]. Futalan et al. [24] have conducted studies to evaluate the adsorptive capacity of chitosan-kaolinite composite to remove Cu(II) using a fixed bed adsorption system. Regeneration studies were also carried out using HCl and NaOH eluents. The results demonstrated high adsorption capacities, which could be due to the higher surface area of composite material.

Magnetic kaolinite-chitosan nanocomposites have been tested successfully for the removal of various inorganic and organic pollutants from wastewater. Magnetic nanocomposites can be easily recovered from the solutions for regeneration/desorption studies. However, tedious procedures involved in synthesizing nanocomposites, pre-treatment of the clays, and dissolution of chitosan in a suitable solvent and crosslinking or precipitation make their use restricted. The synthesis of clay-biopolymer nanocomposites through these multiple steps increases the cost of production, production time and higher use of chemicals. Thus, there is a need for the development of single-step procedures for the synthesis of clay-biopolymer nanocomposites [44,45,46]

3.4 Desalination applications

In the current century, demand for pure water has been increased exponentially due to the exhaust of water resources. Scarcity of water resources and meeting pure water demand are severe global challenges for human beings. Usage of renewable water resources and reuse of the already utilized water is the only solution for this. Desalination using membranes is considered a promising water purification technique. Numerous studies have been focused on the development of advanced membranes for pure water generation through desalination techniques. Biopolymer membranes have attracted many researchers due to their excellent biocompatibility and competitiveness in energy-saving and eco-friendly operation [47]. Biopolymers exhibit excellent mechanical and chemical stability and show good temperature resistance over a wide range of temperatures. Recently biopolymer membranes have attracted many industries and academia since they are synthesized from renewable raw materials [48].

3.5   Food and packaging applications

Polythene and plastic materials are widely used in the food packaging industry. However, with recent awareness of environmental protection, there is a great demand for innovative alternatives. Regular packaging materials are produced from fossil fuels, and their impact on environmental damage is very much concerned. Whereas kaolinite-chitosan nanocomposite-based packaging materials are naturally available biopolymers, low cost, and provide eco-friendly packaging. These materials are biodegradable, and hence no toxic chemicals are disposed of into the environment. Film coatings are critical in active packaging. It protects against microbial attacks, preserves the food and extends the shelf life of food products. Chitosan is widely used in food preservation and packing applications due to its non-toxic nature, excellent film-forming characteristics, biocompatible and biostability properties. However, films and coatings made with chitosan alone show brittleness, high oxygen permeability and low bioactive functionality [49]. Modifying the chitosan with Kaolinite provides an improvement in mechanical and antioxidant properties [50].

Chitosan-kaolinite nanocomposite exhibits good antimicrobial, antioxidant and good thermal stability properties and provides protection through packaging. In the food packaging industry, diffusion of oxygen, carbon dioxide, moisture, and other flavours is to be controlled to keep the food hygiene and taste. Decreasing the diffusion/permeability of these compounds is critical in active packaging; Film coating made with kaolinite-chitosan nanocomposites decreases the permeability characteristics and oxygen, water vapour, etc. Thus, gaining importance in the food industry.

Conclusion

Production of biopolymers in large quantities requires the farming of energy crops in huge quantities in competition with food security crops. Thus, developments should be done in the utilization of industrial wastes and by-products for biopolymer production. Chitosan obtained from the marine and food processing industries as a waste material represents a good equilibrium between food consumption and value addition to the industrial waste materials.

This chapter presented the potential use of kaolinite-chitosan nanocomposites in various industries. In recent research studies, it has been proved that kaolinite-chitosan-based nanomaterials have superior biological activity and other Physico-chemical properties than the bulk or microparticles. Recent studies display the various conventional applications and new applications of kaolinite-chitosan nanocomposites in the field of biosensors, biomedical devices, and pharmaceutical industries. The use of nanocomposites technology is still in its early stages; more research should be placed to develop new processes and technologies to prepare these nanocomposites in an economical, environment-friendly manner. More studies should be conducted to improvise the development of compatible filler polymer systems, better processing techniques and design methods. Kaolinite-chitosan nanocomposites may advance the research in various applications in future.

References

[1] S.C. Dey, Md. Al-Amin, T.U. Rashid, Md. Ashaduzzaman, S. Md. Shamsuddin, pH Induced Fabrication of Kaolinite-Chitosan Biocomposite. Int. Lett. Chem. 68 (2016) 1-9. https://doi.org/10.18052/www.scipress.com/ILCPA.68.1

[2] M.S. Chiou, H.Y. Li, Equilibrium and kinetic modeling of adsorption of reactive dye on cross-linked chitosan beads, J. Hazard. Mater. 93 (2002) 233-248. https://doi.org/10.1016/S0304-3894(02)00030-4

[3] S. Hasan, T.K. Ghosh, D.S. Viswanath, V.M. Boddu, Dispersion of chitosan on perlite for enhancement of copper (II) adsorption capacity, J. Hazard. Mater. 152 (2008) 826-837. https://doi.org/10.1016/j.jhazmat.2007.07.078

[4] S.R. Popuri, Y. Vijaya, V.M. Boddu, K. Abburi, Adsorptive removal of copper and nickel ions from water using chitosan coated PVC beads, Bioresour. Technol. 100 (2009) 194-199. https://doi.org/10.1016/j.biortech.2008.05.041

[5] S. Biswas, T.U. Rashid, A.K. Mallik, Md.M. Islam, M.N. Khan, P. Haque, M. Khan, Md.M. Rahman, Facile Preparation of Biocomposite from Prawn Shell Derived Chitosan and Kaolinite-Rich Locally Available Clay, Int. J. Polym. Sci. 2017 (2017) 8 https://doi.org/10.1155/2017/6472131

[6] H.H. Murray, Kaolin applications, Applied Clay Mineralogy. Occurrences, Processing and Application of Kaolins, Bentonites, Palygorskite Sepiolite, and Common Clays, in: H.H. Murray (Ed.), Developments in Clay Science, vol. 2, Elsevier, Amsterdam, 2006, pp. 85-109. https://doi.org/10.1016/S1572-4352(06)02005-8

[7] M. Dabbaghianamiri, S. Das, G.W. Beall, Improvement approach for gas barrier behavior of polymer/clay nanocomposites films, MRS Adv. 2 (2017) 3547-3552. https://doi.org/10.1557/adv.2017.458

[8] C. Pagano, F. Marmottini, M. Nocchetti, D. Ramella, L. Perioli, Effects of different milling techniques on the layered double hydroxides final properties, Appl. Clay Sci. 151 (2018) 124-133. https://doi.org/10.1016/j.clay.2017.10.030

[9] Amina Baccour Neji, Mourad Jridi, Hela kchaou, Moncef Nasri, Rym Dhouib Sahnoun, Preparation, characterization, mechanical and barrier properties investigation

of chitosan-kaolinite nanocomposite, Polym. Test. 84 (2020) 106380
https://doi.org/10.1016/j.polymertesting.2020.106380

[10] A. Laaraibi, F. Moughaoui, F. Damiri, A. Ouakit, I. Charhouf, S. Hamdouch, A. Jaafari, A. Abourriche, N. Knouzi, A. Bennamara, Md. Berrada, Chitosan-Clay Based (CS-NaBNT) Biodegradable Nanocomposite Films for Potential Utility in Food and Environment, in: R.S. Dongre (Eds), Chitin-Chitosan - Myriad Functionalities in Science and Technology, Intechopen, 2018, pp. 1-8, doi:10.277./intechopen.98468. https://doi.org/10.5772/intechopen.76498

[11] Y. Bréchet, J.Y. Cavaillé, E. Chabert, L. Chazeau, R. Dendievel, L. Flandin, C. Gauthier, Polymer based nanocomposites: Effect of filler-filler and filler-matrix interactions. Adv. Eng. Mater. 3 (2001) 571-577. https://doi.org/10.1002/1527-2648(200108)3:8<571::AID-ADEM571>3.0.CO;2-M

[12] D. Paul, L. Robeson, Polymer nanotechnology: Nanocomposites, Polym. 49 (2008) 3187-3204. https://doi.org/10.1016/j.polymer.2008.04.017

[13] V. Mittal, Polymer Layered Silicate Nanocomposites: A Review, Mater. 2 (2009) 992-1057. https://doi.org/10.3390/ma2030992

[14] V. Kanchana, T. Gomathi, V. Geetha, P. Sudha, Adsorption analysis of Pb (II) by nanocomposites of chitosan with methyl cellulose and clay, Der Pharm. Lett. 4 (2012) 1071-1079.

[15] H.Y. Zhu, R. Jiang, L. Xiao, Adsorption of an anionic azo dye by chitosan/kaolin/γ-Fe2O3 composites. Appl. Clay Sci 48 (2010) 522-526. https://doi.org/10.1016/j.clay.2010.02.003

[16] I.P. Chen I.P, C.C Kan, C.M. Futalan, M.J.C. Calagui, S.S. Lin, W.C. Tsai, M.W. Wan, Batch and fixed bed studies: Removal of copper (II) using chitosan-coated kaolinite beads from aqueous solution, Sustain. Environ. Res. 25 (2015) 73-81.

[17] A.K. Pradhan, P.K. Rana, P.K. Sahoo, Biodegradability and swelling capacity of kaolin based chitosan-g-PHEMA nanocomposite hydrogel, Int. J. Biol. Macromol. 74 (2015) 620-626. https://doi.org/10.1016/j.ijbiomac.2014.12.024

[18] S. Mahmoudabadi, M.R. Farahpour, S. Jafarirad, Effectiveness of Green Synthesis of Silver/Kaolinite Nanocomposite Using Quercus infectoria Galls Aqueous Extract and Its Chitosan-Capped Derivative on the Healing of Infected Wound, in IEEE Transactions on NanoBioscience, 20 (2021) 530-542 https://doi.org/10.1109/TNB.2021.3105356

[19] H.S. Vardikar, B.A. Bhanvase, A.P. Rathod, S.H. Sonawane, Sonochemical synthesis, characterization and sorption study of Kaolin-Chitosan-TiO2 ternary

nanocomposite: Advantage over conventional method, Mater. Chem. Phys. 217 (2018) 457-467 https://doi.org/10.1016/j.matchemphys.2018.07.014

[20] R. Jiang, H. Zhu, Y. Fu, Equilibrium and Kinetic studies on adsorption of methyl orange from aqueous solution on chitosan/kaolin/γ-Fe2O3 nanocomposite, International Conference on Remote Sensing, Environ. Transp. Eng. (2011)7565-7568 https://doi.org/10.1109/RSETE.2011.5966122

[21] S. Biswas, T.U. Rashid, T. Debnath, P. Haque, M.M. Rahman, Application of Chitosan-Clay Biocomposite Beads for Removal of Heavy Metal and Dye from Industrial Effluent, J. Compos. Sci. 4 (2020) 16. https://doi.org/10.3390/jcs4010016

[22] M. Elsabahyd, M.A Hamad, Sashiwa Hitoshi, Design and Preclinical Evaluation of Chitosan/Kaolin Nanocomposites with Enhanced Hemostatic Efficiency, Mar. Drugs. 19 (2021) 50-50. https://doi.org/10.3390/md19020050

[23] A.H. Jawad, A.S. Abdulhameed, A.N. Najwa, A. Malek, Z.A. Alothman, Statistical optimization and modeling for color removal and COD reduction of reactive blue 19 dye by mesoporous chitosan-epichlorohydrin/kaolin clay composite, Int. J. Biol. Macromol. 164 (2020) 4218-4230, https://doi.org/10.1016/j.ijbiomac.2020.08.201

[24] C.M. Futalan, J.H. Yang, P. Phatai, Fixed-bed adsorption of copper from aqueous media using chitosan-coated bentonite, chitosan-coated sand, and chitosan-coated kaolinite, Environ. Sci. Pollut. Res. 27 (2020) 24659-24670. https://doi.org/10.1007/s11356-019-06083-0

[25] N. Khatoon, M.Q. Chu, C.H. Zhou, Nanoclay-based drug delivery systems and their therapeutic potentials, J. Mater. Chem. B, 8 (2020) 7335-7351 https://doi.org/10.1039/D0TB01031F

[26] J.B. Glick, R.R. Kaur, D.Siegel, Achieving hemostasis in dermatology-Part II: Topical hemostatic agents, Indian Dermatol. Online J., 4 (2013) 172. https://doi.org/10.4103/2229-5178.115509

[27] M.E. Chávez-Delgado, C.V. Kishi-Sutto, X.N. Albores de la-Riva, M. Rosales-Cortes, P. Gamboa-Sánchez, Topic usage of kaolin-impregnated gauze as a hemostatic in tonsillectomy, J. Surg. Res. 192 (2014) 678-685. https://doi.org/10.1016/j.jss.2014.05.040

[28] X. Sun, Z. Tang, M. Pan, Z. Wang, H. Yang, H. Liu, Chitosan/kaolin composite porous microspheres with high hemostatic efficacy, Carbohydr. Polym. 177 (2017) 135-143 https://doi.org/10.1016/j.carbpol.2017.08.131

[29] R. Kumar, A.M. Isloor, T. Matsuura, Synthesis and characterization of novel water soluble derivative of chitosan as an additive for polysulfone ultrafiltration membrane, J. Membr. Sci., 440 (2013), 140-147 https://doi.org/10.1016/j.memsci.2013.03.013

[30] I. Hamed, F. Ozogul, J.M. Regenstein, Industrial applications of crustacean byproducts (chitin, chitosan, and chitooligosaccharides): a review, Trends Food Sci. Technol., 48 (2016), 40-50 https://doi.org/10.1016/j.tifs.2015.11.007

[31] D. Zhao, S. Yu, B. Sun, S. Gao, S. Guo, K. Zhao, Biomedical Applications of Chitosan and Its Derivative Nanoparticles, Polymers (Basel).10 (2018) 462, https://doi.org/10.3390/polym10040462

[32] W. Ma, J. Dai, X. Dai, Y. Yan, Preparation and Characterization of Chitosan/Kaolin/Fe3O4 Magnetic Microspheres and Their Application for the Removal of Ciprofloxacin. Adsorp. Sci. Technol. 32 (2014) 775-790. https://doi.org/10.1260/0263-6174.32.10.775

[33] K.V.N. Saibaba, R.V. Kandisa, Adsorption isotherm studies on methylene blue dye removal using naturally available biosorbent, Rasayan J. Chem. 12 (2019) 2176-2182. https://doi.org/10.31788/RJC.2019.1245478

[34] K.V.N. Saibaba, P. King, R. Gopinadh, D.K.N. Lakshmi, Response surface optimization for the decolorization of crystal violet dye from aqueous solutions by waste crab shells, Int. J. Appl. Environ. Sci. 7 (2012) 149-154.

[35] K. Veerabhadram, K.V.N. Saibaba, K. Sattibabu, Efficient removal of hexavalent chromium from industrial automobile solid waste using bioremediation technique, Indian J. Environ. Prot. 40 (2020) 203-206

[36] M.N.V.R. Kumar, A review of chitin and chitosan applications. React. Functional Polym. 46 (2000) 1-27. https://doi.org/10.1016/S1381-5148(00)00038-9

[37] Q. Tang, X.W. Tang, Z.Z. Li, Y.M. Chen, N.Y. Kou, Z.F. Sun, Adsorption and desorption behaviour of Pb(II) on a natural kaolin: equilibrium, kinetic and thermodynamic studies, J. Chem. Technol. Biotechnol., 84 (2009) 1371-1380. https://doi.org/10.1002/jctb.2192

[38] W.H. Cheung, Y.S. Szeto, G. McKay, Enhancing the adsorption capacities of acid dyes by chitosan nano particles, Bioresour. Technol. 100 (2009) 1143-1148. https://doi.org/10.1016/j.biortech.2008.07.071

[39] G. Crini, P.M. Badot, Application of Chitosan, a Natural Aminopolysaccharide, for Dye Removal from Aqueous Solutions by Adsorption Processes Using Batch Studies: A Review of Recent Literature. Polym. Sci. 33 (2008) 399-447. https://doi.org/10.1016/j.progpolymsci.2007.11.001

[40] H. Gecol, P. Miakatsindila, E. Ergican, Biopolymer coated clay particles for the adsorption of tungsten from water, Desalination, 197 (2006) 165-178. https://doi.org/10.1016/j.desal.2006.01.016

[41] S.SD. Elanchezhiyan, P. Karthikeyan, R. Karthik, M.H. Farzana, C.M. Park, Magnetic kaolinite immobilized chitosan beads for the removal of Pb(II) and Cd(II) ions from an aqueous environment, Carbohydr. Polym. 261 (2021) 117892 https://doi.org/10.1016/j.carbpol.2021.117892

[42] S.C. Dey, M. Moztahida, M. Sarker, M. Ashaduzzaman, S.M. Shamsuddin, pH-Triggered Interfacial Interaction of Kaolinite/Chitosan Nanocomposites with Anionic Azo Dye. J. Compos. Sci. 3 (2019) 39-50. https://doi.org/10.3390/jcs3020039

[43] X.H. Wang, Y. Zheng, A.Q. Wang, Fast removal of copper ions from aqueous solution by chitosan-g-poly(acrylic acid)/attapulgite composites. J. Hazard. Mater. 168 (2009) 970-977. https://doi.org/10.1016/j.jhazmat.2009.02.120

[44] V. Arya, L. Philip, Adsorption of pharmaceuticals in water using Fe3O4 coated polymer clay composite, Microporous Mesoporous Mater., 232 (2016), 273-280 https://doi.org/10.1016/j.micromeso.2016.06.033

[45] M.M. Sobeih, M.F. El-Shahat, A. Osman, M.A. Zaid, M.Y. Nassar, Glauconite clay-functionalized chitosan nanocomposites for efficient adsorptive removal of fluoride ions from polluted aqueous solutions, RSC Adv., 10 (2020) 25567-25585 https://doi.org/10.1039/D0RA02340J

[46] R. Rusmin, B. Sarkar, R. Mukhopadhyay, T. Tsuzuki, Y. Liu, R. Naidu, Facile one pot preparation of magnetic chitosan-palygorskite nanocomposite for efficient removal of lead from water, J. Colloid Interface Sci. 608 (2022) 575-587 https://doi.org/10.1016/j.jcis.2021.09.109

[47] M.T. Ravanchi, T. Kaghazchi, A. Kargari, Application of membrane separation processes in petrochemical industry: a review, Desalination, 235 (2009) 199-244 https://doi.org/10.1016/j.desal.2007.10.042

[48] S.B. Rekik, S. Gassara, J. Bouaziz, A. Deratani, S. Baklouti, Development and characterization of porous membranes based on kaolin/chitosan composite, Appl. Clay Sci. 143 (2017) 1-9 https://doi.org/10.1016/j.clay.2017.03.008

[49] M. Mujtaba, R.E. Morsi, G. Kerch, M. Elsabee, M. Kaya, J. Labidi, K.M. Khawar, Current advancements in chitosan-based film production for food technology; A review, Int. J. Biol.Macromol. 121 (2019) 889-904 https://doi.org/10.1016/j.ijbiomac.2018.10.109

[50] B. Qu, Y. Luo, A review on the preparation and characterization of chitosan-clay nanocomposite films and coatings for food packaging applications, Carbohydr. Polym. Technol. Appl. 2 (2021) 100102-100109, https://doi.org/10.1016/j.carpta.2021.100102

Chapter 5

Chitosan-Halloysite Nano-Composite for Scaffolds for Tissue Engineering

Krishnapur M. Pragna[1], N. Sushma[1], K. Bhanu Revathi[1], K. Shinomol George[2,*]

[1]Department of Biotechnology, Dayananda Sagar College of Engineering, Bengaluru, India

[2]Faculty Biochemistry, Department of Life Sciences, Kristu Jayanti College (Autonomous), Bengaluru, India

*shinojesu@gmail.com; shinomol@kristujayanti.com

Abstract

Repair mechanism maintains the integrity and function of damaged tissues. However, natural repair slows down with age and diseases. Generation of synthetic materials for tissue renewal adopts technique from molecular biology as well as structural and molecular engineering. Nanomaterials possess superior strength that outdoes the relative characteristics of conventional materials. Nano scaffolds along with growth factors used in organ regeneration; hasten wound healing process. Recently, halloysites nanotubes are replacing Carbon Nanotubes, and doping with chitosan enhances biocompatibility and mechanical strength; also lowering cytotoxicity. Thus, exploring economical and superior performance Halloysite Nanotubes incorporated into chitosan is essential in cost effective biomedical applications and regeneration of tissues.

Keywords

Chitosan, Halloysite, Scaffolds, Nanotechnology, Tissue Engineering

Abbreviations

3D - Three Dimensional
ECM - Extracellular matrix
TKA – Total knee arthroplasty
HAP – Hydroxyapatite
TCP – Tricalcium phosphate
BG – Bioactive glass
CS – Calcium silicate

CaP – Calcium phosphate
DNA – Deoxyribonucleic acid
RNA – Ribonucleic acid
PCL – Polycaprolactone
CNT – Carbon nanotube
MNP – Magnetic nanoparticles
MSC – Mesenchymal stem cells
Mag - TE – Magnetic force-based tissue engineering
MCL – Magnetite cationic liposomes
HAEC – Human aortic endothelial cells
BG – Bioactive glass
HNT – Halloysite nanotubes
1D – One Dimensional

Contents

Adv. App. of Micro and Nano Clay – Biopolymer-based Composites   Materials Research Forum LLC
Materials Research Foundations **125** (2022) 103-123   https://doi.org/10.21741/9781644901915-5

1.   Introduction

Repair in an organism is a very important procedure that helps in preserving and restoring the integrity of the injured organs, tissues etc., [1]. The tissues of liver, skin, and bone, can be rehabilitated post the infection, injury or trauma by natural mechanism [2,3]. Although, this can be time consuming, resulting in slower recovery as with age or type of the disease [4]. Natural living organisms can adapt to their environment because of their genetic potential. Conversely, artificial materials cannot be adjusted or adapted. In this approach, to create synthetic materials for the revival of novel group of cells, tissue engineering adopts methods derived from structural engineering and molecular biology [5].

Tissue engineering has endured to emerge as one of the exhilarating and diverse area amid the 1980s, aimed at improving the elements that replace, restore or regenerate defective tissue [6,7]. It is an interdisciplinary field which utilizes the engineering concepts along with health science for the development of replacements which restore, and enhances the functioning of group of cells and group of tissues [8]. It can be used to make complex replacement objects from the bench to regenerate, preserve, or boost the functions of tissues and organs [9,10]. The essential elements of the engineered tissue are combination of cells, scaffold, and signals constitute the tissue engineering triad (Fig. 1).

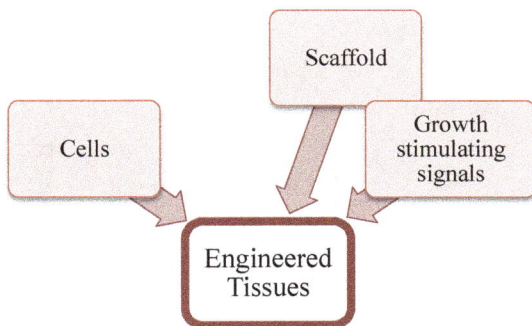

Figure 1: Tissue engineering triad

Tissue Engineering involves three main methods (Fig. 2): (1) transplantation of cells isolated from a healthy part to an injured tissue, (2) injecting the factors (growth factors, differentiation factors, polysaccharides, and peptides) induce tissue regeneration to the desired site and (3) seeding of cells together with growth factors on a three-dimensional (3D) matrix, often known as a 'scaffold', which usually acts as a short lived framework, helps in adhering and differentiation of cells in *vitro* before implantation *in vivo* [8].

Two main methods are used to manufacture novel tissues. The initial technique of scaffolding acts as an aid for the cell tool. The process begins with transfer of the cells *in vitro* on top layer of scaffold. Thereafter, cells are allowed to settle down the matrix. In due time the foundations for reconstituted tissue starts producing. The latter technique, comprises the employment of scaffolds as a tool or growth factor for drug delivery, In this technique the scaffold is made to be adhered, to the growing material. Hence, during the implantation, cells are recruited to the scaffold site from body, thus constituting the tissues. The two above mentioned techniques are not consistent and could be collaborated effortlessly [11].

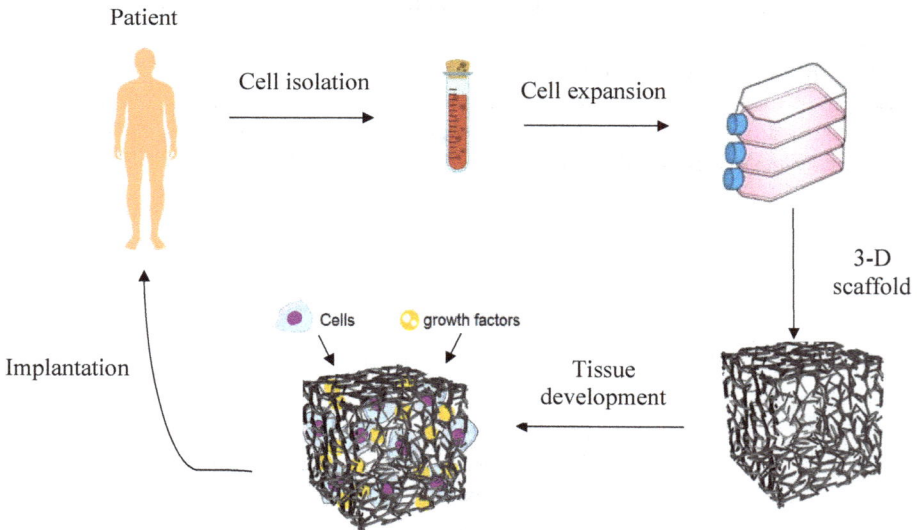

Figure 2: Schematic representation of scaffold-based tissue engineering method.

Scaffolds are often considered as a model. This further assist in the reconstruction of flaws. It also assists in attachment, growth of cells, regeneration of extracellular matrix, restoration of nerves, muscles, bones, etc. Scaffolds are biomaterials which can be porous, fibrous, or permeable. They try to ensure that body liquids and gases are transported, encourage cell interaction and ECM deposition low rate of inflammation and toxicity [12]. Simultaneously, 3D scaffolds are used as tissue, which duplicates the formation of living tissue. Thus, for the identical cause, macro, micro, and nano-architecture of scaffolds and biomaterials are used.

2.   Scaffolds

To meet the demand for various aspects of tissue engineering; different types of scaffolds of various compositions were developed. Naturally, the finest scaffold for the tissue regeneration will be target tissue's matrix in its original state. Yet, the numerous tasks, composition, and vital identity of ECM in original tissues turns out to be troublesome. Hence, the current idea of scaffolding involves mimicking functions at minimum of original ECM partly. [13].

2.1   Scaffold requirements

The significant factors for an ideal scaffold (Fig. 3) for tissue engineering are [13]:

2.1.1 Architecture

To aid the incorporation of host tissue attachment, scaffolds should offer vacant place for vascularization, fabrication of novel tissues and remodelling. Bio materials must be refined, thus providing a design for transportation of nutrients as well as metabolites, without trading off the scaffold firmness. In addition, upon implantation, biomaterials should be degradable that matches the pace of the fabrication of the novel matrix.

2.1.2 Compatibility of cells and tissues

Scaffolds are anticipated to provide support for both externally and interior cells for attachment, growth and differentiation throughout *in vivo* grafting and culturing via *in vitro*. Biomaterials used for manufacturing the scaffolds are required to maintain affinity with interior cells in host along with cellular units of constructed tissues.

2.1.3 Bioactivity

To facilitate and regulate the activities of engineered tissues, scaffolds are made to interact with their cellular components. To enhance the attachment biological cues like cell adhesive ligands are used. Similarly, to influence the cell alignment and morphology

Adv. App. of Micro and Nano Clay – Biopolymer-based Composites   Materials Research Forum LLC
Materials Research Foundations **125** (2022) 103-123   https://doi.org/10.21741/9781644901915-5

physical cues extensively like topography are included. Thus, biomaterials are required to keep up affinity with bio molecules with preserved bio activity. To mention, production of hydrogels by crosslinking aids in the capturing of proteins as well as releasing proteins by hydrogels through controlled mechanism.

2.1.4 Mechanical characteristics

Scaffolds offer stability to the tissue shortcomings. The biomaterials' inherent properties in use for scaffolding should correspond to the host. Studies have emphasized significance on the mechanical properties of implanted cells [14]. The mechanosensitivity studies had exhibited for mesenchymal stem cells differentiation [15], determining the stiffness of the agarose gel differentiation tendency.

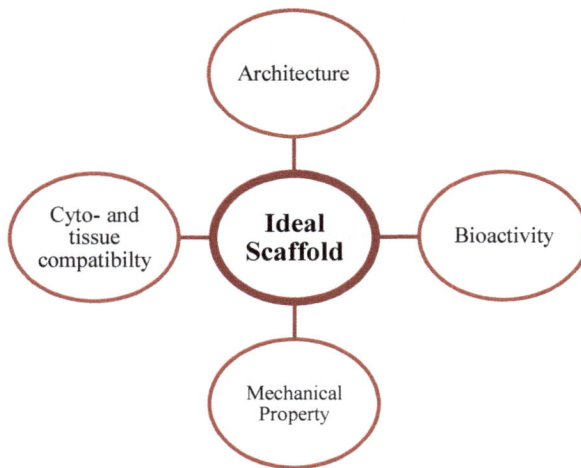

Figure 3: Ideal characteristics of a scaffold.

2.2 Types of scaffolds

2.2.1 Metal based scaffolds

Researchers have developed personalized metallic scaffolds. This aids for supporting a wide variety of purposes in dental and medical field. Changes in surface of proven biologically compatible metals are one of the primary specifications for generation of tissues. This is mostly for the reason that surface of the metals should be restricted to bring about the biomolecules adsorption as well as attachment and growth of cells [16]. They possess various mechanical properties like ductility, resistance to the corrosion, to name a few [17]. Very benefit of the metals is due to their bioinert nature making it easy to produce finished goods. Metal-based scaffolds like Stainless steel, Tantalum, Magnesium, Titanium, and its alloys, Copper, and their alloys [16] are widely employed in dental and orthopaedic surgeries. The metallic substance has various refining demands as well as the various level of techniques for the formation of scaffold.

Porous tantalum is one of the unique biomaterials with high-surface porosity (>80%) that are highly interconnected, to permit safe and swift bone development [18]. It also possesses a elasticity modulus that is very comparable to bone, that helps in minimizing stress shielding. With its various unique properties, porous tantalum can be a substitute metal for initial and revised total knee arthroplasty (TKA). Magnesium and Magnesium alloys have been used for specific surgical applications. Lately, Mg-Ca alloys were tested *in vitro* and *in vivo* for the biodegradability of biomaterials for use in implants [19]. Titanium is found to be well tolerated in the human body environment. Titanium's toughness can be improved when it is alloyed with Aluminium, Molybdenum, Manganese, Nickel, Niobium, Tantalum, Cobalt, or Vanadium [20]. The generally familiar Titanium alloy, Ti6Al4V, is commonly used in bone reconstruction. This is simply owing to their ability to advance the binding of osteoblasts, in comparison to regular substance. For any bone reconstructions, there should be increased amounts of calcium deposits, which are formed due to the advancements in adhesion [20]. When compared to steel and alloys of Cr-Co, alloys of titanium show superior biocompatibility.

2.2.2 Ceramic based scaffolds

In the past few decades, bioactive ceramics have captured immense attention due to its achievement in inducing cell division, differentiation, and bone regeneration. They tend to usually react and chemically join to form chemical bonds with various types of cells. Some bioactive ceramic materials due to their ability to bond directly with the bone in the living system following implantation in bone shortcomings, such as Hydroxyapatite (HAP), Phosphate and silicate of Calcium and bioactive glass (BG) were abundantly considered as scaffold materials [21]. HAP is a key component of natural bone. It is found to coalesce with tissues by means of chemical bonds. This helps to regenerate new bone tissue post implantation [22]. Tricalcium Phosphate (TCP) has good biological activity, biodegradability and is compatible in living systems. It is seen to guide bone regeneration and also boost stem cell proliferation capacity [23]. BG promotes the expression of osteocalcin gene [24]. CS possesses the ability to bind to bone along with the soft tissue along with excellent bioactivity.

Due to their performance in supporting the bone formation, BG and TCP is employed in restoration therapy. TCP nanoparticles possess the capability to improve the synthesis of

Adv. App. of Micro and Nano Clay – Biopolymer-based Composites   Materials Research Forum LLC
Materials Research Foundations **125** (2022) 103-123   https://doi.org/10.21741/9781644901915-5

hard tissue. The main advantages of the TCP based material being that it can be synthesized in huge measure, comparatively lesser in cost and stable, high likeness to bone mineral, biocompatibility and biodegradability under moderate acidic conditions [25].

2.2.3 Polymer based scaffolds

Owing to their exceptional attributes like elevated surface-to-volume ratio and porosity along with very small pore size, biodegradability, and mechanical strength polymeric scaffolds tend to gain great attention. Natural polymers are considered to be the earliest biodegradable biomaterials, that have been used in clinical pratices [26]. Since natural substances are proven to be bioactive, they exhibit improved cellular interactions that permit them to performance better in cells. Natural materials are classified into proteins, polysaccharides, and polynucleotides (Fig. 5). Proteins comprise silk, actin, elastin, keratin, fibrinogen, collagen and myosin, whereas polysaccharides comprise dextran, chitin, cellulose and glycosaminoglycans, and amongst the polynucleotides are DNA, RNA [27].

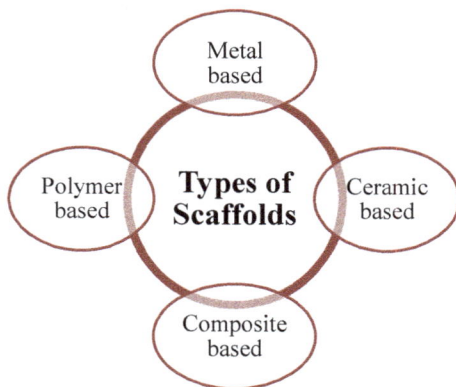

Figure4: Types of Scaffolds

Different procedures have been employed for the manufacture of porous scaffolds with the help of synthetic biopolymers. Synthetic biopolymers are usually considered to be cheaper than biological scaffolds and correspond to the leading collection of biodegradable polymers. They have longer shelf time and can be produced in huge quantities under controlled settings. Most of the manmade polymers available in the market exhibit enhanced mechanical properties, when compared to biological tissues [28].

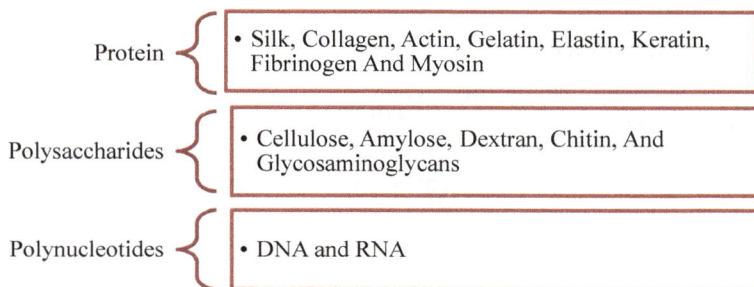

Protein	• Silk, Collagen, Actin, Gelatin, Elastin, Keratin, Fibrinogen And Myosin
Polysaccharides	• Cellulose, Amylose, Dextran, Chitin, And Glycosaminoglycans
Polynucleotides	• DNA and RNA

Figure5: Natural Polymers

The most widely utilized artificial polymers are polyglycolic acid (PGA), polylactic acid (PLA) and polylactic acid-co-glycolic acid (PLGA) copolymers [29]. Polyhydroxyalkanoates (PHA) are considered to be a group of microbial derived polyesters and well considered for tissue engineering [30]. Polycaprolactone (PCL) degrades at a considerably milder pace when compared to the PHAs. The time-consuming degradation renders PCL more appealing in long-term implants and controlled release application but then less attractive for biomedical applications [29, 31].

2.2.4 Composite based scaffolds

The amalgamation of two or more substances which differ in composition, to facilitate certain chemical, physical and mechanical properties on a macroscopic scale is generally referred to as 'composite material'. The benefit is that the subsequent composite material might have a blend of the finest attributes of their constituents, and other noteworthy properties that are not quite often shown by the single constituents [32]. Composite based scaffolds are considered to represent good bioactivity, better mechanical properties, flexibility and structural integrity.

Bioactive ceramics like HAP and CaPs (as mentioned in section 2.2.2) have shown suitable bone conductivity and biocompatibility for bone repair [33], owing to the chemical and structural similarity to native bone but then again, they possess poor shape ability. By adding the HAP or CaP, they tend to increase the surface roughness which in turn enhances the cellular adhesion and promotes a fine distributed morphology. Bioactive glass composite scaffolds that mimics hypoxia, promotes angiogenesis and bone regenration and control the issue of insufficient vascularization of grafts usually observed in tissue engineering [5].

3.   Role of Nanotechnology in scaffolding

Present scaffold systems are used for grafting, mostly in degenerative diseases for tissue regeneration. These are intended to provide functions to stem cells that are transplanted and as well as guide them towards the desired tissue. Hence, to deliver an appropriate niche for stem cells, nanotechnology has a critical role to play in engineering the scaffolds for growth of cells and its function [34].

Nanotechnology is adopted in tissue engineering to gain enhanced mechanical and biological performances. When compared to the relative features of established materials, nanomaterials possess exceptional strength and the reason is shortcomings in materials decline as the size of crystal-like grain decreases into nanometres [35]. Nanomaterials correspondingly have exceptional tensile modulus, that attributes to minor grain size, strong interfacial interactions, also large surface area [36]. The conductivity of gold nanoparticles, the power of silver nanoparticles restricting the microbial growth, and distinctive properties of CNTs are the reason why nanomaterials are advantageous in various appliances of tissue engineering. The use of nanofibrous scaffolds is widely investigated, due to their identical phenomenal apperance to protein nanofibers in ECM [37]. On other hand, nanocomposite-based scaffolds like nano-hydroxyapatite are gaining more popularity for the reconstruction of bone tissues [38].

Several novel methods have been designed that integrate growth factors into the nanocarrier systems [39]. To preclude the scarcity of growth factors, integration of the growth nutrients is loaded with nanocarriers inside scaffolds. This in turn hints on likelihood and tolerance of continued discharge of growth factors, which in turn stops death of cells [40,41]. Various approaches have been adopted to bring growth factors to planted cells by utilising micro as well as nanocarrier systems.

Recently, magnetic nanoparticles (MNPs) have been proven to be useful for delivery of cell gene, control on cell pattern, mechano-transduction, and fabrication of dense 3D tissues. Due to the distinctive dimensions as well as extraordinary magnetic property MNPs are trying to gain lots of consideration. They vary 20–300 nm in size, and at room or biological temperature, they support thermal fluctuations which is due to the directions of their magnetization.

Adv. App. of Micro and Nano Clay – Biopolymer-based Composites   Materials Research Forum LLC
Materials Research Foundations **125** (2022) 103-123   https://doi.org/10.21741/9781644901915-5

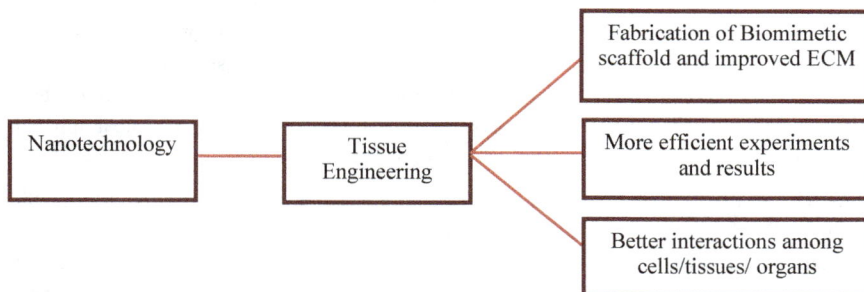

Figure 6: Nanotechnology in tissue engineering

3.1 Nanoscaffolds for skin regeneration

Nanoscaffolds along with growth factors fasten the wound healing process [42]. However, electrospun nanofibers demonstrate the similarity with the ECM. But they are connected with challenges are related to density, hence compositenanofibrous scaffolds obtained from electrospinning are in demand for various applications [43]. Recent studies have been focused on the manufacture of collagen fusion structures solely based on mimicking in the field of skin tissue engineering [44]. By utilizing different combinations various hybrid scaffolds have been manufactured. The benefit of these scaffolds is proven to have great capacity designed for linkage and development of MSCs derived fat tissues. The reason is the existence of microsponges of collagen along with solid support offered by PCL-Poly (L-lactic acid) [44].

In another trial by Rheinwald and Green, various sheets of keratinocyte cellular layers were constructed. These were found to be significantly effective for transplantation and improvement. The hydrogel films along with magnet of neodymium were bound together and was utilised to harvest keratinocytes that were magnetically labeled. The results showed that devoid of the magnet or MCLs, the keratinocytes refused to attach to surface's culture [45]. Likewise, the sheets that contained the cells were undifferentiated, which may be added advantage in curing the lesions.

3.2 Nanoscaffolds for liver regeneration

Various researchers have confirmed that magnetic force-based tissue engineering is capable to be employed as layers of cells for manufacturing liver tissues. It was discovered that the cell interaction was close and tight, which was caused by the magnetic force. This led to the deposition of cytokines between the layers and an ECM, thus enhancing the function of the liver. The hepatocytes were cocultured along with endothelial cells using Mag-TE. The human aortic endothelial cells were categorised using MCLs and was assigned on a sheet of hepatic cells of rat. Multiple layers of HAECs were designed and made sure that they remain attached to hepatic cells film in a uniform style. The amplified quantity of secreted albumin was preserved by both the cultures for 8 days [46].

In another study, it was observed that HepG2 which was magnetically labeled along with NIH3T3 cells remained to be attracted to the magnet at the ultra-low attachment. They were magnetised towards the bottom. In addition they were seen to form strong sheet-like structures. The cells got detached and were seen to be disengaged without disruption when the magnet was used [47].

3.3   Nanoscaffolds for bone regeneration

Nanoscaffolds serve as an alternative to complex surgical treatment. Biomaterials are used for fabricating the composition of bone-like Titanium, Calcium phosphate, Carbon nanotubes, Graphene oxide, and Nanohydroxyapatite [48]. By endorsing the regeneration, shape-memory nanocomposite nanoscaffolds were designed to repair bone defects. These scaffolds were made of chemically cross-linked PCL and HAP, these studies were proven by *in vivo* studies. The studies were conducted on rabbit mandibular bone defects [49]. While the *in vitro* studies were conducted on MG63 osteoblast electrospun nanocomposite scaffold (PCL/Gel(50/50)/BG) exhibited improved growth. This was because of the existence of 5% bioactive glass (BG) in the scaffold [50]. These compositions are quite similar to the bone or cartilage composition. The combination of organic and inorganic particle scaffolds which were developed demonstrated better biocompatibility and mechanical strength in bone tissue engineering [51].

4.   Use of chitosan – halloysite nanocomposite scaffolds

Chitosan is a cationic linear polysaccharide constituting β-(1,4)-linked units of N-acetyl-glucosamine (Fig. 7). It is also a naturally obtained hydrogel, which is generally obtained by deacetylation in alkaline conditions. It is distinguished by the larger population of NH_2 free moiety. Chitosan is usually soluble in aqueous solvents of pH more than 7. Nevertheless, in diluted acids of pH 6, protonation occurs among the unbound amino groups of chitosan thereby making it water soluble. Depending on not only the source but also the preparation methods, the molecular weight of chitosan was found to be 300 kDa to more than 1,000 kDa. Chitosan is proven to be a biocompatible, decomposable, and a biopolymer that does not elicit toxicity, that can be used inside or outside the human body safely. It also has the ability to assemble a tissue-specific ECM [52]. It has been extensively applied in several biomedical purposes (Fig. 8).

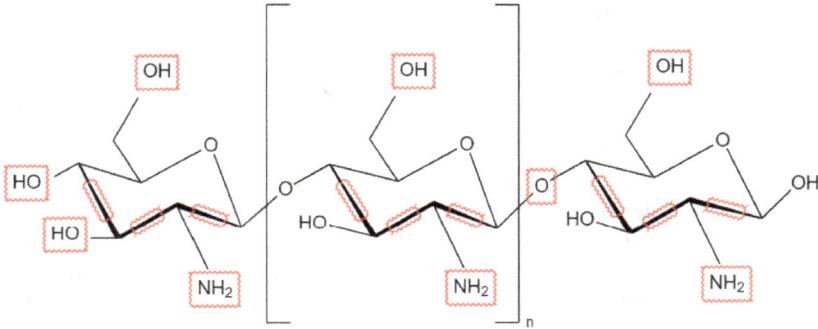

Figure 7: Chemical structure of Chitosan

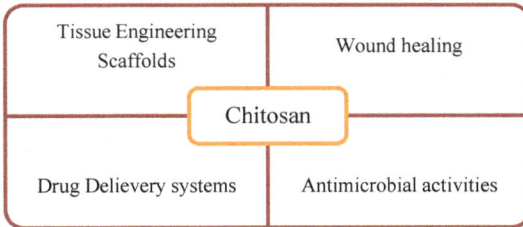

Figure 8: Chitosan and its biomedical applications

Electrospun chitosan nanofibers are being elaborately employed in regeneration of skin tissue. They can be directed towards crosslinkage by means of physical or chemical processes. Heat treatment, electron beam or gamma irradiation, and complexation with ions include the physical methods. Chemical methods consist of reactions involving crosslinking agents like glutaraldehyde and glyoxal [53].

Halloysite clay $(Al_2Si_2O_5(OH)_4 \cdot nH_2O)$ is a natural aluminosilicate tubule material that is formed by rolled kaolin sheets of 0.7nm of thickness to which it is chemically identical. It resembles the CNTs. On the other hand, halloysite nanotubes (HNT) were found to be Aluminosilicate nanostructures with extreme biocompatibility and low toxicity [54]. The

Adv. App. of Micro and Nano Clay – Biopolymer-based Composites   Materials Research Forum LLC
Materials Research Foundations 125 (2022) 103-123   https://doi.org/10.21741/9781644901915-5

aforesaid nanotubes demonstrate rodlike surface without any curving or aggregation. Halloysite is considered to be a secure and biocompatible nanomaterial, which is 1D with an outer diameter of 50-200 nm, 5-30 nm lumenal diameter, and length of 0.5-2 μm [55]. The chemical attributes of HNTs external morphology are very comparable to those of silicon dioxide, whereas the properties of the inner cylindrical core are alike to that of aluminium oxide. Moreover, current studies depict the biocompatibility and low cytotoxicity of HNTs, which makes them one of the best choices in the development of new drugs, as agents of gene delivery, as scaffolds used in tissue engineering, wound healing, for tumor cell isolation, and for improved adhesion of human cells [57].

Many studies have demonstrated the amalgamation of chitosan with various types of nanoparticles. Recently, these nanocomposites have proven to be biocompatible, antimicrobial, mucoadhesive, and could be engineered into several forms which include films, coatings, hydrogels, and membranes. Moreover, upon the addition of HNTs, chitosan proves to suggestively enhance physical strength, tensile modulus, rigidity, and toughness [57]. Chitosan – HNTs complex might be formed by electrostatic attractions when HNTs are added to chitosan in a dilute acid. Correspondingly, on the surfaces of HNTs, the Al-O-H and silanols may form hydrogen bonds involving hydroxyl and amino groups of chitosan. Thus, it is likely for chitosan to have adequate interfacial compatibility with the HNTs that may be disseminated in chitosan without any difficulty [58].

Liu and team [57], developed chitosan/HNTs bionanocomposite films via the solution casting technique. NIH3T3 mouse cell lines were studied to assess the cytotoxicity of the nanocomposites. The results demonstrated the adherence of 3T3 cells and proliferation on chitosan/HNTs nanomaterial. It was observed that when the culturing time was increased, the absorbance for all the samples was also increased, thus suggesting the good growth of the cells. The nanocomposite showed good cytocompatibility and was noncytotoxic in nature. The results from this study offered a better indulgence of the interaction between chitosan and HNTs in applications such as scaffold resources in tissue engineering.

In another study, chitosan– HNTs nanomaterial scaffolds were developed using combination of solution-mixing and freeze-drying procedures [59]. The chitosan–HNTs scaffolds displayed significantly enhanced strength, compressive modulus, and thermal stability in comparison to whole chitosan scaffold when *in vitro* studies were conducted on NIH3T3 cells. Overall, the nanocomposite scaffolds shown enhanced physicochemical and biological properties along with good cytocompatibility.

Naumenko et al., [60], designed a porous chitosan-gelatine-agarose hydrogel synthesized at 3-5 wt% HNTs via freeze-drying process. The resultant scaffolds exhibited the advancement of physical strength, enhanced water uptake, and enhanced thermal properties. The nanotubes were seen to show uniform dispersal in the matrix. The fabricated material were seen to uphold cell adhesion and proliferation without any deviations in viability and cytoskeleton formation. *In vivo* studies were performed on rats to evaluate biocompatibility as well as biodegradability. It was observed that scaffolds promoted the development of novel blood circulation surrounding the implantation sites. The scaffolds also showed enhanced resorption just within 6 weeks of implantation in rats, thus reiterating biocompatibility of scaffolds doped with halloysite.

Sandri et al., [61], developed HNTs/chitosan oligosaccharide nanocomposites for wound dressing. These nanomaterials exhibited biocompatibility with fibroblasts of human origin along with elevated cell proliferation as shown in an *in vitro* wound healing test. *In vivo* tests on wound therapy exhibited an enhanced extent of revascularization and restoration of hair follicles. Hence, this therapy might be employed in treating skin burns.

Recently, Koosha et al., [53], reinforced glyoxal crosslinked chitosan/poly (vinyl alcohol) (PVA) hydrogel nanofibers along with HNT and were organized through electrospinning. The mats were proven to show cell adhesion and propagation after 3 days of culture, indicating that integration of HNT certainly attracted the fibroblast cells to latch on and divide to a larger degree on the nanocomposite surfaces. The studies also suggested that crosslink with glyoxal did not elicit any side effects on the biocompatibility of the nanocomposites.

Conclusions

It is a known fact that the scaffold's architecture in tissue engineering from macroscale to nanoscale is critical for apt cellular interactions, physical stability, configuration of functional tissue, etc. In recent times, a wide number of 3D cell culture methods and resources are developed. In spite of advancements in composite and scaffold design for 3D culture of cells, sometimes it could be complicated to attain biosimiliarity of the tissues in the scaffolds, which changes the consequential tissue regeneration by the means of ECM environment. The present chapter has attempted to focus on the applications of several varieties of tissue engineering scaffolds. As an alternative for CNTs, HNTs are available in abundance, and doping them with chitosan are proven to be biocompatible, less cytotoxic, and are also physically sturdy. Thus, experimenting on the use of HNTs on chitosan is very much essential for their lesser cost and better performance for tissue engineering, drug delivery, in wound therapy and many more.

References

[1] R Madonna, G Novo, CR Balistreri, Cellular and molecular basis of the imbalance between vascular damage and repair in ageing and age-related diseases: As biomarkers and targets for new treatments, Mech Ageing Dev. 159 (2016) 22-30. https://doi.org/10.1016/j.mad.2016.03.005

[2] R. M Raftery, D.P Walsh, I. M. Castaño, A Heise, G. P. Duffy, S. A. Cryan, & F. J. O'Brien, Delivering Nucleic-Acid Based Nanomedicines on Biomaterial Scaffolds for Orthopedic Tissue Repair: Challenges, Progress and Future Perspectives, Adv Mater. 28 (2016) 5447-5469. https://doi.org/10.1002/adma.201505088

[3] ES Place, ND Evans, MM Stevens, Complexity in biomaterials for tissue engineering, Nature Mater. 8 (2009) 457-70. https://doi.org/10.1038/nmat2441

[4] PP Spicer, JD Kretlow, S Young, JA Jansen, FK Kasper, AG Mikos, Evaluation of bone regeneration using the rat critical size calvarial defect, Nat Protoc.7 (2012) 1918-1929. https://doi.org/10.1038/nprot.2012.113

[5] M Okamoto. The role of scaffolds in tissue engineering,Mozafari, Masoud, Farshid Sefat, and Anthony Atala (Eds.) Handbook of Tissue Engineering Scaffolds: Volume One, Cambridge, England: Woodhead Publishing, 2019, pp. 23-49. https://doi.org/10.1016/B978-0-08-102563-5.00002-2

[6] JM Karp, R Langer, Development and therapeutic applications of advanced biomaterials, Curr Opin Biotechnol. 18 (2007) 454-459. https://doi.org/10.1016/j.copbio.2007.09.008

[7] R Langer, DA Tirrell, Designing materials for biology and medicine, Nature. 428 (2004) 487-492. https://doi.org/10.1038/nature02388

[8] R Langer, JP Vacanti, Tissue engineering, Science. 260 (1993) 920-926. https://doi.org/10.1126/science.8493529

[9] Wang, Yao & Zhao, Qiang & Luo, Yiyang & Xu, Zejun & Zhang, He & Yang, Sheng & Wei, Yen & Jia, Xinru, A High Stiffness Bio-inspired Hydrogel from the Combination of Poly (amido amine) Dendrimer with DOPA. Chemical communications (Cambridge, England). 51 (2015) 16786-16789. https://doi.org/10.1039/C5CC05643H

[10] A Atala, FK Kasper, AG Mikos, Engineering complex tissues, Sci Transl Med. 4 (160) (2012). https://doi.org/10.1126/scitranslmed.3004890

[11] D Howard, L. D. Buttery, K. M. Shakesheff, &S. J. Roberts, Tissue engineering: strategies, stem cells and scaffolds, J Anat. 213 (2012) 66-72. https://doi.org/10.1111/j.1469-7580.2008.00878.x

[12] M. P. Nikolova &M. S. Chavali, Recent advances in biomaterials for 3D scaffolds: A review, BioactMater. 4 (2019) 271-292. https://doi.org/10.1016/j.bioactmat.2019.10.005

[13]B P Chan and K W Leong, Scaffolding in tissue engineering: general approaches and tissue-specific considerations. European spine journal: official publication of the European Spine Society, the European Spinal Deformity Society, and the European Section of the Cervical Spine Research Society. 17 (2008)467-79. https://doi.org/10.1007/s00586-008-0745-3

[14] DE Discher, P Janmey, YL Wang, Tissue cells feel and respond to the stiffness of their substrate, Science. 310 (2005) 1139-1143. https://doi.org/10.1126/science.1116995

[15] AJ Engler, S Sen, HL Sweeney, DE Discher, Matrix elasticity directs stem cell lineage specification, Cell.126 (2006) 677-89. https://doi.org/10.1016/j.cell.2006.06.044

[16] K Alvarez,& H Nakajima,Metallic Scaffolds for BoneRegeneration, Materials. 2 (2009) 790-832. https://doi.org/10.3390/ma2030790

[17] F Matassi, A Botti, L Sirleo, C Carulli, & M Innocenti, Porous metal for orthopedics implants, Clin Cases MinerBone Metab. 10 (2013) 111-115.

[18] J. D. Bobyn, G. J. Stackpool,S. A. Hacking, M Tanzer,&J. J. Krygier, Characteristics of bone ingrowth and interface mechanics of a new porous tantalum biomaterial. The Journal of bone and joint surgery. British volume, 81 (1999) 907-914. https://doi.org/10.1302/0301-620X.81B5.0810907

[19] Li, Z., Gu, X., Lou, S., &Y Zheng, The development of binary Mg-Ca alloys for use as biodegradable materials within bone. Biomaterials, 29 (2008) 1329-1344. https://doi.org/10.1016/j.biomaterials.2007.12.021

[20] Z.J. Wally, W. Van Grunsven, F. Claeyssens, R. Goodall, G.C. Reilly, Porous titanium for dental implant applications, Metals. 5 (2015) 1902-1920. https://doi.org/10.3390/met5041902

[21] Liu, FH, Synthesis of bioceramic scaffolds for bone tissue engineering by rapid prototyping technique, J Sol-Gel Sci Technol. 64 (2012) 704-710. https://doi.org/10.1007/s10971-012-2905-5

[22] H Zhou,J Lee, Nanoscale hydroxyapatite particles for bone tissue engineering, Acta Biomater. 7 (2011) 2769-2781. https://doi.org/10.1016/j.actbio.2011.03.019

[23] A. A Mirtchi,J. Lemaitre,N. Terao, Calcium phosphate cements: Study of the β-tricalcium phosphate-monocalcium phosphate system, Biomaterials 10 (1989) 475-480. https://doi.org/10.1016/0142-9612(89)90089-6

[24] L. L. Hench, The story of Bioglass®, J. Mater. Sci. 17 (2006) 967-978. https://doi.org/10.1007/s10856-006-0432-z

[25] W. Habraken, P. Habibovic, M. Epple, M. Bohner, Calcium phosphates in biomedical applications: materials for the future? Mater Today. 19 (2015) 69-87. https://doi.org/10.1016/j.mattod.2015.10.008

[26] L. S. Nair and C. T. Laurencin,Biodegradable polymers as biomaterials,Progress in Polymer Science. 32 (2007) 762-798. https://doi.org/10.1016/j.progpolymsci.2007.05.017

[27] I. V. Yannas, Classes of materials used in medicine: natural materials. B. D. Ratner, A. S. Hoffman, F. J. Schoen, and J. Lemons (Eds.) Biomaterials Science-An Introduction to Materials in Medicine,Elsevier Academic Press, San Diego, Calif, USA, 2004, pp. 127-136.

[28] P. Gunatillake, R. Mayadunne, and R. Adhikari, Recent developments in biodegradable synthetic polymers,BiotechnolAnnu Rev. 12 (2006) 301-347. https://doi.org/10.1016/S1387-2656(06)12009-8

[29] P. X. Ma, Scaffolds for tissue fabrication. Mater Today, 7 (2004) 30-40. https://doi.org/10.1016/S1369-7021(04)00233-0

[30] LJ Chen, M Wang, Production and evaluation of biodegradable composites based on PHB-PHV copolymer,Biomaterials, 23 (2002) 2631-2639. https://doi.org/10.1016/S0142-9612(01)00394-5

[31] W. He, T. Yong, Z.W. Ma, R. Inai, W.E. Teo, S. Ramakrishna, Biodegradable poly¬mer nanofiber mesh to maintain functions of endothelial cells, Tissue Eng 12 (2006) 2457-2466. https://doi.org/10.1089/ten.2006.12.2457

[32] A Gloria, R De Santis, &L Ambrosio, Polymer-based composite scaffolds for tissue engineering, J Appl BiomaterBiomech. 8 (2010) 57-67.

[33] H.W. Kim, J.C. Knowles, H.E. Kim, Hydroxyapatite/poly(epsilon)-caprolactone) com¬posite coating on hydroxyapatite porous bone scaffold for drug delivery, Biomaterials 25 (2004) 1279-1287. https://doi.org/10.1016/j.biomaterials.2003.07.003

[34] LM Montaser, SM Fawzy, NANO scaffolds and stem cell therapy in liver tissue engineering, SPIE 9550 (2015) 8. https://doi.org/10.1117/12.2188342

[35] M.A. Meyers, A. Mishra, Benson, D.J, Mechanical properties of nanocrystalline materials, Prog. Mater. Sci. 51 (2006) 427-556. https://doi.org/10.1016/j.pmatsci.2005.08.003

[36] J.C. Martinez-Garcia,A. Serraïma-Ferrer, A Lopeandía-Fernández,M. Lattuada, J. Sapkota, J. Rodríguez-Viejo,A Generalized Approach for Evaluating the Mechanical Properties of Polymer Nanocomposites Reinforced with Spherical Fillers, Nanomaterials 11 (2021)830. https://doi.org/10.3390/nano11040830

[37] M Goldberg, R Langer, XJia, Nanostructured materials for applications in drug delivery and tissue engineering, J Biomater Sci Polym Ed. 18(3) (2007) 241-268. https://doi.org/10.1163/156856207779996931

[38] R. Murugan, And S Ramakrishna, Development of Nanocomposites for Bone Grafting, Compos Sci Technol, 65 (2005) 2385-2406. https://doi.org/10.1016/j.compscitech.2005.07.022

[39] PA Sharma, R Maheshwari, M Tekade, RK Tekade. Nanomaterial Based Approaches for the Diagnosis and Therapy of Cardiovascular Diseases, Curr Pharm Des. 21 (2015) 4465-4478. https://doi.org/10.2174/1381612821666150910113031

[40] N Monteiro, A Martins, RL Reis, NM Neves, Nanoparticle-based bioactive agent release systems for bone and cartilage tissue engineering, Regen Ther. 1 (2015) 109-118. https://doi.org/10.1016/j.reth.2015.05.004

[41] M. M.Silva, L.A. Cyster, J. J. Barry, X. B. Yang, R. O. Oreffo, D. M. Grant, C. A. Scotchford, S. M. Howdle, K. M. Shakesheff,&F. R. Rose,The effect of anisotropic architecture on cell and tissue infiltration into tissue engineering scaffolds, Biomaterials. 27 (2006) 5909-5917. https://doi.org/10.1016/j.biomaterials.2006.08.010

[42] D SundaramurthI, UM Krishnan, S Sethuraman, Electrospun nanofibers as scaffolds for skin tissue engineering, Polym Rev. 54 (2014) 348-376. https://doi.org/10.1080/15583724.2014.881374

[43] Zhu, Caihong, Chengwei Wang, Ruihua Chen, and Changhai Ru., A Novel Composite and Suspended Nanofibrous Scaffold for Skin Tissue Engineering, EMBEC & NBC. (2017) 1-4. https://doi.org/10.1007/978-981-10-5122-7_1

[44] ARD Bakhshayesh, E Mostafavi, Alizadeh, N Asadi, A Akbarzadeh, S Davaran, Fabrication of three-dimensional scaffolds based on nanobiomimetic collagen hybrid

constructs for skin tissue engineering, ACS Omega.3 (2018) 8605-8611. https://doi.org/10.1021/acsomega.8b01219

[45] J Rheinwald, H Green, Serial cultivation of strains of human epidermal keratinocytes in defined clonal and serum-free culture, J Invest Dermatol. 6 (1975) 331-342.

[46] A Ito, Y Takizawa, H Honda, K Hata, H Kagami, M Ueda&T Kobayashi,Tissue engineering using magnetite nanoparticles and magnetic force: heterotypic layers of cocultured hepatocytes and endothelial cells. Tissue Eng, 10 (2004) 833-840. https://doi.org/10.1089/1076327041348301

[47] A Ito, H Jitsunobu, Y Kawabe, M Kamihira. Construction of heterotypic cell sheets by magnetic force-based 3-D coculture of HepG2 and NIH3T3 cells, J Biosci Bioeng. 104 (2004) 371-378. https://doi.org/10.1263/jbb.104.371

[48] Pan, Su, Hongmei Yu, Xiao-Yu Yang, Xiaohong Yang, Y. Wang, Qin-yi Liu, Liliang Jin and Yudan Yang, Application of Nanomaterials in Stem Cell Regenerative Medicine of Orthopedic Surgery, J Nanomater (2017): 1-12. https://doi.org/10.1155/2017/1985942

[49]K. Zhang, S. Wang, C. Zhou, L. Cheng, X. Gao,X. Xie, J. Sun, H. Wang, M. D. Weir, M. N. Reynolds, N. Zhang,Y. Bai&H. Xu,Advanced smart biomaterials and constructs for hard tissue engineering and regeneration, Bone research. 6 (2018) 31. https://doi.org/10.1038/s41413-018-0032-9

[50] Shirani, Keyvan, Mohammad Sadegh Nourbakhsh, and Mohammad Rafienia, Electrospun Polycaprolactone/Gelatin/Bioactive Glass Nanoscaffold for Bone Tissue Engineering, Int J Polym Mater. 68 (2019) 607-15. https://doi.org/10.1080/00914037.2018.1482461

[51] G Funda, S Taschieri, GA Bruno, E Grecchi, S Paolo, D Girolamo, MD Fabbro,Nanotechnology scaffolds for alveolar bone regeneration, Materials. 13 (2020) 201. https://doi.org/10.3390/ma13010201

[52] M. Rinaudo,Chitin and chitosan: Properties and applications, Prog Polym Sci. 31 (2006) 603-63. https://doi.org/10.1016/j.progpolymsci.2006.06.001

[53] M. Koosha, M. Raoufi&H. Moravvej,One-pot reactiveelectrospinning of chitosan/PVA hydrogel nanofibers reinforced by halloysitenanotubes with enhanced fibroblast cell attachment for skin tissue regeneration, Colloids Surf B Biointerfaces. 179 (2019) 270-279. https://doi.org/10.1016/j.colsurfb.2019.03.054

[54] V. Vergaro, E. Abdullayev, Y.M. Lvov, A. Zeitoun, R. Cingolani, R. Rinaldi, S. Leporatti, Cytocompatibility and uptake of halloysite clay nanotubes, Biomacromolecules. 11 (2010) 820-826. https://doi.org/10.1021/bm9014446

[55] Y. Luo&D. K. Mills, The Effect of Halloysite Addition on the Material Properties of Chitosan-Halloysite Hydrogel Composites, Gels (Basel,Switzerland). 5 (2019) 40. https://doi.org/10.3390/gels5030040

[56] Satish, Swathi & Tharmavaram, Maithri, Deepak, Halloysite nanotubes as a nature's boon for biomedical applications, Nanobiomedicine. 6 (2019) 184954351986362. https://doi.org/10.1177/1849543519863625

[57] M. Liu, Y. Zhang, C. Wu, S. Xiong, C. Zhou, Chitosan/halloysite nanotubes bionanocomposites: structure, mechanical properties and biocompatibility, Int J Biol Macromol.51 (2012) 566-75. https://doi.org/10.1016/j.ijbiomac.2012.06.022

[58] X. Sun, Y. Zhang, H. Shen, & N. Jia, Direct electrochemistry and electrocatalysis of horseradish peroxidase based on halloysite nanotubes/chitosan nanocomposite film, Electrochimica Acta. 56 (2010) 700-705. https://doi.org/10.1016/j.electacta.2010.09.095

[59] Liu, Mingxian, Chongchao Wu, Yanpeng Jiao, Sheng Xiong, and Changren Zhou, Chitosan-Halloysite Nanotubes Nanocomposite Scaffolds for Tissue Engineering, J Mater Chem B. 15 (2013) 2078-89. https://doi.org/10.1039/c3tb20084a

[60]E. A. Naumenko, I. D. Guryanov, R. Yendluri, Y. M. Lvov&R. F. Fakhrullin,Clay nanotube-biopolymer composite scaffolds for tissue engineering, Nanoscale. 8 (2016) 7257-7271. https://doi.org/10.1039/C6NR00641H

[61] G. Sandri, C. Aguzzi, S. Rossi, M. C. Bonferoni, G. Bruni, C. Boselli, A. I. Cornaglia, F. Riva, C. Viseras, C. Caramella&F. Ferrari,Halloysite and chitosan oligosaccharide nanocomposite for wound healing, Acta biomaterialia. 57 (2017) 216-224. https://doi.org/10.1016/j.actbio.2017.05.032

Adv. App. of Micro and Nano Clay – Biopolymer-based Composites   Materials Research Forum LLC
Materials Research Foundations **125** (2022) 124-151   https://doi.org/10.21741/9781644901915-6

Chapter 6

Vermiculite Starch-based Nanocomposites and Applications

Nadia Akram[1,*], Khawaja Taimoor Rashid[1], Tanzeel Munawar[1], Muhammad Usman[1], Muhammad Saeed[1]

[1]Department of Chemistry, Government College University Faisalabad, Faisalabad-38000, Pakistan

* nadiaakram@gcuf.edu.pk

Abstract

Nowadays, the demand of biodegradable as well as biocompatible nanocomposites-based products is growing all over the world. In this perspective, a little effort is done in terms of exploring the effectiveness of environment friendly vermiculite starch-based nanocomposites. Both the constituents are naturally occurring so they are considered as low-cost materials. Different methods such as solution casting, melt solution blending as well as in-situ polymerization approaches are used for the fabrication of vermiculite starch-based nanocomposites. Functional characterization of nanocomposites can be investigated in terms of XRD, SEM, FTIR, TGA, DSC as well as by some mechanical testing methods. The resultant nanocomposites give attractive solutions in the subject of wastewater treatment and food packaging additionally may also exhibits improved fire properties that are resistive.

Keywords

Starch, Expanded Vermiculite, Silicates Sheets, Nanofillers, Biodegradable

Contents

1.   Introduction

Petrochemical waste has created a lot of environmental problems in the last decade. So, the researcher community showed more attention towards the reduction or elimination of petrochemical based products. Nowadays the use of biodegradable polymers is considered to be the replacement of petrochemical products to obtain environment responsive constituents. The biodegradable nature of starch in soil and water as well as its low cost and easiness in availability at commercial and domestic level attracts the more attention of researchers in terms of fabricating the starch clay-based nanocomposites [1-4].

The main disadvantage of producing the biodegradable or environment friendly material is the poor mechanical performance and high absorption rate of water [5]. Therefore, the modification is necessary with regards to the technical and physical properties for the nanocomposites. So the great potential has been found in the starch-clay based nanocomposites in terms of achieving the improved mechanized and physical attributes [6].

Starch matrix can be modified by adding the low amounts of clay for the fabrication and exhibiting admirable properties of nanocomposites [7]. Usually, clay in the form of layered silicates such as vermiculite and montmorillonite can be used as a filler or reinforcement for synthesizing the starch-clay based nanocomposites [8, 9].

2.   Starch based clay nanocomposites

Nowadays, nano-clay exhibits a huge amount of impact i.e., 70% as a nanomaterial in the commercial international market of the world. On the basis of the chemistry involve in the structure of clay, it can be divided into five different types, i.e., chlorite, montmorillonite (MMT)/smectite, illite, kaolinite and vermiculite as well.

Generally, MMT is considered to be the most beneficial nanomaterial for the fabrication of various types of clay-based nanocomposites. The highly attractive and well accepted properties of layered clay such as improved surface reactivity as well as large surface area makes the MMT more reliable and advantageous. So, these biopolymers which are all-natural such as starch based layered silicates greatly improve the properties of nanocomposites especially in the world of food packaging applications in order to prevent the food from microbial attack.

Starch which is well known polysaccharide and has the potential to play vital role due to its low price, naturally as well as easily extracted material from different crops in the fabrication of organic clay-based nanocomposites [10]. Normally the starch is present in various forms such as wheat starch (WS), corn starch (CS), rice starch (RS), potato starch that could be easily extracted from grains [11].

Starch is principally consists of amylase and amylopectin that are combined with α- D-(1–4) and/or α-D-(1–6) linkages. Amylase and amylopectin show linear and branched framework respectively. In the composition of the starch amylopectin is the major contributor while the amylose only contributes between 15 to 20% [12].

For different applications, the starch which is present in raw form contains high extent of intermolecular hydrogen bonds in its polymeric chain which is not suitable for the paper industry. On the other hand, raw starch also exhibits higher melting temperature as compared to its degradation temperature [13].

In order to reduce the limitation of starch, some structural changes in its structure by adding some plasticizers or organic silicates in the form of clay makes the glass transition

Adv. App. of Micro and Nano Clay – Biopolymer-based Composites   Materials Research Forum LLC
Materials Research Foundations **125** (2022) 124-151   https://doi.org/10.21741/9781644901915-6

temperature lower [14]. The low molecular weight of starch can be improved by the addition of plasticizers that makes the resulting polymeric films more flexible and reliable [15].

Plasticizers also reveals the change with certainly considerable mobility of polymer string in terms of decrease in hydrogen bonding ability. Most frequently used solvent such as water also diminishes the granules of starch. In aqueous medium, starch is insoluble at room temperature. On the other hand, the irreversible swelling nature of starch at high temperature showed gelatinization as well as degradation of amylase and amylopectin [16].

Nevertheless, the use of plasticizers and water as a solvent mainly rely on storage condition. So, by the accumulation of non-volatile plasticizers such as glycerol, this weakness can be resolved. The sample shows physical ageing as well as distinction in mechanical properties and poor water resistance, if they are secured at below glass transition temperature [17].

In the fabrication of many synthetic materials, starch exhibits various advantages. But owing to its deprived mechanical properties, more attention is still required to achieve the desired expectations [18]. In order to achieve the desired expectations from starch, nano fillers are incorporated into the starch that improves both the structural and physiochemical properties of the starch.

The biological activities of the starch can also be improved by the incorporation of nanofillers by showing enhancement in its particle size. In the subsequent section, we have summarized some applications of starch-based nanocomposites with special consideration to starch/ clay nanocomposites [19]. The applications of starch based composites is shown in Fig 1.

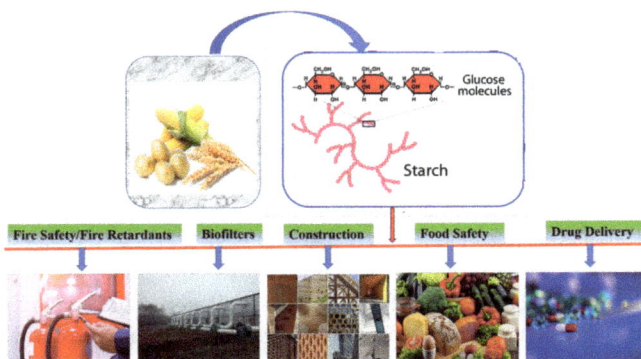

Fig: 1 Different application of starch-based nanocomposites

3.   Properties of vermiculite

Some of the most attractive properties of vermiculite includes excellent thermal and chemical stability, better flame retardancy, environmentally responsive nature and inertness as well. It also shows better thermal resistance as compared to other organic clay silicates [20]. It shows expansion up to 20% to 30% as compared to its original size, when vermiculite is thermally treated [21]. The minerals of vermiculite have been shown in Fig 2.

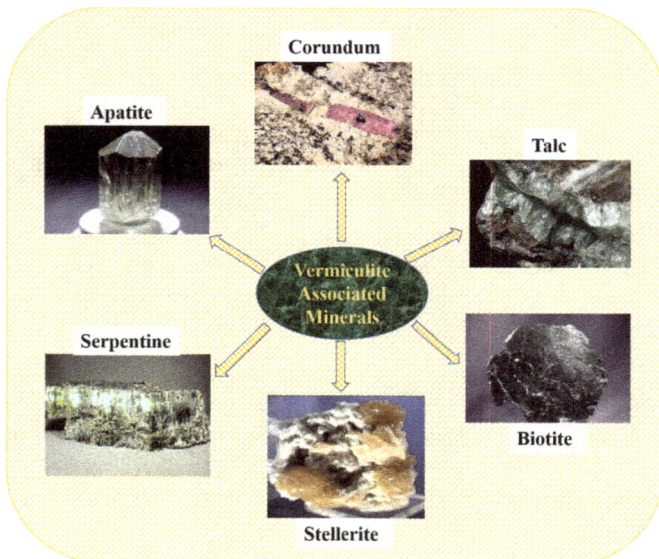

Fig: 2 Vermiculite associated minerals

Due to expansion in its size, vermiculite showed curved appearance. The expanded layers of vermiculite show higher absorbency, huge particular surface, reduced thickness, and reduced thermal conductivity as compared to vermiculite that is unexpanded and these expanded layers are separated from each other as well [22]. On the other hand, some additional properties of vermiculites include active surface groups in its structure that shows many useful applications including surface assimilation of oil, organic solvent (OS), gas, ions of metallic element, for the splitting of microorganisms, wastewater remedies, sound, as well as the fabrication of some thermal and electrical insulated materials.

Vermiculite also exhibits good fire resistance and improved gas barrier characteristics as well [23].

In the past, not too much researcher showed the attention towards the synthesis of starch-vermiculite based nanocomposites due to lack of literature work. But nowadays vermiculite is considered to be the effective and low-cost nanomaterial as a filler or reinforcement for biopolymer-based nanocomposites [24]. Vermiculite is really a mica, a hydrated, magnesium–iron–aluminum trioctahedral sheet silicate of mixed structure. Its elementary component is two tetrahedral silicate sheets that contains aluminum and iron interlayered by an octahedral sheet [25].

The two basic layers of silicate sheets in vermiculite are joined together by electrostatic and van der Waals forces. Silanol (Si–O–H) groups on the surface of vermiculite allows the biopolymer such as starch as well as chitosan to develop strong hydrogen bond connections with hydroxyl group of starch and chitosan [9].

Lu et al., (2012) prepared the starch–VER nanocomposites films by adopting the solution casting method by varying the concentration of vermiculite nanofiller as a reinforcement into the starch matrix. Improvements in water resistance characteristics of starch-vermiculite nanocomposites films have been found as well as starch can be protected from recrystallization during storage and transportation process.

The resultant starch-vermiculite based nanocomposites films can be inspected by diverse characterization practices such as X-ray diffraction (XRD), Fourier transform infrared spectroscopy (FTIR), and scanning electron microscope (SEM). Moreover, thermal effects on starch based varied vermiculite concentration nanocomposites has also shown improvements as compared to their individual phases of constituents [24].

The attractive properties of plastic based products at domestic as well as on commercial scale due to its low prices, light weight and good stability at room temperature exhibits a huge impact on the environment as well as bad impacts on human health.

In the recent past, scientists of different labs in the whole world analyzed that plastic-based products not only produce bad impacts on environment but plastic based packed foods create panic in terms of plastic eating. So nowadays the main focus of researchers is to find those sources that are biodegradable and biocompatible as well as environment friendly [26].

3.1 Vermiculite-modified bio nanocomposites

Although, single constituent based plastic products lacks the various environment friendly properties in terms of biodegradable nature, lower viscosity as well as thermal instability which prevents to allow pure plastic-based products in food packaging industry [27] So, vermiculite as a nanofiller provides various facilities in terms of using plastic-vermiculite based products due to well-known biodegradable nature of vermiculite clay [28]. Starch and vermiculite induced nanocomposite's preparation is shown in Fig. 3. Reis *et al*., (2017) synthesized bio composites films of all-natural occurring vermiculite and checked the properties of vermiculite by melt intercalation technique. Investigation of thermal behavior of nanocomposites films found that vermiculite-based nanocomposites films have well thermal stability and high biodegradability rate [29]. They are also effective in terms of food packaging applications and shows inertness as well in terms of interaction with food. So, these properties make them vermiculite important material [30]. Oliveira *et al*., (2015) studied the polyester-starch matrix-based vermiculite biodegradable nanocomposites by the melt intercalation method by using injection molding as a processing technique. Naturally occurring vermiculite and chemically treated vermiculite with tetra butyl phosphonium bromide are used as a nanomaterial filler in each nanocomposite film composition [31]. During the degradation behavior of nanocomposite films dark spots were appeared on the samples due to the action of microorganisms. By scanning electron microscopy measurements, it was noticed that after 15 to 30 days, lighter regions were appeared on the samples of polyester-starch based vermiculite nanocomposites films owing to presence of microorganisms

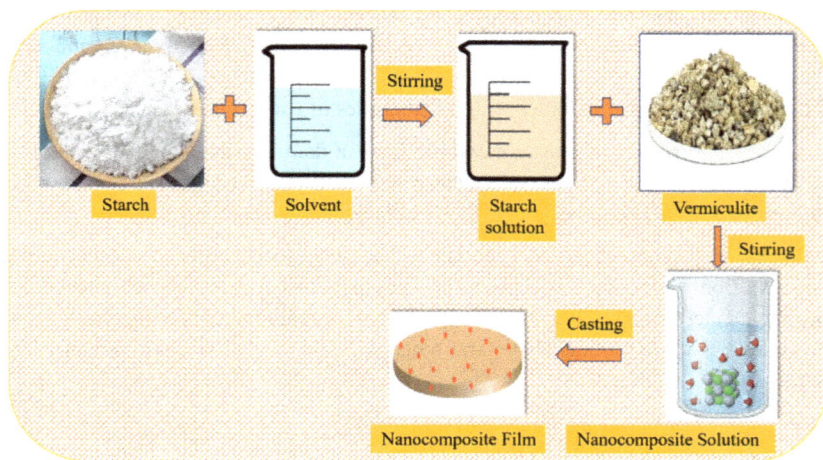

Fig. 3 Preparation of Starch-Vermiculite nanocomposite Film

The crystallization temperature (TC) and crystal melting temperature (TM) increases by inclusion of VMT, tributylphosphonium bromide (P-VMT) to starch blends of polybutylene adipate-terephthalate [32]. Fernandez *et al.*, investigated the attrition mechanism of polymer matrix and reinforcement of clay minerals. The category and circulation of clay nutrients notably disturbs the degradation portion of PLLA for vermiculite L-lactic acid (PLLA) based nanocomposites [33].

Biopolymer-clay nanoparticles when prepared with low clay to biopolymer ratio are modern materials which have huge potential for the improvement of its mechanical structure. For the improvement there are challenges in which nanoscale dispersion of clay is a major one. The common example of this behavior can be seen in Montmorillonite clay-starch nanoparticles which are being used in many nano systems. Reports showed that it faces low clay to biopolymer dispersion [3]

3.2 Basis and division of vermiculite

Clay mineral system can easily be classified as illite group, kaolinite group and smectite-vermiculite group. Vermiculite which belongs to smectite-vermiculite group is collected through removal of potassium from biotite or muscovite [34].

Vermiculite has been known since 1861 and means worm-like according to English dictionary owing towards the better expansion that it is thermal. Vermiculite is found in America, China and Zimbabwe. China is considered to be richer in case of vermiculite reserves, it contains vast majority (one-sixth) of the world's overall supplies having more than 100 million tons in the Xinjiang Province [35].

3.3 Morphologic and material characteristics of vermiculite

Vermiculite belongs to phyllosilicate material or silicates sheets. Vermiculite is a 2:1 type layered aluminosilicate mineral, embracing of octahedral sheet among two tetrahedral layers [23].

Variations in vermiculites can be found in the nature including macroscopic vermiculite and vermiculite clays. The vermiculites are usually in brown color with bronze arrival. Essentially, the color is dependent on the amount of the components. Depending on type and percentage of the components, color change from black to dark brown and even to green, yellow and yellowish-brown [34].

The modifications between the structures of minerals disturb not only the color but also some other properties as well. The variation in the particle size of vermiculite and expanded vermiculite can be seen from Figure and also optical images of expanded vermiculite are given in Fig. 4.

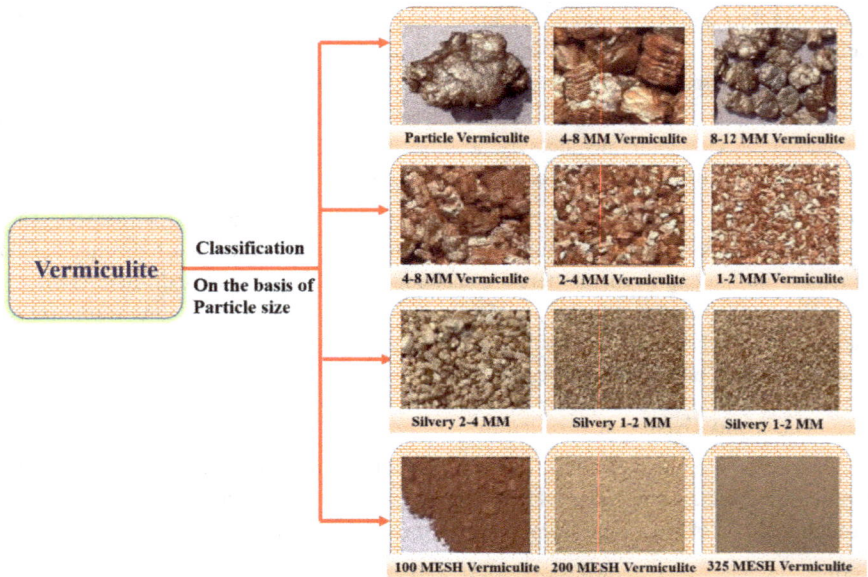

Figure: 4 Classification of vermiculite on the basis of particle size

The elemental percentage composition of vermiculite is given in table 1. The substitution belongs to isomorphism that brings (-) charge on the surface of vermiculite as a whole. These negative signs on atoms can be adjusted by interlayered transferable water containing cations such as Calcium (II), Magnesium (II), Copper (II) respectively [36].

The vermiculites can be expanded by using thermal, chemical and microwave treatment from 20% to 30% as compared to their original size. The exfoliation occurs in vermiculite is owing to release of water being existed in the layers of vermiculite [37]. The resultant expanded vermiculite exhibits very useful distinctive properties in the forms of porous structure, low thermal conductivity as well as showed resistance against the high temperature. In addition, the characteristics of expanded vermiculite, it also showed good ion exchange properties that gives useful results in terms of removal of unsought species from water [38].

Table 1: % composition of vermiculite minerals

Component	% by weight
SiO_2	38.00-46.00
Al_2O_3	10.00-16.00
MgO	16.00-35.00
CaO	1.00-5.00
K_2O	1.00-6.00
Fe_2O_3	6.00-13.00
TiO_2	1.00-3.00
H_2O	8.00-6.00
Other	0.200-1.200

The spacing or space involving the layers of vermiculite is entirely reliant on the size associated with the cations and on the incident of interlaminar water. The rate of hydration of cations may be diverse due to spacing regarding the layers that are vermiculite [39].

Normally all types of organic clays showed common properties like cation substitution, surface assimilation, chemical durability and energetic surface groups. But the vermiculite showed some additional properties in terms of expansion competence and frost resistance [40].

4.   Varieties and ways of processing of clay nanocomposites

Three different method that could explain the interactions between the constituents of nanocomposites such as polymer matrix and reinforcement as shown in fig 5.

- Intercalated design of nanocomposites
- Exfoliated design of nanocomposites
- Flocculated design of nanocomposites [41-43].

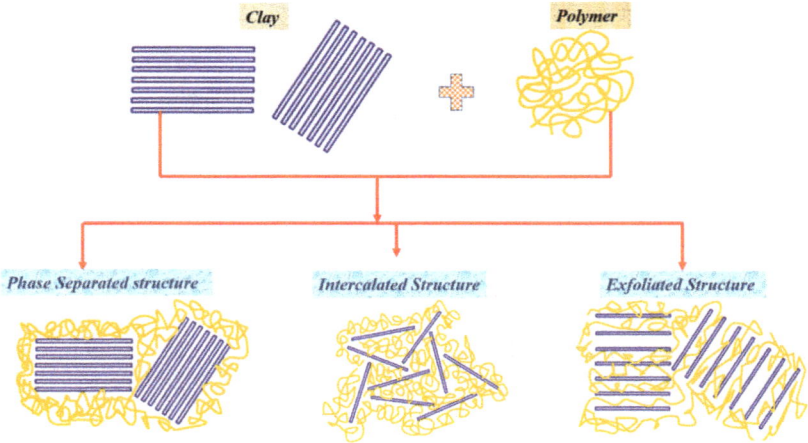

Fig: 5 Different structures of clay-based nanocomposites

4.1   Preparation techniques for polymer clay nanocomposites

Generally speaking, three straight ways that are different are used for the fabrication of polymer- clay nanocomposites:

4.2   Solution-blending method

In this valuable sort of process, first of all both constituents that can be used in the preparation of nanocomposites have to be dissolved in a suitable solvent that could well mix both the constituent such as matrix and reinforcement in the form of nanomaterial or nanofiller. After the solvation, the resultant homogenous solution is allowed to evaporate at room temperature until the solution dry completely. The overall procedure of this process is shown in fig 6 A [44-46]

4.3   Melt-blending method

This is the second method of preparation of clay-based nanocomposites films. In this method, mixing of both constituents such as reinforcement and matrix could be achieved by the solution-blending technique or by a melt-integration approach. This would allow the mixture of polymer and clay to mix more properly. The overall procedure of this process is shown in fig 6B [47, 48].

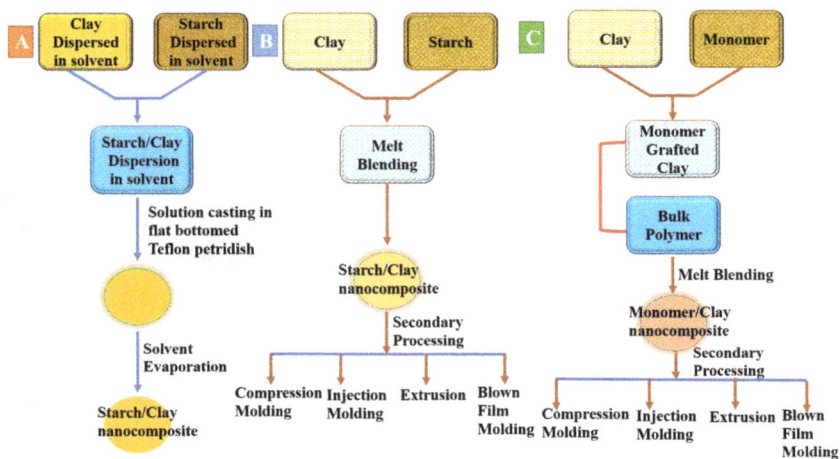

Figure:6 Distinctive strategies for the processing of starch-clay nanocomposites: (A)
solution casting; (B) melt blending; (C) in-situ polymerization

4.4   In-situ polymerization method

This approach helps to improve the rate of mixing concerning polymer and furthermore subatomic particle associated with clay-based nature. Certainly one of the main features of this method is the distribution that is uniformity of particles in the polymer matrix. Since the particles of clay are dispersed within the solution comprising monomer, which will be then followed by polymerization.

The grafting of clay specks is possible by the addition of reinforcement in the shape of thin layer of polymer to the bulk polymer matrix. So, the mixture is allowed for supplementary handling methods such as for instance compression and extrusion molding. The overall procedure of this process is shown in fig 6C [49].

5.   Processing of vermiculite based composites

Vermiculite is naturally present in particle or powder form. To get the three-dimensional form of vermiculite usually it is dispersed in a binder. The 3D form vermiculite could be easily obtained by preparing the nanocomposites of vermiculite [13]. The nanocomposites that could be prepared from vermiculite would be designated as intercalated or exfoliated composites. Methods for fabrication includes melt mixing, in situ polymerization, solution casting, or a mixture of solution casting and compression molding [16].

Vermiculite polymeric biocomposites is specified as also elastomeric composites. Elastomers displays various properties that are helpful in flexibility or they may be flexible and warped at various degree plus they may also regain their shape easily [15]. Their unique qualities are modified simply by the type associated with elastomeric solid and density that is cross-linking. In order to develop elastomers which are efficient varieties involving fillers can be merged to the framework. Vermiculite had been additionally utilized being a period that is reinforcing different types of elastomers [15].

6.   Vermiculite-modified polymer nanocomposites

The nanocomposites that could be modified by vermiculite and polymer matrix are fabricated through melt infiltrate, in situ form of polymerization as well as solution molding phenomena [50]. The interactions can be enhanced between the two components such as vermiculite and polymer in terms of treating the organic vermiculite clay through cation exchange as well as by cationic surfactant. Nowadays, a large variety of vermiculite-based nanocomposites could be fabricated synthetically [51].

The modification of vermiculite could be well achieved by various alkali metal ions, organic modifiers as well as some other organic molecules [52]. So, the attractive package of vermiculite properties makes them competent, inexpensive, and biologically responsive raw material to synthesize reinforcing nanomaterial filler, packing materials, energy storage materials, gas barrier properties and for the fabrication of thermal insulator [53].

7.   Mechanistic and thermic properties of nanocomposites

Polymeric and biopolymeric based derived products can be characterized by thermal and mechanical properties. Incorporation of vermiculite into the biopolymers significantly improves the thermal and mechanical properties of the resultant materials [54]. On the other hand, the tensile strength and impact strength of the resultant material could also be improved in case of nanocomposites. In vermiculite, spacing as well as mechanical and thermal properties can be enhanced by the presence of surfactants.

Nonetheless, the progress that is supplementary of nanocomposites is mainly controlled because of the agglomerate and low-quality dispersion of polymer matrix with bigger percentage of VMT [55].

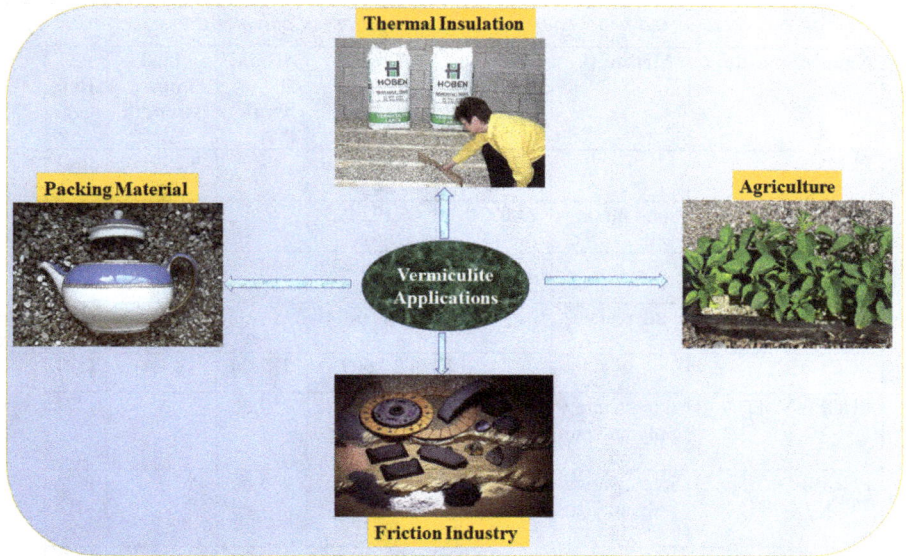

Figure:7 Applications of vermiculite-based nanocomposites

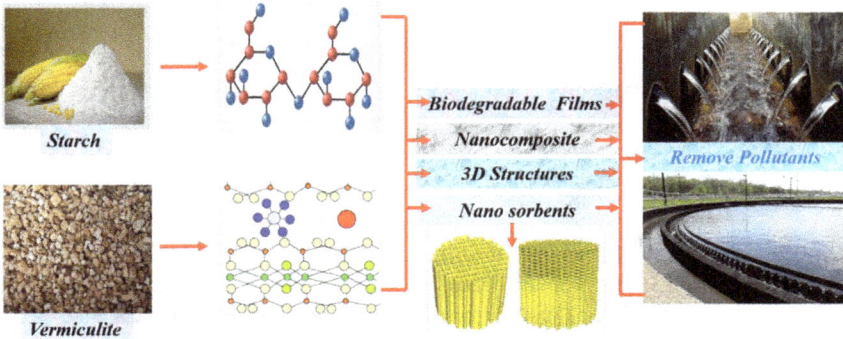

Figure:8 Applications of wastewater treatment of vermiculite-starch based nanocomposites

Table 2: Specialized properties of Vermiculite-adapted polymer nanocomposites

Nanocomposites	Methods	Tensile modulus (GPa)	Tensile strength (MPa)	Strain at break (%)	Izod impact strength	Ref.
PLLA	-	1.700±0.0200	73.544±2.045	-	9.299±0.257	[56]
PLLA/VMT10	melt mixing	2.009±0.133	66.192±1.169	-	11.593±1.400	[56]
EA	-	-	7.299±0.402	15.00±0.5	5.50±0.504	[51]
2BTPC-VMT	intercalate	-	13.501±0.600	9.00±0.500	9.10±0.40	[51]
PUCPB	-	-	17.609	337.00	-	[52]
PUCPB/VMT	intercalating polymerization	-	19.790	359.00	-	[52]
PUCPB/OVMT	intercalating polymerization	-	26.799	443.00	-	[52]
PVB	-	~1.601	-	-	-	[57]
2Vk+PVB	Solution molding method	~2.813	-	-	-	[57]
PMMA	Ultrasonic in-situ polymerization	0.966±0.253	56.00±2.00	8.320±0.660	-	[54]
PBTP-S3	Ultrasonic in-situ polymerization	1.188±0.029	65.00±4.00	14.00±1.799	-	[54]

8.   Characterization of nanocomposites

The nanocomposites properties are dependent on the nature of constituents and various parameters such as the dimensions, microstructure and homogeneity of the dispersed phase.

Numerous characterization techniques are being established that are used for examining the microstructure and effectiveness of nanocomposites including x-ray diffraction (XRD), scanning electron microscopy (SEM), transmission electron microscopy (TEM), atomic force microscopy (AFM), Fourier transform infrared (FTIR) spectroscopy, nuclear magnetic resonance (NMR) spectroscopy, Thermogravimetric analysis (TGA),

Differential scanning calorimetry (DSC), etc. A more descriptive classification tools laterally with their actual applications are described in Table 3 [58].

Table 3. Typical characterization techniques for clay-based nanocomposites.

Sr. no	Characterization Technique	Characteristic Property of Clay-based Nanocomposite
1	XRD	Interlayer distance of clays Degree of clay platelets dispersion Nanocomposite morphology (e.g. intercalated vs. exfoliated)
2	TEM	Microstructure and spatial distribution Structural defects
3	SEM	Surface morphology Degree of dispersion of clay particles
4	FTIR	Composition analysis
5	ASM	Crystallization analysis of polymer Surface morphology Particle size and distribution
6	TGA	Thermal stability
7	NMR	Local dynamics of organic polymer chains Surface chemistry
8	DSC	Melting and crystallization analysis Local dynamics of polymer chains
9	Rheometry Mechanical test	Young's modulus Tensile strength Viscoelastic properties
10	Cone Calorimetry	Flame retardancy Thermal stability

9. Applications

9.1 Deletion of pollutants from water/waste water from by starch-vermiculite based nanocomposites

Starch is a natural bio-compatible and bio-degradable biopolymer which commonly present in roots, and crop seeds. Starch present in granules with 3D shaped are characterized by a crystallinity between 15% and 45% and by D-glucose units with bio-macromolecules. Starch have potential to remove of various pollutants are summarized in Table 4 [59].

Table 4. Potential of starch in pollutant removal.

Target Pollutant	Starch	References
Dye	Crosslinked porous sorbents	[60-66]
Metals	Crosslinked copolymers	[67, 68]
Organic pollutants	Composite photocatalysts	[59, 69, 76]
	Magnetic nanocatalysts	[77]
	Crosslinked starches adsorbents	[78]
	Crosslinked cyclodextrins sorbents/gels	[79,84]
Biomass Harvesting	Cationic crosslinked starch	[85,86]

Vermiculite and starch-based nano-composites are provided as a potential alternative choice for the deletion of numerous toxins from aqueous environment. These nano-composite are offered many auspicious benefits such as the great capability of ion sorption, grater definite surface area, less expansive, high stability of chemical and mechanical and large swelling capability by the conventional methods [59].

For deletion of substantial metals from aqueous solution, clay materials and clay-polymer nano-composites have shown the high capacity significantly. Recently a novel approach for water remediation have been revealed where the combination of adsorption and bio-degradation are involved. Explicitly, adsorption of pollutants took place on clay nano-composites and then these pollutants are decomposed in less toxic constituent in the presence of specific microorganisms [87].

9.2   Packaging applications of starch-vermiculite based nanocomposites

In the advancement of a lot more eco-friendly and safe food providing items starch-based biodegradable materials play a major role in the possible future. One of the recent disadvantages is the retro gradation of starch causing poor mechanical properties. Different techniques such as physical alteration of starch, enzymatic, chemical and the inclusion of some other biopolymers and efficient additives to the films were used [88].

Though numerous techniques had been directed to perfections in film properties but still a huge effort is essential to generate starch-based films which are synthesized from synthetic polymers having similar mechanical, optical and barrier properties. Different factors such

Adv. App. of Micro and Nano Clay – Biopolymer-based Composites   Materials Research Forum LLC
Materials Research Foundations **125** (2022) 124-151   https://doi.org/10.21741/9781644901915-6

as temperature, type of starch, time during formation of film, symbiotic polymers, plasticizers and storage conditions. The light packing of Vermiculite reduces the extra costs of shipping rates. The transformed vermiculite does not break down, decay or release odors [89].

Exfoliated vermiculite provides a cushion against shock during handling and transportation as it is easily poured around irregular shaped objects. It acts as a good adsorbent which makes it an ideal solution to contain leakage that might take place while shipping liquids. This is mainly important for particular chemicals containing flammable types. It is a commonly used material due to these properties. Expanded vermiculite is widely used as packaging material weight in the shipping industry due to its lightweight. It is easy to pour and clean objects inside boxes. It is safe to use as it protects products against mishandling and impact through the shipping process.

Vermiculite is commonly used when shipping dangerous materials because of its effectiveness as an absorbent. Vermiculite is non-flammable so it is usually used to ship lithium batteries, which sometimes catch or explode fire when inadequately packaged.

As an inorganic material, this ensures that, if a fire occurs it will not spread but it will be available in package. Whereas other materials such as styrofoam packing peanuts can take eras to break down in landfills but vermiculite degrades naturally which makes it less dangerous for the environment.

9.3   Flame-retardant applications of starch-vermiculite based nanocomposites

Vermiculite is physical and chemical mineral that is highly stable and it cannot break under a fire of up to 1200 °C. Consequently, vermiculite can be used in the sense of filler for developing flame retardant nano-composite. The flame-retardant nano-composites need the modification by using the continuous phase (polymer matrix) with vermiculite or can develop the innovative film with nanosheets of matrixes. The impact of flame-retardant nanocomposites structured on polymers had been primarily examined by LOI (limited oxygen index). [55]. Wang *et al.*, (2016) prepared flame-retardant coating that are used for surface treatment. The results revealed that the treated prepared composite increases limited oxygen index value and reduces the thermal degradation energy [90].

Cheong *et al.*, (2015) prepared the vermiculite–PEG (polyethylene glycol) hybrid film. The results of scanning electron microscopy confirmed the cross-sectional view of the vermiculite–PEG based hybrid nanocomposites films that showed approximately same thickness. The synthesized films were burned off in the absence of vermiculite because the films re only dependent upon the polymers. By adding the suitable concentration of vermiculite i.e., 75%, the nanocomposites-based films showed fire resistive property for more than one minute. So, this type of films provides efficient and controllable properties of materials with approximate thickness [91].

Conclusion

The naturally occurring and low-cost nature of both vermiculite and starch make them more attractive and useful in the field of nanocomposites. Vermiculite and starch-based nano-composites have provided a potential alternative choice for the removal of various pollutants from aqueous environment. These nano-composite offer many auspicious benefits such as the great capability of ion sorption, greater definite surface area, less expansive, high stability of chemical and mechanical and large swelling capability by the conventional methods. In nanocomposites form, they exhibit improved thermal and mechanical properties such as young's modulus and tensile strength when compared with the starch matrix. The light packing of vermiculite reduces the extra costs of shipping rates. The transformed vermiculite does not break down, decay or release odors. In shipping industry, vermiculite and starch-based nano-composites are also effective due to its light weight.

References

[1] F. Chivrac, E. Pollet, L. Averous, Progress in nano-biocomposites based on polysaccharides and nanoclays, Mater. Sci. Eng. 67 (2009) 1-17. https://doi.org/10.1016/j.mser.2009.09.002

[2] F. Chivrac, E. Pollet, M. Schmutz, L. Avérous, New approach to elaborate exfoliated starch-based nanobiocomposites, Int. J. Biol. Macromol. 9 (2008) 896-900. https://doi.org/10.1021/bm7012668

[3] Y.L. Chung, S. Ansari, L. Estevez, S. Hayrapetyan, E.P. Giannelis, H.M. Lai, Preparation and properties of biodegradable starch-clay nanocomposites, Carbohydr. Polym. 79 (2010) 391-396. https://doi.org/10.1016/j.carbpol.2009.08.021

[4] M. Huang, J. Yu, Structure and properties of thermoplastic corn starch/montmorillonite biodegradable composites, J. Appl. Polym. Sci. 99 (2006) 170-176. https://doi.org/10.1002/app.22046

[5] Y.L. Chung, H.M. Lai, Preparation and properties of biodegradable starch-layered double hydroxide nanocomposites, Carbohydr. Polym. 80 (2010) 525-532. https://doi.org/10.1016/j.carbpol.2009.12.020

[6] M.F. Huang, J.G. Yu, X.F. Ma, P. Jin, High performance biodegradable thermoplastic starch-EMMT nanoplastics, Polym. J. 46 (2005) 3157-3162. https://doi.org/10.1016/j.polymer.2005.01.090

[7] H.M. Park, W.K. Lee, C.Y. Park, W.J. Cho, C.S. Ha, Environmentally friendly polymer hybrids Part I Mechanical, thermal, and barrier properties of thermoplastic starch/clay nanocomposites, J. Mater. Sci. 38 (2003) 909-915. https://doi.org/10.1023/A:1022308705231

[8] C. Zeppa, F. Gouanvé, E. Espuche, Effect of a plasticizer on the structure of biodegradable starch/clay nanocomposites: Thermal, water-sorption, and oxygen-barrier properties, J. Appl. Polym. Sci. 112 (2009) 2044-2056. https://doi.org/10.1002/app.29588

[9] K. Zhang, J. Xu, K. Wang, L. Cheng, J. Wang, B. Liu, Preparation and characterization of chitosan nanocomposites with vermiculite of different modification, Polym. Degrad. Stab. 94 (2009) 2121-2127. https://doi.org/10.1016/j.polymdegradstab.2009.10.002

[10] F. Zia, K.M. Zia, M. Zuber, S. Kamal, N. Aslam, Starch based polyurethanes: A critical review updating recent literature, Carbohydr. Polym. 134 (2015) 784-798. https://doi.org/10.1016/j.carbpol.2015.08.034

[11] N.J. Vickers, Animal communication: when i'm calling you, will you answer too?, Curr. Biol. 27 (2017) 713-715. https://doi.org/10.1016/j.cub.2017.05.064

[12] Y. Lu, L. Weng, X. Cao, Morphological, thermal and mechanical properties of ramie crystallites-reinforced plasticized starch biocomposites, Carbohydr. Polym. 63 (2006) 198-204. https://doi.org/10.1016/j.carbpol.2005.08.027

[13] M.N. Anges, A. Dufresne, Plasticized starch/tunicin whiskers nanocomposites. 1. Structural analysis, Int. J. Biol. Macromol. 33 (2000) 8344-8353. https://doi.org/10.1021/ma0008701

[14] N.M. Hansen, D. Plackett, Sustainable films and coatings from hemicelluloses: a review, Int. J. Biol. Macromol. 9 (2008) 1493-1505. https://doi.org/10.1021/bm800053z

[15] T. Galicia-García, F. Martínez-Bustos, O. Jiménez-Arévalo, D. Arencón, J. Gámez-Pérez, A.B. Martínez, Films of native and modified starch reinforced with fiber: Influence of some extrusion variables using response surface methodology, J. Appl. Polym. Sci. 126 (2012) 327-336. https://doi.org/10.1002/app.36982

[16] A.P. Mathew, A. Dufresne, Plasticized waxy maize starch: effect of polyols and relative humidity on material properties, Int. J. Biol. Macromol. 3 (2002) 1101-1108. https://doi.org/10.1021/bm020065p

[17] H.M. Park, X. Li, C.Z. Jin, C.Y. Park, W.J. Cho, C.S. Ha, Preparation and properties of biodegradable thermoplastic starch/clay hybrids, Macromol. Mater. Eng. 287 (2002) 553-558. https://doi.org/10.1002/1439-2054(20020801)287:8<553::AID-MAME553>3.0.CO;2-3

[18] J.K. Pandey, R.P. Singh, Green nanocomposites from renewable resources: effect of plasticizer on the structure and material properties of clay-filled starch, Starke. 57 (2005) 8-15. https://doi.org/10.1002/star.200400313

[19] B. Chen, J.R. Evans, Thermoplastic starch-clay nanocomposites and their characteristics, Carbohydr. Polym. 61 (2005) 455-463. https://doi.org/10.1016/j.carbpol.2005.06.020

[20] S. Guggenheim, J.M. Adams, D.C. Bain, F. Bergaya, M.F. Brigatti, V.A. Drits, M.L. Formoso, E. Galán, T. Kogure, H. Stanjek, Summary of recommendations of nomenclature committees relevant to clay mineralogy: report of the Association Internationale pour l'Etude des Argiles (AIPEA) Nomenclature Committee for 2006, Clays Clay Miner. 54 (2006) 761-772. https://doi.org/10.1346/CCMN.2006.0540610

[21] A.F. Lopez, M.V. Martinez, T. Arbeloa, A.I. Lopez, Crystal Structures of Clay Minerals and their X-ray Identification, J. Photochem. Photobiol. 8 (2007) 85-108. https://doi.org/10.1016/j.jphotochemrev.2007.03.003

[22] A.M. Mathieson, G.F. Walker, Crystal structure of magnesium-vermiculite, Am. Mineral. 39 (1954) 231-255.

[23] M.J. Fernández, M.D. Fernández, I.J. Aranburu, Effect of clay surface modification and organoclay purity on microstructure and thermal properties of poly (l-lactic acid)/vermiculite nanocomposites, Appl. Clay Sci. 80 (2013) 372-381. https://doi.org/10.1016/j.clay.2013.06.034

[24] P. Lu, M. Zhang, Y. Liu, J. Li, M. Xin, Characteristics of vermiculite-reinforced thermoplastic starch composite films, J. Appl. Polym. Sci. 126 (2012) 116-122. https://doi.org/10.1002/app.36342

[25] H.H Murray, Applied clay mineralogy: occurrences, processing and applications of kaolins, bentonites, palygorskitesepiolite, and common clays, Elsevier, Netherlands, 2006, pp.101-105. https://doi.org/10.1016/S1572-4352(06)02008-3

[26] I. Ahmad, F. Ali, F. Rahim, Clay Based Nanocomposites and Their Environmental Applications, Nanomaterials for Environmental Applications their Fascinating Attributes, Bentham Science Publishers, Sharjah U.A.E., 2018, pp. 166. https://doi.org/10.2174/9781681086453118020006

[27] P.J.P. de Mesquita, R.D.J. Araujo, L.H. de Carvalho, T.S. Alves, R. Barbosa, Thermal evaluation of PHB/PP-g-MA blends and PHB/PP-g-MA/vermiculite bionanocomposites after biodegradation test, Polym. Eng. Sci. 56 (2016) 555-560. https://doi.org/10.1002/pen.24279

[28] Y. Guo, M. Chen, J. Li, G. Gao, Effect of vermiculite dispersion in poly (lactic acid) preparation and its biodegradability, Polym. Sci. Ser. B. 58 (2016) 47-53. https://doi.org/10.1134/S1560090416010024

[29] D.C. da Costa Reis, T.A. de Oliveira, L.H. de Carvalho, T. Soares Alves, R. Barbosa, Biodegradability of and interaction in the packaging of poly (3-hydroxybutyrate-co-3-hydroxyvalerate)-vermiculite bionanocomposites, J. Appl. Polym. Sci. 134 (2017). https://doi.org/10.1002/app.44700

[30] D.C.D.C. Reis, T.A.D. Oliveira, L.H.D. Carvalho, T.S. Alves, R. Barbosa, The influence of natural clay and organoclay vermiculite on the formation process of bionanocomposites with poly (3-hydroxybutyrate-co-3-hydroxyvalerate), Mater. Jpn. 22 (2017). https://doi.org/10.1590/s1517-707620170004.0220

[31] M.F. Oliveira, A.L. China, M.G. Oliveira, M.C. Leite, Biocomposites based on Ecobras matrix and vermiculite, Mater. Lett. 158 (2015) 25-28. https://doi.org/10.1016/j.matlet.2015.05.030

[32] M.F. Oliveira, F.C. Braga, M.C. Leite, M.G. Oliveira, Evaluation of thermal properties of nanocomposites based on ecobras matrix and vermiculite modified with alkylphosphonium salt, Macromol. Symp. 367 (2016) 42-48. https://doi.org/10.1002/masy.201600014

[33] M.D. Fernández, M.J. Fernández, Vermiculite/poly (lactic acid) composites: Effect of nature of vermiculite on hydrolytic degradation in alkaline medium, Appl. Clay Sci. 143 (2017) 29-38. https://doi.org/10.1016/j.clay.2017.03.010

[34] M. Valaskova, G.S. Martynkova, Clay minerals in nature: their characterization, modification and application, BoD-Books on Demand Publisher, 2012, pp. 209-238. https://doi.org/10.5772/2708

[35] S. Malamis, E. Katsou, A review on zinc and nickel adsorption on natural and modified zeolite, bentonite and vermiculite: examination of process parameters,

kinetics and isotherms, J. Hazard. Mater. 252 (2013) 428-461.
https://doi.org/10.1016/j.jhazmat.2013.03.024

[36] M. Valášková, J. Tokarský, M. Hundáková, J. Zdrálková, B. Smetana, Role of vermiculite and zirconium-vermiculite on the formation of zircon-cordierite nanocomposites, Appl. Clay Sci. 75 (2013) 100-108.
https://doi.org/10.1016/j.clay.2013.02.015

[37] M. Sutcu, Influence of expanded vermiculite on physical properties and thermal conductivity of clay bricks, Ceram. Int. 41 (2015) 2819-2827.
https://doi.org/10.1016/j.ceramint.2014.10.102

[38] A.M. Rashad, Vermiculite as a construction material-A short guide for Civil Engineer, Constr. Build. Mater. 125 (2016) 53-62.
https://doi.org/10.1016/j.conbuildmat.2016.08.019

[39] N. Xie, J. Luo, Z. Li, Z. Huang, X. Gao, Y. Fang, Z.J.S.E.M. Zhang, S. Cells, Salt hydrate/expanded vermiculite composite as a form-stable phase change material for building energy storage, Sol. Energy Mater Sol. Cells. 189 (2019) 33-42.
https://doi.org/10.1016/j.solmat.2018.09.016

[40] L. Chmielarz, P. Kuśtrowski, Z. Piwowarska, B. Dudek, B. Gil, M. Michalik, Montmorillonite, vermiculite and saponite based porous clay heterostructures modified with transition metals as catalysts for the DeNOx process, Appl. Catal. B. 88 (2009) 331-340. https://doi.org/10.1016/j.apcatb.2008.11.001

[41] S.Y. Lee, H. Chen, M. Hanna, Preparation and characterization of tapioca starch-poly (lactic acid) nanocomposite foams by melt intercalation based on clay type, Ind. Crops. Prod. 28 (2008) 95-106. https://doi.org/10.1016/j.indcrop.2008.01.009

[42] K.M. Dean, M.D. Do, E. Petinakis, L. Yu, Key interactions in biodegradable thermoplastic starch/poly (vinyl alcohol)/montmorillonite micro-and nanocomposites, Compos. Sci. Technol. 68 (2008) 1453-1462.
https://doi.org/10.1016/j.compscitech.2007.10.037

[43] S.A. McGlashan, P.J. Halley, Preparation and characterisation of biodegradable starch-based nanocomposite materials, Polym. Int. 52 (2003) 1767-1773.
https://doi.org/10.1002/pi.1287

[44] J.W. Rhim, S.I. Hong, C.S. Ha, Tensile, water vapor barrier and antimicrobial properties of PLA/nanoclay composite films, LWT - Food Sci. Technol. 42 (2009) 612-617. https://doi.org/10.1016/j.lwt.2008.02.015

[45] M. Jamshidian, E.A. Tehrany, M. Imran, M.J. Akhtar, F. Cleymand, S.J. Desobry, Structural, mechanical and barrier properties of active PLA-antioxidant films, J. Food Eng. 110 (2012) 380-389. https://doi.org/10.1016/j.jfoodeng.2011.12.034

[46] X. Gong, L. Pan, C.Y. Tang, L. Chen, Z. Hao, W.C. Law, X. Wang, C.P. Tsui, C. Wu, Preparation, optical and thermal properties of CdSe-ZnS/poly (lactic acid)(PLA) nanocomposites, Compos. B. Eng. 66 (2014) 494-499. https://doi.org/10.1016/j.compositesb.2014.06.016

[47] N. Najafi, M. Heuzey, P.J. Carreau, Polylactide (PLA)-clay nanocomposites prepared by melt compounding in the presence of a chain extender, Compos. Sci. Technol. 72 (2012) 608-615. https://doi.org/10.1016/j.compscitech.2012.01.005

[48] I.S. Tawakkal, M.J. Cran, S.W. Bigger, Effect of kenaf fibre loading and thymol concentration on the mechanical and thermal properties of PLA/kenaf/thymol composites, Ind. Crops. Prod. 61 (2014) 74-83. https://doi.org/10.1016/j.indcrop.2014.06.032

[49] W. Song, Z. Zheng, W. Tang, X. Wang, A facile approach to covalently functionalized carbon nanotubes with biocompatible polymer, Polym. J. 48 (2007) 3658-3663. https://doi.org/10.1016/j.polymer.2007.04.071

[50] H.M. Ye, K. Hou, Q. Zhou, Improve the thermal and mechanical properties of poly (L-lactide) by forming nanocomposites with pristine vermiculite, Chin. J. Polym. Sci. 34 (2016) 1-12. https://doi.org/10.1007/s10118-016-1724-5

[51] H. Liao, D. Ma, Z. Jiao, Y. Xie, S. Tan, X. Cai, L. Huang, Fabrication of quaternary phosphonium-intercalated vermiculite for reinforcing UV-curable epoxy acrylate coatings, J. Adhes. Sci. Technol. 29 (2015) 171-184. https://doi.org/10.1080/01694243.2014.980615

[52] T. Zhang, F. Zhang, S. Dai, Z. Li, B. Wang, H. Quan, Z. Huang, Polyurethane/organic vermiculite composites with enhanced mechanical properties, J. Appl. Polym. Sci. 133 (2016). https://doi.org/10.1002/app.43219

[53] Y. Wan, Y. Fan, J. Dan, C. Hong, S. Yang, F. Yu, A review of recent advances in two-dimensional natural clay vermiculite-based nanomaterials, Mater. Res. Express. 6 (2019). https://doi.org/10.1088/2053-1591/ab3c9e

[54] J. Zhang, H. Liu, Z. Wu, W. Xiang, S. Wen, X. Cai, S. Tan, T. Wu, Rubber, Composites, Preparation of organic vermiculite/polymethylmethacrylate nanocomposite featuring excellent mechanical and thermal properties via ultrasonic in

situ polymerisation, Plastics, Plast. Rubber Compos. 46 (2017) 333-340.
https://doi.org/10.1080/14658011.2017.1356053

[55] J. Wang, F. Wang, Z. Gao, M. Zheng, J. Sun, Flame retardant medium-density
fiberboard with expanded vermiculite, BioResources 11 (2016) 6940-6947.
https://doi.org/10.15376/biores.11.3.6940-6947

[56] H.M. Ye, K. Hou, Q. Zhou, Improve the thermal and mechanical properties of poly
(L-lactide) by forming nanocomposites with pristine vermiculite, Chin. J. Polym. Sci.
34 (2016) 1-12. https://doi.org/10.1007/s10118-016-1724-5

[57] S. İşçi, Y. İşçi, Characterization and comparison of thermal & mechanical properties
of vermiculite polyvinylbutyral nanocomposites synthesized by solution casting
method, Appl. Clay Sci. 151 (2018) 189-193.
https://doi.org/10.1016/j.clay.2017.10.009

[58] A.D. Chandio, I.A. Channa, M. Rizwan, S. Akram, M.S. Javed, S.H. Siyal, M.
Saleem, M.A. Makhdoom, T. Ashfaq, S. Khan, S. Hussain, Polyvinyl Alcohol and
Nano-Clay Based Solution Processed Packaging Coatings, Coatings. 11 (2021) 942.
https://doi.org/10.3390/coatings11080942

[59] T. Russo, P. Fucile, R. Giacometti, F. Sannino, Sustainable Removal of
Contaminants by Biopolymers: A Novel Approach for Wastewater Treatment. Current
State and Future Perspectives, Process. 9 (2021) 719.
https://doi.org/10.3390/pr9040719

[60] M. Nasrollahzadeh, M. Sajjadi, S. Iravani, R.S. Varma, Starch, cellulose, pectin,
gum, alginate, chitin and chitosan derived (nano) materials for sustainable water
treatment: A review, Carbohydr. Polym. 251 (2021) 116986.
https://doi.org/10.1016/j.carbpol.2020.116986

[61] L. Guo, G. Li, J. Liu, Y. Meng, Y. Tang, Adsorptive decolorization of methylene
blue by crosslinked porous starch, Carbohydr. Polym. 93 (2013) 374-379.
https://doi.org/10.1016/j.carbpol.2012.12.019

[62] S. Ahmad, B.A. Palvasha, B.B.K. Abbasi, M.S. Nazir, M.N. Akhtar, Z. Tahir, M.A.
Abdullah, Preparation and Applications of Polysaccharide-Based Composites,
Polysaccharides: Properties and Applications, Scrivener publishing, John Wiley &
Sons, New York, 2021. https://doi.org/10.1002/9781119711414.ch26

[63] A. Pourjavadi, A. Abedin-Moghanaki, A. Tavakoli, Efficient removal of cationic
dyes using a new magnetic nanocomposite based on starch-g-poly (vinylalcohol) and

functionalized with sulfate groups, RSC. Adv. 6 (2016) 38042-38051. https://doi.org/10.1039/C6RA02517J

[64] J. Guo, J. Wang, G. Zheng, X. Jiang, P. Research, A TiO 2/crosslinked carboxymethyl starch composite for high-efficiency adsorption and photodegradation of cationic golden yellow X-GL dye, Environ. Sci. Pollut. Res. 26 (2019) 24395-24406. https://doi.org/10.1007/s11356-019-05685-y

[65] G. Gong, F. Zhang, Z. Cheng, L. Zhou, Facile fabrication of magnetic carboxymethyl starch/poly (vinyl alcohol) composite gel for methylene blue removal, Int. J. Biol. Macromol. 81 (2015) 205-211. https://doi.org/10.1016/j.ijbiomac.2015.07.061

[66] F. Delval, G. Crini, N. Morin, J. Vebrel, S. Bertini, G. Torri, The sorption of several types of dye on crosslinked polysaccharides derivatives, Dyes Pigm. 53 (2002) 79-92. https://doi.org/10.1016/S0143-7208(02)00004-9

[67] L.M. Zhang, D.Q. Chen, An investigation of adsorption of lead (II) and copper (II) ions by water-insoluble starch graft copolymers, Colloids Surf. A. Physicochem. Eng. Asp. 205 (2002) 231-236. https://doi.org/10.1016/S0927-7757(02)00039-0

[68] B. Kim, S.T. Lim, Removal of heavy metal ions from water by cross-linked carboxymethyl corn starch, Carbohydr. Polym. 39 (1999) 217-223. https://doi.org/10.1016/S0144-8617(99)00011-9

[69] N.D. Suzaimi, P.S. Goh, N.A. Nizam Nik Malek, B.C. Ng, A.F. Ismail, Nano-Adsorbents in Wastewater Treatment for Phosphate and Nitrate Removal, J. Environ. Nanotechnol. 5 (2021) 339-370. https://doi.org/10.1007/978-3-030-73010-9_10

[70] A. Malathi, A.J. Singh, Antimicrobial activity of rice starch based film reinforced with titanium dioxide (TiO2) nanoparticles, Agric. Res. 56 (2019) 111. https://doi.org/10.5958/2395-146X.2019.00017.6

[71] Z. Hejri, A.A. Seifkordi, A. Ahmadpour, S.M. Zebarjad, A. Maskooki, Biodegradable starch/poly (vinyl alcohol) film reinforced with titanium dioxide nanoparticles, Int. J. Miner. Metall. Mater. 20 (2013) 1001-1011. https://doi.org/10.1007/s12613-013-0827-z

[72] D.K. Ban, S.K. Pratihar, S.J. Paul, An investigation of optical properties of zinc oxide nanoparticle synthesized by starch mediated assembly and its application in photocatalytic bleaching of methyl green and rhodamine-B, Mater. Sci. Semicond. Process. 39 (2015) 691-701. https://doi.org/10.1016/j.mssp.2015.05.038

[73] K. Vidhya, M. Saravanan, G. Bhoopathi, V. Devarajan, S. Subanya, Structural and optical characterization of pure and starch-capped ZnO quantum dots and their photocatalytic activity, Appl. Nanosci. 5 (2015) 235-243. https://doi.org/10.1007/s13204-014-0312-7

[74] J. Ma, W. Zhu, Y. Tian, Z. Wang, Preparation of zinc oxide-starch nanocomposite and its application on coating, Nanoscale Res. Lett. 11 (2016) 1-9. https://doi.org/10.1186/s11671-015-1209-4

[75] S.T. Lin, M. Thirumavalavan, T.Y. Jiang, J.F. Lee, Synthesis of ZnO/Zn nano photocatalyst using modified polysaccharides for photodegradation of dyes, Carbohydr. Polym. 105 (2014) 1-9. https://doi.org/10.1016/j.carbpol.2014.01.017

[76] J.H. Lee, H.S. Kim, E.T. Yun, S.Y. Ham, J.H. Park, C.H. Ahn, S.H. Lee, H.D. Park, Vertically Aligned Carbon Nanotube Membranes: Water Purification and Beyond, J. Membr. Sci. 10 (2020) 273. https://doi.org/10.3390/membranes10100273

[77] H. Sharma, M. Bhardwaj, M. Kour, S. Paul, Highly efficient magnetic Pd (0) nanoparticles stabilized by amine functionalized starch for organic transformations under mild conditions, Mol. Catal. 435 (2017) 58-68. https://doi.org/10.1016/j.mcat.2017.03.019

[78] F. Delval, G. Crini, J. Vebrel, M. Knorr, G. Sauvin, E. Conte, Starch-modified filters used for the removal of dyes from waste water, Macromol. Symp. 203 (2003) 165-172. https://doi.org/10.1002/masy.200351315

[79] G. Crini, Studies on adsorption of dyes on beta-cyclodextrin polymer, Bioresour. Technol. 90 (2003) 193-198. https://doi.org/10.1016/S0960-8524(03)00111-1

[80] G. Crini, N. Morin, J.C. Rouland, L. Janus, M. Morcellet, S. Bertini, Adsorption de béta-naphtol sur des gels de cyclodextrine-carboxyméthylcellulose réticulés, Eur. Polym. J. 38 (2002) 1095-1103. https://doi.org/10.1016/S0014-3057(01)00298-1

[81] G. Crini, Recent developments in polysaccharide-based materials used as adsorbents in wastewater treatment, Prog. Polym. Sci. 30 (2005) 38-70. https://doi.org/10.1016/j.progpolymsci.2004.11.002

[82] L. Janus, B. Carbonnier, A. Deratani, M. Bacquet, G. Crini, J. Laureyns, M. Morcellet, New HPLC stationary phases based on (methacryloyloxypropyl-β-cyclodextrin-co-N-vinylpyrrolidone) copolymers coated on silica. Preparation and characterisation, New J. Chem. 27 (2003) 307-312. https://doi.org/10.1039/b206718h

[83] C.Y. Jimmy, Z.T. Jiang, H.Y. Liu, J. Yu, L. Zhang, β-Cyclodextrin epichlorohydrin copolymer as a solid-phase extraction adsorbent for aromatic compounds in water

samples, Anal. Chim. Acta .477 (2003) 93-101. https://doi.org/10.1016/S0003-2670(02)01411-3

[84] M. Kitaoka, K. Hayashi, Adsorption of bisphenol A by cross-linked β-cyclodextrin polymer, J. Incl. Phenom. Macrocycl. Chem. 44 (2002) 429-431. https://doi.org/10.1023/A:1023024004103

[85] A. Mohseni, L. Fan, F. Roddick, H. Li, Y. Gao, Z. Liu, Cationic starch: an effective flocculant for separating algal biomass from wastewater RO concentrate treated by microalgae, J. Appl. Soc. Psychol. 33 (2021) 917-928. https://doi.org/10.1007/s10811-020-02348-1

[86] J.P. Wang, S.J. Yuan, Y. Wang, H. Yu, Synthesis, characterization and application of a novel starch-based flocculant with high flocculation and dewatering properties, Water Res. 47 (2013) 2643-2648. https://doi.org/10.1016/j.watres.2013.01.050

[87] M. Mohd Amin, S. Heijman, L. Rietveld, Clay-starch combination for micropollutants removal from wastewater treatment plant effluent, Water Sci. Technol. 73 (2016) 1719-1727. https://doi.org/10.2166/wst.2016.001

[88] H. Cheng, L. Chen, D.J. McClements, T. Yang, Z. Zhang, F. Ren, M. Miao, Y. Tian, Z. Jin, Technology, Starch-based biodegradable packaging materials: A review of their preparation, characterization and diverse applications in the food industry, Trends Food Sci. Technol. (2021). https://doi.org/10.1016/j.tifs.2021.05.017

[89] M. Aghazadeh, R. Karim, R.A. Rahman, M.T. Sultan, M. Paykary, S. Johnson, Effect of glycerol on the physicochemical properties of cereal starch films, Czech J. Food Sci. 36 (2018) 403-409. https://doi.org/10.17221/41/2017-CJFS

[90] F. Wang, Z. Gao, M. Zheng, J. Sun, Thermal degradation and fire performance of plywood treated with expanded vermiculite, Fire Mater. 40 (2016) 427-433. https://doi.org/10.1002/fam.2297

[91] J.Y. Cheong, J. Ahn, M. Seo, Y.S. Nam, Flame-retardant, flexible vermiculite-polymer hybrid film, RSC Adv. 5 (2015) 61768-61774. https://doi.org/10.1039/C5RA08382F

Adv. App. of Micro and Nano Clay – Biopolymer-based Composites   Materials Research Forum LLC
Materials Research Foundations **125** (2022) 152-171   https://doi.org/10.21741/9781644901915-7

Chapter 7

Halloysite-Starch based Nano-Composites and Applications

Farrukh Rafiq Ahmed[1], Muhammad Sikandar[1], Muhammad Harris Shoaib[1*], Rabia Ismail Yousuf[1], Kamran Ahmed[1] *

[1]Department of Pharmaceutics, Faculty of Pharmacy and Pharmaceutical Sciences, University of Karachi, Sindh, Pakistan

* harrisshoaib2000@yahoo.com; mhshoaib@uok.edu.pk

Abstract

Halloysite is a novel mineral belonging to the kaolinite family of clays. It consists of largely cylindrical particles in the size range of few hundred to few micrometers in length. The negatively charged Si-O-Si functional groups at the surface and positively charged $Al_2(OH)_4$ at the luminal side offer unique chemistry to this clay mineral. Biopolymers such as starch are considered biodegradable and non-toxic in nature. But their higher water permeability, poor mechanical strength, and rigid characteristics limit their applications in many fields. Halloysite and starch hybrid materials or composites have been demonstrated to improve on these properties and at the same time remain natural. They have a wide variety of applications such as tissue engineering, drug delivery and food packaging materials. Besides this, they have also been used as catalyst and flame retardant materials.

Keywords

Halloysite Nanotubes, Starch, Nanocomposite, Nano-Clay, Nano-Biocomposite, Applications

Contents

1.   Introduction

Natural clay minerals have received a wide-scale attention for their various applications in the fields ranging from environmental and materials sciences to biomedical and pharmaceutical sciences. Halloysite is one of such clay minerals, which, owing to its low cost, high availability, biocompatibility and environment friendly nature, has received increased attention. [1-3]. The clay mineral was first named as 'halloysite' by a French geologist, Pierre Berthier in the year 1826. It was named in honor of a Belgian geologist, Juan Baptiste Julien d'Omalius d'Halloy, who is believed to have discovered this clay (Figure 1) [4][7][5]. It is chemically categorized as a morphological type of kaolin clays [6]. But, unlike other kaolin clay minerals, it has tubular shaped concentric multilayered particles, in size range of a few hundred nanometers to few micrometers; various bioactive agents or molecules can be encapsulated [7]. The special geometry of such clay is believed to have formed naturally by the surface weathering of aluminosilicate kaolin with rolling-over phenomena, specifically during volcanic activity [8]. Based on origin, it is found in different arrangements particularly large tubes, short tubes, and spheroids [9]. It is resourced in abundant quantities from various diversified regions and countries such as Australia, Brazil, China, South Korea, Russia, and the United States of America [10]. The structure of these nanotubes is composed of concentric multilayer hollow tubules formed by rolling of the silica and alumina sheets having chemical formula $Al_2(OH)_4Si_2O_5(2H_2O)$, that is similar to kaolinite. The Al−OH groups are dispersed on their inner luminal surfaces of the tubules while the Si-O-Si are distributed on the outer surface (Figure 2) [11-13]. The outer and inner surfaces of the nanotubes are negatively and positively charged [14]. This clay material is morphologically present in white form to slightly red form due to presence of high content of iron oxide. The mineral stones are usually crushed into powder form and the particulate structure is usually comprised of 15-20 layers [6, 15, 16].

Figure 1. Representative scanning electron microscopic (SEM) image of halloysite with nanotubes protruding from the bulk materials

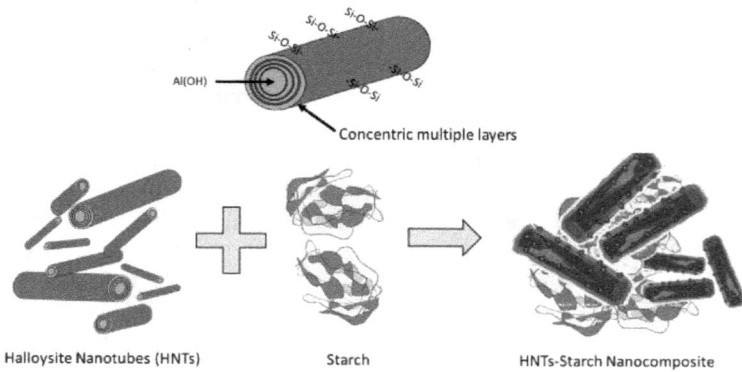

Figure 2. Structure of halloysite nanotubes and halloysite based starch nanocomposites.

Adv. App. of Micro and Nano Clay – Biopolymer-based Composites   Materials Research Forum LLC
Materials Research Foundations **125** (2022) 152-171   https://doi.org/10.21741/9781644901915-7

The halloysite nanotubes (HNTs) are used commonly for the reinforcement of polymers and are generally in the range of 300 to 1500 nm in length with an external and internal diameter of 40 to120 nm and 15 to 100 nm [3]. The interlayer spacing of hydrated nanotubes is measured to be 1 nm, while this spacing becomes narrow upon irreversible dehydration at temperatures in between 30 °C to 110 °C [3, 6]. Moreover, this mineral with a nanotubular structure is considered relatively non-toxic and biocompatible [6, 17, 18]. Halloysite nanotubes also possess chemical and physical stability, high surface area and porosity, sufficient hydroxyl groups for conjugation, large aspect ratio, good dispersion, and lower density compared to other kaolinites [3, 6, 9, 17, 18]. These features make the nanotubes favorable nanofiller for polymer materials to enhance their properties [3, 19].

Biopolymers that are generally biodegradable and categorized as renewable materials are considered superior to synthetic polymers [20]. The polysaccharides include alginate, chitosan, and starch, have been categorized as natural biopolymer adsorbents. They are biodegradable and non-toxic in nature. But, the high water permeability, poor mechanical strength, and rigid characteristics limit their application on a practical scale [17]. With the advent of an era of nanotechnology, the functional properties of these materials have been rescaled, and several biocomposite materials have been developed to improve on their properties [21]. The polymeric nano-biocomposites are made up of biopolymer and an inorganic nanomaterial, usually a nanoclay such as HNTs (Fig. 2) [22-26]. The biopolymer acts as a matrix while the nanoclays as a filler improve its mechanical, physical or thermal properties [19, 26]. The composites of starch (among the biopolymers) and HNTs (among the nanofillers) are considered attractive due to their large natural availability, biocompatibility, renewability, and biodegradability. They are moreover preferred over traditional fossil-based non-biodegradable and non-renewable polymers [17, 27]. The preparations of HNTs-Starch nanocomposites are widely reported in the literature and are summarized here in Table 1. Various field specific applications of the halloysite and starch nanocomposites are discussed in the proceeding sections of the chapter.

Adv. App. of Micro and Nano Clay – Biopolymer-based Composites   Materials Research Forum LLC
Materials Research Foundations **125** (2022) 152-171   https://doi.org/10.21741/9781644901915-7

Table 1. Summary of the various studies on the development of halloysite nanotubes-starch nanocomposite films.

Nanocomposite Materials	Preparation Methods	Prepared nanocomposites Properties	References
HNTs, β-cyclodextrin and Folic acid	Casting method	The developed material was suggested as a highly promising tool in isolating circulating tumor cells (CTCs) for the diagnosis and treatment of cancer	[15]
Non-treated HNTs, quaternary ammonium salt with benzoalkonium chloride treated HNTs, Wheat starch and Glycerol	Melt-extrusion method	Enhanced thermal stability and improved mechanical properties of starch. Composite with treated HNTs were found superior in properties than non-treated HNTs	[19]
HNTs, Corn starch and Chitosan	Solution casting process	The water vapor transmission rate, water solubility and water absorption capacity of the materials were observed improved. Moreover, the prepared composite was found impermeable to bacteria.	[22]
HNTs, Potato starch, glycerol and sorbitol	Melt blending method	The mechanical and thermal stability was observed improved while the moisture absorption capability of the starch was reduced	[27]
HNTs, Potato starch, Sorbitol and glycerol	Casting method	The mechanical and barrier properties of potato starch film were found improved	[51]
HNTs and Corn starch	Solvent exchange method	The adsorption capacity was perceived improved	[53]
Poly (glycidyl methacrylate) modified HNTs and β-cyclodextrin	A facile method	The composite could be used as a adsorbent for methylene blue and a catalyst in the oxidation reaction of benzyl alcohol	[55]
Chlorinated HNTs, aminated Starch and Pd (II) acetate	Conjugation method	An efficient catalytic activity was observed	[57]
HNTs, Starch, Polyvinyl alcohol (PVA) and herbicide atrazine	Molecule loading method	Slower release of herbicide was observed from the developed system	[59]

1,4-phenylenebisdiboronic acid-modified halloysite nanotubes (HNTs-BO), compressible starch and Pentoxifylline	Hydrogel formulated by coupling of materials	A complete release of drug was accomplished with H2O2, indicating development of a successful H_2O_2-responsive composite	[60]
HNTs, Starch and Mexican oregano Essential Oil (EO)	Casting method	A change in morphology of the film structure was shown with EO. The pore density was observed decreased while the pore size was found increased. Moreover, a better physical and satisfactory antifungal properties especially against *A. niger* were found.	[61]
Acid modified HNTs, Sweet potato, β-cyclodextrin and Urea	Free radical copolymerization and casting method	The release of fertilizer was effective controlled from the system	[62]
HNTs, Starch and polyurethane-urea (PUU)	Casting method	The material is considered a favorable top coat constituent in human-body conditions	[63]
HNTs, Thermoplastic starch and poly(butylene adipate-*co*-terephthalate)	Melt blending method	The mechanical properties of the thermoplastic starch was observed improved	[64]
HNTs, Thermoplastic sago starch (TPSS), glycerol and Chloramphenicol	Casting method	An efficient antimicrobial activity with the inhibition zone sizes 67.3 ± 0.5 for TPSS/HNT/0.25mL Chloramphenicol and 100 ± 0.3 mm^2 for TPSS/HNT/0.5mL Chloramphenicol were observed	[65]
HNTs, Starch, PVA and glycerol	Casting method	The thermal stability of polymer was observed improved	[66]
HNTs and thermoplastic starch	Melt-processing method	A material having excellent mechanical properties with increased Young's modulus and tensile strength and improved thermal stability was obtained	[67]
HNTs, Starch, and Nisin	Melt blending method	The developed system was found effective against *Staphylococcus aureus*, *Clostridium perfringens* and *Listeria monocytogenes*	[68]

HNTs, TPSS and glycerol	Solvent casting process	The mechanical of the TPSS was observed increased while the water absorption capacity decreased	[69]
HNTs, Low density polyethylene and thermoplastic starch	Extrusion and injection molding	The mechanical properties of the nanocomposites was found enhanced	[70]
HNTs, Polypropylene and thermoplastic starch	Melt blending method	The thermal stability and biodegradability of the polymer was improved and moisture uptake capability was reduced	[71]
HNTs, PVA, Starch and Glycerol	Solution casting method	Water solubility and absorption capacity of the polymers were observed reduced by almost 50% but the water contact angle was found increased. The developed nanocomposites were claimed suitable for food packaging due to adequate transparency, good biodegradability and efficient water resistance properties	[72]

2. Biomedical applications

Nano-biocomposites of halloysite nanotubes with glycerol plasticized-starch sufficiently improves the thermal resistance, tensile strength and decrease moisture permeation, and thereby have diverse applications in paper, textile, cosmetic and bio-medical industries [28]. Moreover, polyethylene glycol mediated dispersed hydrogels of halloysite nanotubes have also been reported with prime applications in biomedical implants and tissue scaffolds [29]. The composites have been suggested as useful in several fields of biomedical and pharmaceuticals such as tissue engineering, drug carrier delivery system, biosensor, wound dressing as well as in food packaging [30].

2.1 Drug/molecular carrier

A molecular carrier system is based on the delivery of molecules to the required site [31]. Recently, the advances in nanotechnology applications related to the delivery of molecules using various types of nanoparticles as carriers have become a topic of interest. Similarly, special consideration has been paid to the use of porous starch (PS) as a carrier system for the encapsulation of biologically active compounds. PS is a modified form of natural polysaccharide starch that plays an important role as an adsorbent in biotechnological and drug carrier systems [32]. PS composites with enhanced adsorption properties have been produced with the combination of HNTs.

The composites have been demonstrated as a carrier system for controlled, sustained, or extended-release drug delivery [29]. The effective controlled-release rate of the molecules has been achieved by coating the starch onto the drug-loaded HNTs [29]. Furthermore, the development of PS-HNT carrier composites using solvent exchange process has been reported and suggested appropriate for the oral delivery of sensitive or volatile and water-insoluble substances. It is believed that the porous structure of the PS in the composites acts as adsorbent and promotes controlled-release of the adsorbed substance from the system [33]. Several studies based on successful controlled delivery of active substances such as antimicrobial peptide and herbicide atrazine using HNT-starch composite films as a carrier have been successfully developed [32, 34, 35].

2.2   Tissue engineering

The rapidly developing field of tissue engineering aims to restore, regenerate, and improve the damaged tissues instead of replacing them by developing highly porous biomaterials as scaffolds [29]. The scaffolds are considered important in tissue repair or regeneration as they provide an appropriate environment for cell interactions and adhesions. They are further found helpful in the transportation of nutrients and regulatory factors needed for cell growth and its proliferation [36]. Tissue engineering scaffolds are usually sponge-like structures having excellent biocompatibility, thermal stability, and mechanical properties [29]. Starch is considered a promising scaffold biomaterial for numerous applications in the developing field of tissue engineering. But its use is limited due to its poor mechanical properties. The materials have accomplished the improvement of mechanical features of starch scaffolds with nano hydroxyapatite and bioactive glasses. However, these fillers require treatment prior to being utilized as fillers in the starch matrix, which again limits their practical applications as tissue engineering scaffolds. HNTs have been considered as a promising alternate for making low cost, biocompatible, mechanical strong and biodegradable starch based scaffolds [37]. HNTs based starch composites have been developed to be used as tissue engineering scaffolds [29, 38].

2.3   Wound dressing

The purpose of wound dressings is to promote effective healing and reduce the duration of the healing process [39]. Different types of wound dressings are available such as traditional, biomaterial-based, bioactive, and interactive dressings [29]. Polysaccharides (starch/chitosan etc.) are considered ideal biomaterials for wound dressings for their nontoxic, biocompatible and biodegradable properties [29]. The use of starch-chitosan composites in wound dressing applications has been reported in several studies, but the little mechanical strength associated with their excessive water solubility limits their practical applications. To overcome the issues, the use of inorganic substances such as nanoclays with biopolymers have gained importance in wound dressing applications. The halloysite nanotubes (HNTs) were shown to improve the mechanical properties of the polymers. It has been claimed that the nanocomposites can be used effectively as wound dressing with appropriate compressive, mechanical and tensile properties in both wet and dry states [40]. Similarly, polyethylene glycol modified HNTs-starch matrix, HNTs-potato starch nanocomposite films, and HNTs-chitosan-starch films have been developed with low toxicity while increased tensile strength, mechanical properties, compatibility, thermal stability, and water solubility. The composite films are suggested as effective wound dressings, provide wound healing by absorbing excess fluid resulting from contraction at the wound site [30, 41, 42].

3.   Food packaging applications

The main purpose of food packaging is to maintain the quality and safety of the food item during storage and transportation by preventing unsuitable conditions such as microbial, chemical and environmental (oxygen, light and moisture etc.) contaminants [43]. Polysaccharides, particularly starches, have been used as food packaging material for their biodegradable, nontoxic and film forming properties [29]. Starch films are either transparent or semitransparent, tasteless, and colorless in nature. The films are considered mechanically stronger than the other polysaccharide based films [44]. But they exhibit less stability, high moisture sensitivity, and poor mechanical properties. However, these properties can be improved by the introduction of plasticizers, either volatile plasticizers such as water or nonvolatile plasticizers like polyols (sorbitol/ glycerol). The plasticized starch is used in agricultural and packaging industries but still cannot fulfill the criteria due to poor moisture sensitivity. These limitations have been successfully addressed by the addition of HNTs into the starch matrix [28, 44]. The use of nano-biocomposites as antimicrobials in food packaging is found significant for increasing the stability of food and protecting it from microbial attacks during storage [38]. Nanocomposites comprising plasticized starch and HNTs have been synthesized by solvent casting and melt extrusion methods [45]. The mechanical properties of films have been found to be enhanced with decreased moisture sensitivity. The films were therefore suggested for their commercial use for food preservation and food packaging application [29].

3.1   As an antimicrobial and antioxidant

The anti-bacterial agents can be classified, based on their mechanism of action, into two groups such as bactericidal (has potential to kill bacteria) and bacteriostatic (effectively inhibits the bacterial cell growth). The antimicrobial activity of the biopolymer films has been found to be improved by the addition of nanoparticles. The nanoparticles are considered more effective against Gram negative than Gram positive bacteria. Similarly, oxygen and moisture permeability are the crucial parameters, cause deterioration of packed food. Starch, among all the biopolymers, is considered the most widely used polysaccharides in the preparation of nanocomposite films used as a preservative [38]. Starch edible films have been investigated for their antimicrobial effect in food packaging and food preservation applications. The stability of the food content was observed to have increased further by the incorporation of essential oils into the composite films [46, 47]. The edible films are claimed to be beneficial in inhibiting the microbial load and thereby increasing the shelf life of the food product [48]. Furthermore, the extracts such as rosemary, peppermint oil, thyme, olive, and ginger extracts have been incorporated into the packaging nanocomposite films as antimicrobial agents [46]. Similarly, antimicrobials such as nisin, pediocin and chloramphenicol have been loaded into HNTs-starch composites. The composites showed adequate antimicrobial activity against several microbes, mainly *L. monocytogenes*, *C. perfringens* with enhanced mechanical and thermal characteristics [29]. The antimicrobial zone of inhibition from nanocomposite films were measured 67.3 ± 0.5 mm for 0.25 mL chloramphenicol and 100 ± 0.3 mm^2 for 0.5 mL chloramphenicol [49].

3.2   As water vapor barrier

Water content for any given product is considered as one of the most important deteriorating factors during its storage, and food items are usually prone to water vapors due to ineffective packaging materials [50]. An effective water vapor barrier properties are preferred for food packaging materials in order to decrease moisture transfer between the materials and the outside environment [51]. Therefore, the resistance of biopolymer or polysaccharides films, used as packaging materials, towards water is deliberated as an essential useful parameter in food packaging. The water acceptability of the polymers is evaluated by water vapor permeability, swelling degree, water content, and water solubility of the films [52]. Although the biopolymers are considered highly biodegradable and generally have high commercial availability, their extraordinary water vapor permeability, poor mechanical properties, and high sensitivity to environmental changes of the biopolymer films make them of a poor choice for use as packaging materials [51, 52].

The incorporation of nanofillers has been proven valuable to reduce these limitations. Such biopolymeric films have been successfully developed and used for the effective storage of several food products, including fish, meat, milk, mushrooms, fruits and vegetables [52]. Additionally, antimicrobial and antioxidant properties have also been observed in the developed nanocomposite films, which are further beneficial in maintaining the food quality during storage [17]. It is believed that the water barrier properties of the nanocomposites films are attributed by the interaction of the nanofillers with the biopolymer through hydrogen bonding. This interaction results in the reduction of the spaces in the polymer matrix and thus decreases the hydrophilicity of the polymer [52]. Accordingly, several nanocomposite films comprised of starch and HNTs have been fabricated for the development of efficient food packaging materials. The mechanical properties and barriers of the developed composites have been considerably improved. Moreover, the decrease in water solubility, water absorption capacity, water vapor, along with gaseous molecules permeability have been perceived in the composites [17, 22, 51, 52]. Consequently, the nanocomposite films have been suggested as effective food packaging systems [17].

4.   Water treatment applications

Carbohydrates, specifically starch and 'porous starch' (PS) have gained importance as biodegradable adsorbents. PS/HNTs carriers have been developed to enhance the adsorbent powers of PS. The sponge-like structure of PS has holes across the surface; hence small molecules can pass through it easily. It is used in the food industry as well as in the pharmaceutical, agriculture, and cosmetics conveniently. The striking features are large surface area, slow-release, and being economically accessible. The adsorbent nature of PS/HNT has been demonstrated useful in dye removal, waste particle removal, and oil removal [53].

Organic pollutants, dyes, and heavy metal ions are some of the most common and essential materials required to be removed from wastewaters. The best method to remove these pollutants from water is adsorption due to its efficiency and low cost compared to alternative methods. HNTs have replaced most traditional adsorbents owing to their unique structure and cost-effectiveness. These benefits are twofold when combined with polysaccharides/carbohydrates. Polysaccharides such as starch with HNTs showed remarkable improvement in adsorption capacity due to characteristic features of both materials. They have been suggested as promising regeneratable adsorbents for different purposes such as wastewater treatment and dye removal techniques [17].

PS showed hydration capacity within the range of 6.36 ± 1.14 %. to 8.50 ± 0.56 %, while PS/HNT enhanced adsorption capacity to 14.30 ± 1.83 % at 0.3 mg/mL [53]. The oil adsorption capacity of PS was reported within a range of 1.29 ± 0.13 % g/g to 4.75 ± 0.37 % g/g. However, PS/HNT showed oil adsorption of 4.42 ± 0.40 g/g. An increase in oil adsorption capacity has been indicated with HNTs up to the dose 0.4 mg/mL but observed decline above 0.5 mg/mL dose. It has also been suggested that high oil adsorption is due to the incorporation of HNTs. However, a high ratio of HNTs seems to block the pores, consequently decreasing the oil adsorption capacity of PS/HNT nanocomposite [53].

HNT has negative Si-O-Si on the outer surface and positive Al-OH on the inner surface that can adsorb both cationic and anionic dyes at the respective sites [17]. The maximum dye adsorption capacity of HNTs has been claimed 84.32 mg/g of methylene blue (MB). PS/HNT had greater dye adsorption capacity compared to PS alone, attributed to the higher surface area provided by HNTs. MB removal was improved from 24.79 to 103.68 mg/g when MB concentration increased from 50 to 500 mg/L [53]. Starch has found its way to get associated with HNTs to exhibit efficient performance as a polysaccharide candidate. Cross-linked porous starch (CPS) has also been used to remove MB from aqueous solutions. When compared with native starch and PS, it was found that CPS has adsorption capacity higher than that of native starch and PS [54]. Cyclodextrin is a natural byproduct of the reaction between novel starch and enzyme cyclodextrins. Cyclodextrin and HNT combination results in novel nanocomposite hybrid formation recognized as HNTs-g-βCD. It has been employed as an environmental treatment and dye removal agent [55].

5.   Applications as a catalysts

Halloysite is an excellent candidate for fabricating hybrid nanocomposites. Halloysite, when fabricated as hybrid materials or composites, exhibits promising catalytic ability like chemical, photochemical and organic transformations as well as hydrogenation reactions. Halloysite nanotubes (HNTs) based composites are reported to be more sophisticated than the constituent materials themselves, leading to the concept that the synergism effect is dominating the process. Hence halloysite hybrids are outstanding catalytic performers. HNTs inner and outer surface modification can also lead to efficient catalytic active forms. The nature of catalyst depends upon the functional units anchored on the surface of halloysites [56]. Starch and halloysite create a unique hybrid by conjugation of amine-functionalized starch with chlorine-HNT and coordination of Pd (II) acetate. This reaction gave a hybrid catalyst. This unique heterogeneous catalyst promotes copper and ligand-free coupling reaction. Pd-HNT-Starch displayed a high catalytic rate and recyclability up to ten times. Comparison of three catalytic species Pd-HNT, Pd-Starch, and Pd-HNT-Starch showed later has superior performance. This has also indicated that synergism exists between starch and halloysite [57]. HNTs-g-βCD i.e. cyclodextrin and halloysite nanocomposites, have been used in catalyst activity, specifically oxidation of benzyl alcohol to aldehyde. This is due to the unique nature and structure of cyclodextrin. The spectral studies have further confirmed the findings of oxidation where increasing intensities of the C=O stretching and the decreasing intensities of the C–O stretching vibration was observed. This approach has made benzyl alcohol oxidation in the water system. It also owns simple methods, cost-effectiveness, high efficiency, and feasible recovery. The HNT-cyclodextrin applications as a catalyst, however, are suggested to be further explored [55].

6.   Flame retardant applications

HNTs have proven to provide thermal stability when combined with starch to fabricate bio nanocomposite films [58]. When mixed with polymers, HNTs give two opposite effects, one being barrier effect which improves thermal stability, while second a promoter effect which increases the rate of thermal decomposition. Nano-biocomposites of potato starch and HNTs combined with plasticizers like glycerol, sorbitol or both formed a new crystalline network. The glass transition temperature of the resultant composite has been observed to have improved as compared to glycerol plasticized matrix, indicating the barrier effect of HNTs on the degradation of polymer. Recent advances into the field of nanocomposite sciences, starch-based nanocomposites are proved to provide superior results. Owning to the larger surface area, nanoclays bring forth an improved version of thermal resistance material and at the same time preserving biodegradability and bio combability of the novel starch matrix [27].

References

[1] J.-H. Choy, S.-J. Choi, J.-M. Oh, T. Park, Clay minerals and layered double hydroxides for novel biological applications, Appl. Clay Sci. 36(1) (2007) 122-132. https://doi.org/10.1016/j.clay.2006.07.007

[2] S. Pavlidou, C. Papaspyrides, A review on polymer-layered silicate nanocomposites, Prog. Polym. Sci. 33(12) (2008) 1119-1198. https://doi.org/10.1016/j.progpolymsci.2008.07.008

[3] M. Du, B. Guo, D. Jia, Newly emerging applications of halloysite nanotubes: a review, Polymer International 59(5) (2010) 574-582. https://doi.org/10.1002/pi.2754

[4] M. Massaro, R. Noto, S.J.M. Riela, Past, present and future perspectives on halloysite clay minerals, 25(20) (2020) 4863. https://doi.org/10.3390/molecules25204863

[5] D.M. MAcEwAN, The nomenclature of the halloysite minerals, Mineral Mag J Mineral Soc 28(196) (1947) 36-44. https://doi.org/10.1180/minmag.1947.028.196.08

[6] M. Liu, Z. Jia, D. Jia, C. Zhou, Recent advance in research on halloysite nanotubes-polymer nanocomposite, Prog. Polym. Sci. 39(8) (2014) 1498-1525. https://doi.org/10.1016/j.progpolymsci.2014.04.004

[7] D. Rawtani, Y. Agrawal, Multifarious applications of halloysite nanotubes: a review, Rev. Adv. Mater. Sci 30(3) (2012) 282-295.

Adv. App. of Micro and Nano Clay – Biopolymer-based Composites   Materials Research Forum LLC
Materials Research Foundations **125** (2022) 152-171   https://doi.org/10.21741/9781644901915-7

[8] M.J. Saif, H.M. Asif, Escalating applications of halloysite nanotubes, J. Chil. Chem. Soc. 60(2) (2015) 2949-2953. https://doi.org/10.4067/S0717-97072015000200019

[9] E. Joussein, S. Petit, J. Churchman, B. Theng, D. Righi, B. Delvaux, Halloysite clay minerals-a review, Clay Miner. 40(4) (2005) 383-426. https://doi.org/10.1180/0009855054040180

[10] C. Cheng, W. Song, Q. Zhao, H. Zhang, Halloysite nanotubes in polymer science: Purification, characterization, modification and applications, Nanotechnol. Rev. 9(1) (2020) 323-344. https://doi.org/10.1515/ntrev-2020-0024

[11] B.H. Bac, N.T. Dung, Finding nanotubular halloysite at Lang Dong kaolin deposit, Phu Tho province, Vietnam J. Earth Sci. 37(4) (2015) 299-306. https://doi.org/10.15625/0866-7187/37/4/8058

[12] L. Guimaraes, A.N. Enyashin, G. Seifert, H.A. Duarte, Structural, electronic, and mechanical properties of single-walled halloysite nanotube models, J. Phys. Chem. C. 114(26) (2010) 11358-11363. https://doi.org/10.1021/jp100902e

[13] H.A. Duarte, M.P. Lourenço, T. Heine, L. Guimarães, Clay mineral nanotubes: stability, structure and properties, Stoichiometry and Materials Science-When Numbers Matter2012, pp. 1-24. https://doi.org/10.5772/34459

[14] P. Pasbakhsh, G.J. Churchman, J.L. Keeling, Characterisation of properties of various halloysites relevant to their use as nanotubes and microfibre fillers, Appl. Clay Sci. 74 (2013) 47-57. https://doi.org/10.1016/j.clay.2012.06.014

[15] X. Li, J. Chen, H. Liu, Z. Deng, J. Li, T. Ren, L. Huang, W. Chen, Y. Yang, S. Zhong, β-Cyclodextrin coated and folic acid conjugated magnetic halloysite nanotubes for targeting and isolating of cancer cells, Colloids Surf B Biointerfaces 181 (2019) 379-388. https://doi.org/10.1016/j.colsurfb.2019.05.068

[16] G. Gorrasi, V. Senatore, G. Vigliotta, S. Belviso, R. Pucciariello, PET-halloysite nanotubes composites for packaging application: preparation, characterization and analysis of physical properties, Eur. polym. j. 61 (2014) 145-156. https://doi.org/10.1016/j.eurpolymj.2014.10.004

[17] Y. Wu, Y. Zhang, J. Ju, H. Yan, X. Huang, Y. Tan, Advances in halloysite nanotubes-polysaccharide nanocomposite preparation and applications, Polymers 11(6) (2019) 987. https://doi.org/10.3390/polym11060987

[18] V. Vergaro, E. Abdullayev, Y.M. Lvov, A. Zeitoun, R. Cingolani, R. Rinaldi, S. Leporatti, Cytocompatibility and uptake of halloysite clay nanotubes, Biomacromolecules 11(3) (2010) 820-826. https://doi.org/10.1021/bm9014446

[19] H. Schmitt, K. Prashantha, J. Soulestin, M. Lacrampe, P. Krawczak, Preparation and properties of novel melt-blended halloysite nanotubes/wheat starch nanocomposites, Carbohydr. Polym. 89(3) (2012) 920-927. https://doi.org/10.1016/j.carbpol.2012.04.037

[20] A.M. Nafchi, R. Nassiri, S. Sheibani, F. Ariffin, A. Karim, Preparation and characterization of bionanocomposite films filled with nanorod-rich zinc oxide, Carbohydr. polym. 96(1) (2013) 233-239. https://doi.org/10.1016/j.carbpol.2013.03.055

[21] Z. Torabi, A. MohammadiNafchi, The effects of SiO2 nanoparticles on mechanical and physicochemical properties of potato starch films, J. Chem. Health Risks 3(1) (2013) 33-42.

[22] N. Devi, J. Dutta, Development and in vitro characterization of chitosan/starch/halloysite nanotubes ternary nanocomposite films, Int. j. biol. macromol. 127 (2019) 222-231. https://doi.org/10.1016/j.ijbiomac.2019.01.047

[23] M. Darder, P. Aranda, E. Ruiz-Hitzky, Bionanocomposites: a new concept of ecological, bioinspired, and functional hybrid materials, Adv. Mater. 19(10) (2007) 1309-1319. https://doi.org/10.1002/adma.200602328

[24] G.A. Ozin, A. Arsenault, Nanochemistry: a chemical approach to nanomaterials, United Kingdom, 2015.

[25] L. Averous, N. Boquillon, Biocomposites based on plasticized starch: thermal and mechanical behaviours, Carbohydr polym 56(2) (2004) 111-122. https://doi.org/10.1016/j.carbpol.2003.11.015

[26] J. George, H. Ishida, A review on the very high nanofiller-content nanocomposites: Their preparation methods and properties with high aspect ratio fillers, Prog. Polym. Sci. 86 (2018) 1-39. https://doi.org/10.1016/j.progpolymsci.2018.07.006

[27] J. Ren, K.M. Dang, E. Pollet, L. Avérous, Preparation and characterization of thermoplastic potato starch/halloysite nano-biocomposites: effect of plasticizer nature and nanoclay content, Polymers 10(8) (2018) 808. https://doi.org/10.3390/polym10080808

[28] J. Ren, K.M. Dang, E. Pollet, L.J.P. Avérous, Preparation and characterization of thermoplastic potato starch/halloysite nano-biocomposites: effect of plasticizer nature and nanoclay content, 10(8) (2018) 808. https://doi.org/10.3390/polym10080808

[29] Y. Wu, Y. Zhang, J. Ju, H. Yan, X. Huang, Y.J.P. Tan, Advances in halloysite nanotubes-polysaccharide nanocomposite preparation and applications, 11(6) (2019) 987. https://doi.org/10.3390/polym11060987

[30] M. Liu, R. He, J. Yang, Z. Long, B. Huang, Y. Liu, C.J.C.m. Zhou, Polysaccharide-halloysite nanotube composites for biomedical applications: a review, 51(3) (2016) 457-467. https://doi.org/10.1180/claymin.2016.051.3.02

[31] S.-H. Lee, H.J.A.d.d.r. Shin, Matrices and scaffolds for delivery of bioactive molecules in bone and cartilage tissue engineering, 59(4-5) (2007) 339-359. https://doi.org/10.1016/j.addr.2007.03.016

[32] B. Zhong, S. Wang, H. Dong, Y. Luo, Z. Jia, X. Zhou, M. Chen, D. Xie, D.J.J.o.a. Jia, f. chemistry, Halloysite tubes as nanocontainers for herbicide and its controlled release in biodegradable poly (vinyl alcohol)/starch film, 65(48) (2017) 10445-10451. https://doi.org/10.1021/acs.jafc.7b04220

[33] L. Liu, W. Shen, W. Zhang, F. Li, Z. Zhu, Porous starch and its applications, Functional Starch and Applications in Food, Springer2018, pp. 91-117. https://doi.org/10.1007/978-981-13-1077-5_4

[34] N. Oliyaei, M. Moosavi-Nasab, A. Tamaddon, M.J.I.j.o.b.m. Fazaeli, Preparation and characterization of porous starch reinforced with halloysite nanotube by solvent exchange method, 123 (2019) 682-690. https://doi.org/10.1016/j.ijbiomac.2018.11.095

[35] F. Liu, L. Bai, H. Zhang, H. Song, L. Hu, Y. Wu, X.J.A.a.m. Ba, interfaces, Smart H2O2-responsive drug delivery system made by halloysite nanotubes and carbohydrate polymers, 9(37) (2017) 31626-31633. https://doi.org/10.1021/acsami.7b10867

[36] B. Dhandayuthapani, Y. Yoshida, T. Maekawa, D.S.J.I.j.o.p.s. Kumar, Polymeric scaffolds in tissue engineering application: a review, 2011 (2011). https://doi.org/10.1155/2011/290602

[37] H. Schmitt, N. Creton, K. Prashantha, J. Soulestin, M.F. Lacrampe, P.J.J.o.A.P.S. Krawczak, Preparation and characterization of plasticized starch/halloysite porous nanocomposites possibly suitable for biomedical applications, 132(4) (2015). https://doi.org/10.1002/app.41341

[38] E. Jamróz, P. Kulawik, P.J.P. Kopel, The effect of nanofillers on the functional properties of biopolymer-based films: A review, 11(4) (2019) 675. https://doi.org/10.3390/polym11040675

[39] L. Guo, G. Li, J. Liu, Y. Meng, Y.J.C.p. Tang, Adsorptive decolorization of methylene blue by crosslinked porous starch, 93(2) (2013) 374-379. https://doi.org/10.1016/j.carbpol.2012.12.019

[40] M. Soheilmoghaddam, M.U. Wahit, S. Mahmoudian, N.A.J.M.C. Hanid, Physics, Regenerated cellulose/halloysite nanotube nanocomposite films prepared with an ionic liquid, 141(2-3) (2013) 936-943. https://doi.org/10.1016/j.matchemphys.2013.06.029

[41] N. Devi, J.J.I.j.o.b.m. Dutta, Development and in vitro characterization of chitosan/starch/halloysite nanotubes ternary nanocomposite films, 127 (2019) 222-231. https://doi.org/10.1016/j.ijbiomac.2019.01.047

[42] N.K. Shrivastava, M.A.A. Saidi, N. Othman, M. Zurina, A. Hassan, Fillers and reinforcements for advanced nanocomposites, Bio-based Polymers and Nanocomposites, Springer, New York City, 2019, pp. 29-48. https://doi.org/10.1007/978-3-030-05825-8_2

[43] J. Sadeghizadeh-Yazdi, M. Habibi, A.A. Kamali, M.J.C.r.i.n. Banaei, f.s. journal, Application of edible and biodegradable starch-based films in food packaging: a systematic review and meta-analysis, 7(3) (2019) 624-637. https://doi.org/10.12944/CRNFSJ.7.3.03

[44] F. Sadegh-Hassani, A.M.J.I.j.o.b.m. Nafchi, Preparation and characterization of bionanocomposite films based on potato starch/halloysite nanoclay, 67 (2014) 458-462. https://doi.org/10.1016/j.ijbiomac.2014.04.009

[45] H. Schmitt, K. Prashantha, J. Soulestin, M. Lacrampe, P.J.C.p. Krawczak, Preparation and properties of novel melt-blended halloysite nanotubes/wheat starch nanocomposites, 89(3) (2012) 920-927. https://doi.org/10.1016/j.carbpol.2012.04.037

[46] R. Aguilar-Sánchez, R. Munguía-Pérez, F. Reyes-Jurado, A.R. Navarro-Cruz, T.S. Cid-Pérez, P. Hernández-Carranza, S.d.C. Beristain-Bauza, C.E. Ochoa-Velasco, R.J.M. Avila-Sosa, Structural, physical, and antifungal characterization of starch edible films added with nanocomposites and Mexican oregano (Lippia berlandieri Schauer) essential oil, 24(12) (2019) 2340. https://doi.org/10.3390/molecules24122340

[47] K. Bisetty, Biocomposites: Biomedical and Environmental Applications, 2018.

[48] H. Wu, Y. Lei, R. Zhu, M. Zhao, J. Lu, D. Xiao, C. Jiao, Z. Zhang, G. Shen, S.J.F.H. Li, Preparation and characterization of bioactive edible packaging films based on pomelo peel flours incorporating tea polyphenol, 90 (2019) 41-49. https://doi.org/10.1016/j.foodhyd.2018.12.016

[49] Z.J.M.J.o.M. Ahmad, Effects of Halloysite Nanotubes (HNT) Structures on Antimicrobial Activity on TPSS Film, 16(2) (2020).

[50] G. Gorrasi, R. Pantani, M. Murariu, P. Dubois, PLA/H alloysite nanocomposite films: water vapor barrier properties and specific key characteristics, Macromol. Mater. Eng. 299(1) (2014) 104-115. https://doi.org/10.1002/mame.201200424

[51] F. Sadegh-Hassani, A.M. Nafchi, Preparation and characterization of bionanocomposite films based on potato starch/halloysite nanoclay, Int. J. Biol. Macromol. 67 (2014) 458-462. https://doi.org/10.1016/j.ijbiomac.2014.04.009

[52] E. Jamróz, P. Kulawik, P. Kopel, The effect of nanofillers on the functional properties of biopolymer-based films: A review, Polym. 11(4) (2019) 675. https://doi.org/10.3390/polym11040675

[53] N. Oliyaei, M. Moosavi-Nasab, A. Tamaddon, M. Fazaeli, Preparation and characterization of porous starch reinforced with halloysite nanotube by solvent exchange method, Inter. J. Biol. Macromol. 123 (2019) 682-690. https://doi.org/10.1016/j.ijbiomac.2018.11.095

[54] L. Guo, G. Li, J. Liu, Y. Meng, Y. Tang, Adsorptive decolorization of methylene blue by crosslinked porous starch, Carbohydr. Polym. 93(2) (2013) 374-379. https://doi.org/10.1016/j.carbpol.2012.12.019

[55] X.T. Cao, A.M. Showkat, D.W. Kim, Y.T. Jeong, J.S. Kim, K.T. Lim, Preparation of β-cyclodextrin multi-decorated halloysite nanotubes as a catalyst and nanoadsorbent for dye removal, J. Nanosci. Nanotechnol. 15(11) (2015) 8617-8621. https://doi.org/10.1166/jnn.2015.11482

[56] S. Sadjadi, Halloysite-based hybrids/composites in catalysis, Appl. Clay Sci. 189 (2020) 105537. https://doi.org/10.1016/j.clay.2020.105537

[57] S. Sadjadi, M. Malmir, M.M. Heravi, F.G. Kahangi, Biocompatible starch-halloysite hybrid: An efficient support for immobilizing Pd species and developing a heterogeneous catalyst for ligand and copper free coupling reactions, Inter. J. Biol. Macromol. 118 (2018) 1903-1911. https://doi.org/10.1016/j.ijbiomac.2018.07.053

[58] X. Qiang, S. Zhou, Z. Zhang, Q. Quan, D. Huang, Synergistic Effect of Halloysite Nanotubes and Glycerol on the Physical Properties of Fish Gelatin Films, Polymers 10(11) (2018) 1258. https://doi.org/10.3390/polym10111258

[59] B. Zhong, S. Wang, H. Dong, Y. Luo, Z. Jia, X. Zhou, M. Chen, D. Xie, D. Jia, Halloysite tubes as nanocontainers for herbicide and its controlled release in

biodegradable poly (vinyl alcohol)/starch film, J. Agric. Food Chem. 65(48) (2017) 10445-10451. https://doi.org/10.1021/acs.jafc.7b04220

[60] F. Liu, L. Bai, H. Zhang, H. Song, L. Hu, Y. Wu, X. Ba, Smart H2O2-responsive drug delivery system made by halloysite nanotubes and carbohydrate polymers, ACS Appl. Mater. Interfaces 9(37) (2017) 31626-31633. https://doi.org/10.1021/acsami.7b10867

[61] R. Aguilar-Sánchez, R. Munguía-Pérez, F. Reyes-Jurado, A.R. Navarro-Cruz, T.S. Cid-Pérez, P. Hernández-Carranza, S.d.C. Beristain-Bauza, C.E. Ochoa-Velasco, R. Avila-Sosa, Structural, physical, and antifungal characterization of starch edible films added with nanocomposites and Mexican oregano (Lippia berlandieri Schauer) essential oil, Molecules 24(12) (2019) 2340. https://doi.org/10.3390/molecules24122340

[62] H. Wei, H. Wang, H. Chu, J. Li, Preparation and characterization of slow-release and water-retention fertilizer based on starch and halloysite, Inter. J. Biol. Macromol. 133 (2019) 1210-1218.

[63] M. Špírková, J. Hodan, R. Konefał, L. Machová, P. Němeček, A. Paruzel, The Influence of Nanofiller Shape and Nature on the Functional Properties of Waterborne Poly (urethane-urea) Nanocomposite Films, Polymers 12(9) (2020) 2001. https://doi.org/10.3390/polym12092001

[64] K.M. Dang, R. Yoksan, E. Pollet, L. Avérous, Morphology and properties of thermoplastic starch blended with biodegradable polyester and filled with halloysite nanoclay, Carbohydr. polym. 242 (2020) 116392. https://doi.org/10.1016/j.carbpol.2020.116392

[65] Z. Ahmad, Effects of Halloysite Nanotubes (HNT) Structures on Antimicrobial Activity on TPSS Film, Malays. J. Microsc 16(2) (2020) 119-127.

[66] Z.W. Abdullah, Y. Dong, Preparation and characterisation of poly (vinyl) alcohol (PVA)/starch (ST)/halloysite nanotube (HNT) nanocomposite films as renewable materials, J Mater Sci 53(5) (2018) 3455-3469. https://doi.org/10.1007/s10853-017-1812-0

[67] K. Prashantha, M. Lacrampe, P. Krawczak, Halloysite Nanotubes-Polymer Nanocomposites:A New Class of Multifaceted Materials, Int. J. Adv. Mat. Manuf. Charac. 3(1) (2013) 1-4.

[68] S.M.M. Meira, G. Zehetmeyer, J.M. Scheibel, J.O. Werner, A. Brandelli, Starch-halloysite nanocomposites containing nisin: Characterization and inhibition of Listeria

monocytogenes in soft cheese, LWT - Food Sci Technol 68 (2016) 226-234. https://doi.org/10.1016/j.lwt.2015.12.006

[69] Z. Ahmad, H.Y. Hermain, N.H. Abdul Razak, Mechanical and physical properties of sago starch/halloysite nanocomposite film, Adv Mat Res 1115 (2015) 394-397. https://doi.org/10.4028/www.scientific.net/AMR.1115.394

[70] A.M. Peres, R. Oréfice, Effect of incorporation of Halloysite nanotubes on the structure and properties of low-density polyethylene/thermoplastic starch blend, J. Polym. Res. 27 (2020) 1-10. https://doi.org/10.1007/s10965-020-02185-3

[71] E. Raee, B. Kaffashi, Biodegradable polypropylene/thermoplastic starch nanocomposites incorporating halloysite nanotubes, J. Appl. Polym. Sci. 135(4) (2018) 45740. https://doi.org/10.1002/app.45740

[72] Z.W. Abdullah, Y. Dong, Biodegradable and Water Resistant Poly(vinyl) Alcohol (PVA)/Starch (ST)/Glycerol (GL)/Halloysite Nanotube (HNT) Nanocomposite Films for Sustainable Food Packaging, Front. Mater. 6(58) (2019) 1-17. https://doi.org/10.3389/fmats.2019.00058

Chapter 8

Montmorillonite-Starch based Nano-Composites and Applications

Milan K Barman[1], Nirmala Tamang[2], Ajaya Bhattarai[2*] and Bidyut Saha[3*]

[1]Department of Chemistry, Sambhu Nath College, The University of Burdwan, Labpur, Birbhum, West-Bengal, India

[2]Department of Chemistry, M.M.A.M.C., Tribhuvan University, Biratnagar 56613, Nepal

[3]Homogeneous catalysis laboratory, Department of Chemistry, The University of Burdwan, Burdwan, West- Bengal, India

* bsaha@chem.buruniv.ac.in & ajaya.bhattarai@mmamc.tu.edu.np

Abstract

Biopolymer nanocomposites are the foremost valuable materials among the existing nanocomposites. Biopolymer nanocomposite compounds are biodegradable, eco-friendly and low in cost. Due to these properties, biopolymer nanocomposites can easily replace petroleum-based nanocomposite in various applications. Compared to pure polymer, clay-polymer nanocomposites exhibit favorable physical, chemical, and mechanical properties since they are dispersed at different sizes and contain improved size dispersion and size distribution. There are several biopolymers on earth, but starch is the most abundant. Moreover, its chemical and physical properties make it an important natural polymer.

Keywords

Nanocomposite, Clay Materials, Biopolymer, Biodegradable, Cost-Effective, Sustainable Materials

Contents

Montmorillonite-Starch based Nano-Composites and Applications..........**172**

1.   Introduction

Most of the Polymers derived from petrochemicals usually damage the environment. They generate many hazardous materials through their highly complicated synthesis process. Due to non-biodegradable plastic wastes, these waste materials are causing a negative impact on the environment. Polymers made from plants are eco-friendly, renewable and low-cost.

Biopolymers derived from agricultural products offer superior properties, such as edible, biocompatible, and degradable characteristics that are not found in synthetic polymers. They can be readily transported and used in a range of disposable plastics, food, and medicine applications [3, 4].

Adv. App. of Micro and Nano Clay – Biopolymer-based Composites   Materials Research Forum LLC
Materials Research Foundations **125** (2022) 172-209   https://doi.org/10.21741/9781644901915-8

Since starch is readily available, renewable, and low-cost, it is of particular interest to all bioengineers [7]. Gluten, lignin, cellulose, chitin, and zein, among others, have been studied by various researchers [5,6]. The semicrystalline polymer has repeating molecules of α-D-glucopyranosyl units. These molecules make up a linear carbohydrate, amylase, and a highly branched polysaccharide, amylopectin. α-D-(1-4) linkages are used to link the repeating units within amylase. Amylopectin has about 5% of α-D-(1-6) linked branches, as well as some α-D-(1-4) linked backbones (Fig.1) [8]. High-level plants primarily store carbohydrates as starches. After appropriate processing, starch is commonly employed in diverse applications like adhesives, papers, textiles, pharmaceutical products, foams, thermoset resin, films and many others [9]. Starch-based films were administered in numerous studies when they were melted or cast from a solution or gel with plasticizers added. Materials made of starch often exhibit highly sensitive mechanical characteristics as a result of humidity, temperature, and pH changes in the environment. To successfully make high performance biomaterials, we need to overcome this problem [10]. In comparison to synthetic polymers, starch films have miserable mechanical properties, are brittle and water sensitive. The hydrophilic property of starch compounds makes it difficult to preserve their behavior. By blending starch with synthetic fibers or synthetic water, it can be enhanced in terms of mechanical properties and water resistance [11] or natural polymers [12, 13] making composite [14, 15] and by crosslinking [16, 17].

Physico-mechanical properties could be enhanced by modern clay-polymer nanocomposites. Clay is added to the biopolymer matrix in small amounts to produce nanocomposites [18]. Clay nanoparticles dispersion within the biopolymer matrix represents the biggest challenge towards preparation of nanocomposites. Because of their small particle size, extremely large surface areas (700-800 m^2/g), high aspect ratio (50-1000) and intercalation abilities, montmorillonite (MMT) minerals have recently received special attention in nanocomposites [19-20]. The MMT consists of silicate layers having a planar thickness 1 nm as well as lateral thickness 200-300 nm. Its standard chemical structure consists of two fused silica tetrahedral sheets sandwiched between a magnesium or aluminium hydroxide octahedral sheet [15]. Among all the polymer nanocomposite systems that are currently in use, MMT is the most commonly used natural clay [21, 22]. Although starch and clay nanocomposite materials have been reported, most of them exhibit poor dispersion [23, 24].

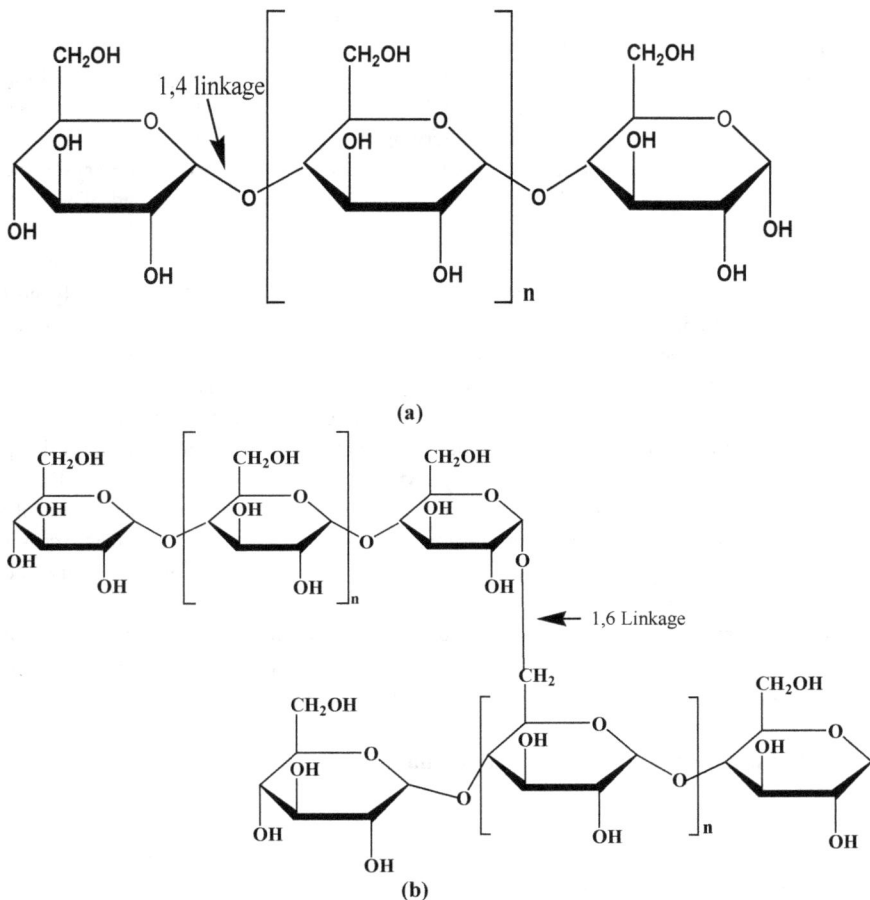

Figure 1. Structures of (a) amylose and (b) amylopectin

Utilizing distearyl dimethyl ammonium chloride, stearyl dihydroxyethyl ammonium chloride, these substances greatly enhance distribution [25] and quaternary ammonium modified starches [26]. The MMT interlayer consists of sodium ions that exchange with organic cations. A well-suited modifier can reduce clay dispersion as much as starch. Nevertheless, miscibility remains a problem, and clay dispersion continues to be a problem.

Adv. App. of Micro and Nano Clay – Biopolymer-based Composites   Materials Research Forum LLC
Materials Research Foundations **125** (2022) 172-209   https://doi.org/10.21741/9781644901915-8

Starch-clay nanocomposite materials also respond to plasticizers, which influence their properties. Nevertheless, the high hydrogen bond fraction and the rigid backbone within the starch molecule make the processing of pure starch extremely challenging. The melting point of the material is therefore higher than the degradation temperature because it has a high transition temperature [27, 28]. When thermoplastic processing of starch is underway, plasticizers are required to induce the creation of hydrogen bonds within the crystalline part of the grain [29]. An examination is conducted and a report is presented of various starch plasticizers, with glycerol being widely because of its efficiency [30]. Xylitol and sorbital are sometimes used as polyol-based plasticizers because of their greater hydroxyl content, but glycerol still provides the best plasticizing properties [31]. Because glycerol plasticizes at a higher rate than larger plasticizers, it is less thermally stable. Various applications are therefore taken into consideration when evaluating metal plasticizer performance and plasticizer volatility. The high hydroxyl content in sorbital of the molecule decreases the efficiency of the starch-plasticizer system, since the hydroxyl groups serve as a hydrophilic cross-link between starch and the sorbitol molecule [29]. Plasticizers are absorbed on the surface of the MMT clay and interfacial layers influence the transference of stress within the polymeric matrix and MMT clay as well as the intensity of the interfacial layer. A plasticizer's primary purpose is to reduce the strong cross-molecular interactions between starch molecules in order to enhance its flexibility and processability. The result of this is that polymeric chains become more mobile, while MMT clay becomes more flexible, extensible and ductile [32]. Here we examine the recent developments in starch-based nanocomposites, focusing on preforming, properties, and application of starch-clay composites, as well as upcoming issues.

2. Structural features of starch and montmorillonite

2.1 Structural features of starch

Starch polymer contains amylase and amylopectin, which are not freely available in nature. These polysaccharides are arranged in granules by a micellar network of related molecules that act as glue holding both polysaccharides together. There are many botanical sources for these granules, which vary in shape, size, and structure. The monomers of glycosyl may have a diameter as small as 0.3 nm, which are the building blocks of starch [33]. Growth rings are concentric shells or layers of concentric starch granules that are either regular or irregular. Optical and electron microscopes can be easily used to observe these growth rings. Amorphous and semicrystalline growth rings alternate with layers of these granules. Starch granules are described as having an onion like structure, equivalent to chemical differences within the layers, density differences in starch molecule deposition and in terms of different crystallinities [34]. The starch structure has been studied extensively in recent years, but a comprehensive model that reflects its complex system is yet to be developed. Multi scale structural model of starch as shown in (Fig. 2).

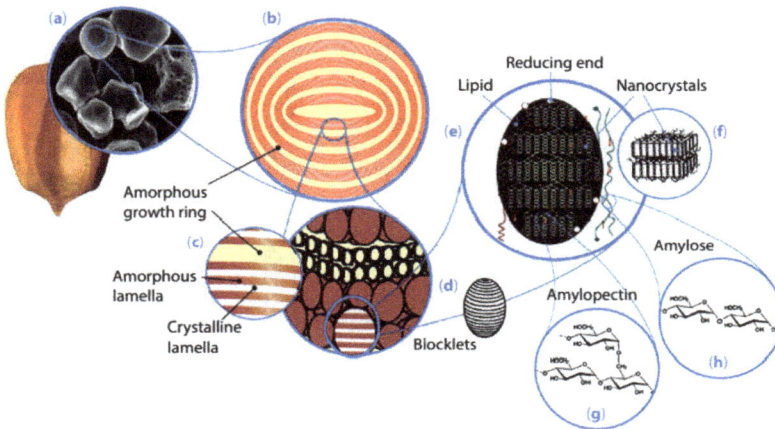

Figure 2 Structure of starch biopolymer. [Reprinted with permission from ref. [38] (Copyright © 2010 American Chemical Society.]

According to this model structure, (a) granules (2-100 μm), (b) growing rings (120-500 nm) (c) d-blocks (20-50 nm), (d) Crystalline and amorphous lamellae (9 nm), (f) amylopectin and (h) amylose chains (0.1-1 nm) [35]. The types of granules, size, shape, and percentage vary depending on the starch source [36]. Table 1 [37-43] show some examples of the starch granule distribution from different sources. There are four types of X-ray dispersion patterns found in original starch granules: A, B, C, and V. A-type polymorphs commonly found in cereals, these structures tend to be closed together. Tuber and root starches tend to be rich in B-type starch. Compared with B-type starches, the A-type is more open. Molecular water is located between each double-helical structure in polymorphs of the A-type and in polymorphs of the B-type, the water is located within the central cavity of the hexagonal, which is constructed of six double-helical structures. (Fig. 3) illustrates types A and B of structures. Acid-resistant B types of starch are better than acid-sensitive A types [44, 45]. The 'C-type' is the result of mixing B and A starches and is found mainly in tubers and legumes. Blocks of the starch polymorphs with B- and C-type structures have been reported to have dimensions of approximately 400-500 nm,

Adv. App. of Micro and Nano Clay – Biopolymer-based Composites   Materials Research Forum LLC
Materials Research Foundations **125** (2022) 172-209   https://doi.org/10.21741/9781644901915-8

considerably larger than the corresponding blocks with A-type structures (25-100 nm). A dispersion pattern characterized by a V-type can be delineated using amylose lipids, iodine as well as alcohol complexes [44, 45]. Starches are crystalline in nature, but can be broken down through a gelatinization process that involves heating the liquid solution that contains starch. Retrogradation is a process where gelatinized starch crystallizes again after being re-collated and stored [46].

Table 1. Starch granule distribution of some starch sources

Starch granules								
	A-type		B-type		C-type		
Starch source	Size (μm)	(%)	Size (μm)	(%)	Size (μm)	(%)	Ref.
Normal barley	8–26	74.7	<8	25.3	ND	ND	37
Waxy barley	8–26	66.4	<8	33.6	ND	ND	37
Zero amylose waxy barley	8–26	43.9	<8	56.1	ND	ND	37
High amylose barley	8-26	19.4	<8	80.6	ND	ND	37
Rye	>62.5	85	≤9.3	10-15	ND	ND	38
Wheat	>62.5	80	≤9.3	20	ND	ND	38
Cassava	>20	20	10.1–20	71.4	5-10	8.6	39
Sweet Potato	>20	8.6	10.1-20	70.5	5-10	20.9	39
Ginger	>20	10.7	10.1-20	82.7	5-10	6.6	39
Yam	>35	67.2	15.1-35	67.2	5-15	19.3	39
Arrowroot	>30	6.8	20.1-30	50.5	5-20	42.7	39
Barley, wheat and triticale	10-35	ND	≈2	ND	ND	ND	40
Wheat	9.48-22.8	75	5.24-9.48	10	<5.24	15	41
Wheat	>15	51-73	5-15	14-37	<5	10.5-15	42
Complete triticale	18-41	80	2-13	20	ND	ND	43
Substituted triticale	8-38	75	0.5-6	25	ND	ND	43

ND = Not determined.

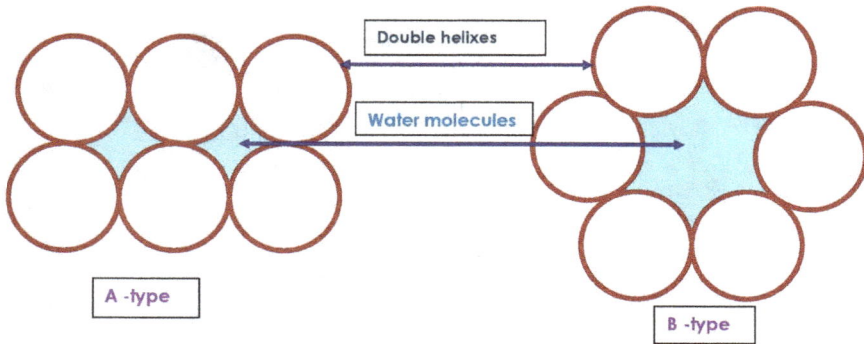

Figure 3. Packing configuration based on crystalline type. [Reprinted with permission from [38] [Copyright © 2010 American Chemical Society.]

2.2   Structural features of montmorillonite

The nanolayered structure of montmorillonite (MMT) makes it a phyllosilicate. Stacks of tetrahedral sheets O-Si-O are sandwiched by octahedral sheets O-Al(Mg)-O in its layered structures. Due to isomorphous substitution in the layer, cations are positively charged during MMT interlayer separation. MTT particles are formed by the electrostatic and van der Waals forces between their neighboring layers. Secondary micrometre- to millimetre-scale particles are then formed from the particles (Fig. 4) [47, 48].

Various strategies are possible in MMT switching and to create hybrids and nanocomposites based on layers. A suitable method of physically and chemically overcoming van der Waals force and electrostatic attraction can prevent layers from adhering [49]. When the aggregates of MMT are dispersed in water they are broken down into MMT particles and individual MMT nanolayer is formed by exfoliation. These MMT nanolayers possess a great aspect ratio and extremely large surface area. Adding nanolayers of MMT to the polymer matrix, clay/polymer nanocomposites (CPN) have improved mechanical properties as well as exhibited thermal stability and flame retardance [50]. According to so many studies, for the formation of MMT/polymer nanocomposites, the crucial step in the enhancement of the component's properties is the exfoliation of MMT nanolayer from the polymer matrix[51].

Adv. App. of Micro and Nano Clay – Biopolymer-based Composites   Materials Research Forum LLC
Materials Research Foundations **125** (2022) 172-209   https://doi.org/10.21741/9781644901915-8

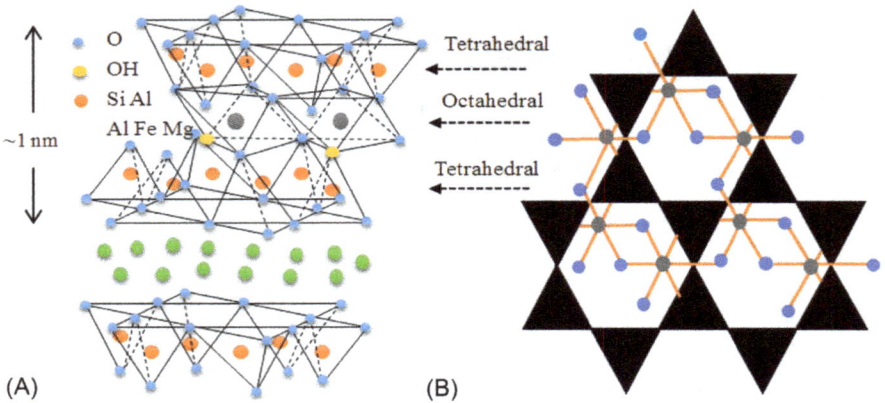

Figure 4. Diagrammatic representation of MTT. (A) Side view: tetrahedrons units of MTT assembled through weak van der Waals and electrostatic forces to form the primary particles, (B) Top view of MTT, hexagonal structure of oxygen and hydroxyl ligands of the octahedral layer. [Reprinted with permission from ref. [40] [Copyright © 2019 Elsevier]

3.   Nanocomposite

Modern technology is centered on composites, which are structured into continuous matrix phase, and non-continuous enforcement phase to give each part the best possible performance [52]. According to the size of reinforcement inside the structure, composites are best categorized into three basic groups: macrocomposites, microcomposites, and nanocomposites. They include nano-composites that have absolutely outstanding features, and are made by embedding reinforcements that are smaller than 100 nm in the composite. Nano-composites also exhibit very high surface to volume ratios due to the excessive interaction between matrix and reinforcement [53]. These multifaceted applications have led to the increased use of nano-composites. There are so many applications in various fields such as engineering, coatings, rubber, plastics, adhesives and electronic as well as optic materials. Nano-composite materials, which decompose slowly in the environment, are also contributing to some environmental problems due to their rapid growth. By preparing nano-composites from renewable materials, these problems can be avoided. Various renewable resources are utilized for the production of nano-composites, including vegetable oil, cellulose, starch, chitin, lignin, natural rubber, and proteins. The nanostructured composite can be split into three main categories on basis of its structure. The reinforcement composites in such composites are obtained from biorenewable sources

Adv. App. of Micro and Nano Clay – Biopolymer-based Composites   Materials Research Forum LLC
Materials Research Foundations **125** (2022) 172-209   https://doi.org/10.21741/9781644901915-8

as the matrix. Therefore, the reinforcement as well as matrix components in composites come from natural resources [54].

In addition to being composed of biorenewable resources, the most frequently expected characteristics of nano-composites include biocompatibility, biodegradability, ease of preparation, low density, and low cost. Additionally to these common characteristics, metal structures are flexible, have a low dielectric constant, and are usually easy to work with [55]. Nano-composite materials are not as robust as petroleum-based nano-composites in terms of thermal, mechanical, and chemical stability, which limits their utility for numerous industrial applications. Research is being conducted in this particular area in order to enhance these properties. The nano-composites produced via renewable resources are environmentally friendly in comparison to those derived from petroleum. Nano-composites are typically used in the food, packaging, and biomedical industries and their use offers wide range of applications, including membranes, sensors, energy storage, optics, automotive, environmental remediation, genetic engineering, flame retardants, and more [56].

3.1   Starch-clay nanocomposites

A lot more interest has been given in recent years to starch-clay owing to their attractive properties which are impossible to achieve by a microfiller. In the recent polymer industry, clay's enhanced properties make it extremely significant as a polymer [57]. A silicate is tetrahedron SiO_4, and minerals contain about 30% silicate. With a positive charge, silicon ions are surrounded by a negatively charged oxygen atom. A silicon ion bonds almost identically with oxygen, making together a cross linkage between silicon ions and continuously bonding between silicon atoms. The different composition of silicate tetrahedron is produced; single units called nesosilicates, double units is called sorosilicastes, chain units is namely inosilicates, sheets (phyllosilicates), rings (cyclosilicates) and framed structure (tectosilicates). For nano-composite preparation, the two most common silicates are layered or sheet silicates and phyllosilicates. The amount of water molecules within the silicate layer of clay makes it distinctive from other mineral phyllosilicates or Layered silicates. Clay has almost all the same chemical and structural properties, but they have different water contents, allowing them to substitute cations more readily [58]. Kaolinite, Smectite, and Illite or mica-clay are among the three major groupings found in the region. These primary groups can alienate clay. Most commonly, starch-clay nanocomposites are reinforced by MMT, hectorite, and saponite. Due to its chemically controlled and easy availability, as well as its low cost, MMT is often used the most [59]. MMT possesses two significant inherent properties that are important in the application of starch-clay nanocomposite to starch-clay interaction. A silicate dispersion is based on its layering ability and a surface chemistry is based upon ion exchange [60]. Despite it being flexible, and compatible with starch, MMT is comprised of ion exchange processes and is readily intercalated. Nanocomposites have greater barrier properties when the clay content is higher. Due to the clay materials dispersion in the starch matrix, tensile strength and dynamic elastic moduli are also enhanced. Even though clay adds significantly to the reduction of water vapor transmission rate, a twist pathway is created through the starch matrix.

3.2 Montmorillonite-starch nanocomposites

The new bioplastic materials class, such as starch montmorillonite nanocomposites, has attracted significant attention recently because of their superior properties as well as an intended alternative to conventional plastics. As a result of these researches, nano-composites were developed as a means of replacing the traditional micro-dimensional fillers with nano-scale particles. During the 1980s, researchers at Toyota company conducted pioneering research on nano-composites that led to the interest in nano-composites today [61]. There has been reported material improvement both in physical and mechanical properties due to the presence of nanoparticles. Throughout this period, nanoparticles played a huge role in improving materials' performance.

Montmorillonite (MMT) particles, which possess unique structure and properties, are used as nano particle component particles. By adding MMT clays to nanocomposites, bioplastics can be made mechanically better, retain moisture better and release plasticizers much more slowly than without them [62, 63]. This accomplishment shows how bioplastic nanocomposites were crucial to fabricating bioplastics whose characteristics can be compared to those of conventional plastics.

Microstructure and the properties of the composite are primarily determined by the dispersion of clay particles as well as the surface interactions between the continuous matrix phase and the reinforcing phases (MMT). Polymer nanocomposites are fabricated by polymer nanotechnology and their dispersion and intercalation of clay particles is affected by factors like nature of the starch, surface modification, and fabrication method. The MMTs, coupled with the starch polymer, can produce four distinct morphologies of the starch composite. These four composites include phase separated composites (microcomposite), flocculated composites, intercalated composites and exfoliated composites (Fig. 5) [64].

Composites that are phase-separation occur when the polymer chains are improbable to penetrate as well as intercalate into the interlayer within the layers of silicates. Incompatibility between these phases results in a poor interaction between polymers and clay particles. The MMT clay plates and the clay filler stacked face to face in a polymeric matrix. It took a large amount of clay loading to achieve the remarkable boost in physical properties of these composites, which were merely performed as a conventional composite (in macro-scale) [65]. Alternatively, layered silicates may be formed by transferring and penetrating polymer chains into the interlayer spacing, which leads to nanomaterials extending in these interlayer spaces. A typical distance separation between clay layers in this structure is about two to five nanometers [66]. Even when the amount of clay added is the same, these composites display better properties than phase-separated composites. In flocculated nano-composites the interactions between hydroxyl groups and clay layers lead to flocculation, which appears as an intercalating structure, though it differs from an intercalating structure.

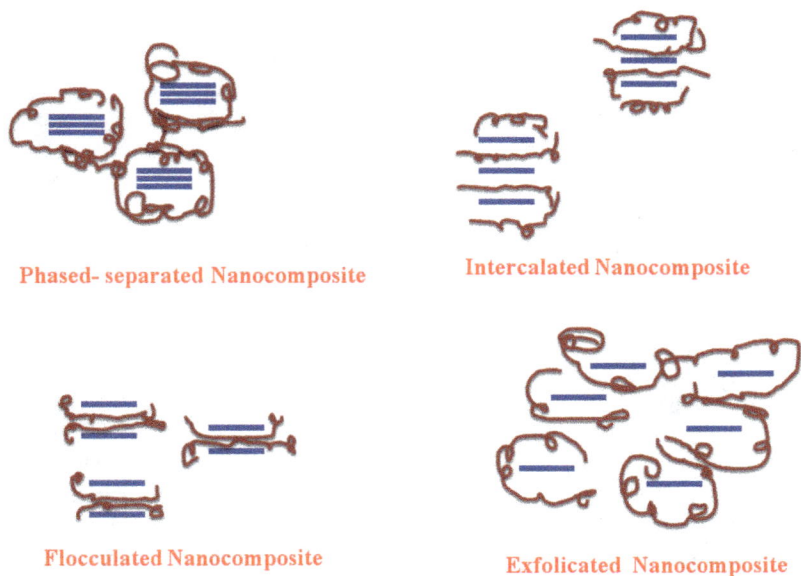

Phased- separated Nanocomposite

Intercalated Nanocomposite

Flocculated Nanocomposite

Exfolicated Nanocomposite

Figure 5: Illustration of four possible polymer-clay morphologies. Reprinted with permission from ref. [56] [Copyright © 2018 Elsevier]

Clay platelets can be separated from one another into large plates that are delaminated or exfoliated to form a nanocomposite in which the periodicities of the clay are completely disoriented. Polymer chains are included in the resulting microstructure, and electrostatic interactions between adjacent lamellar of the MMT are successfully overcome. Clay layers are therefore closely spaced between each other, usually between 5 and 10 nm, or greater. The high aspect ratios as well as strong interaction within the filler phase as well as continuous matrix are responsible for these results. When compared with an intercalated structure, these composites enhance polymer properties in the most favorable manner. In delaminated nanocomposite, clay particles peel off into multiple layer platelets, which indicates continuous organic phase with more reinforcing particles, maximum facilitation and optimized properties of nano-composites also in a little clay loadings [67,68].

4. Methods of polymer-clay nanocomposite synthesis

A number of different methods are available for synthesizing nanocomposite materials, including in situ polymerizing, solution intercalating, and melt intercalating. For polymer-clay nanocomposites synthesis, the most important step is the constant distribution of nanoclay in the polymer matrix. Its poor viscosity and strong agitation power produce similar results to melt intercalation, when it comes to creating good dispersion of nanoclay in polymeric matrix. In any case, melt intercalation is both environmentally and economically feasible, presenting great potential for development [69]. In-situ polymerization methods is one of the most widely applied methods. It provides a uniform dispersion, and can be easily modified by changing the polymerization conditions. Recent advances have guided to the development of nanocomposites with unique properties using diverse synthesis techniques.

4.1 In-situ polymerization process

In this process, the polymerization process occurs with the swelling of the liquid monomer over the organomodified clay layer. As a result, monomers migrate into the interlayer spacing and are inserted forming long organic polymer chains during the polymerization process, as presented in (Fig. 6). Prior to ion exchange reaction, a catalyst or initiator is introduced to begin this polymerization via a heat or radiation source [70]. Toyota's research team invented the in situ polymerization technique and applied it first to creating nylon-6/clay nanocomposites. It is currently widely utilized to synthesize polymer clay nanocomposite due to its potential for production of exfoliated nanocomposite and compatibility with a variety of reactive monomers. The thermosetting based polymer-clay nanocomposite is widely reported to be manufactured by this method. There has been previous research suggesting that layered silicates have high surfaces energies, that maintain the equilibrium between interlayer spacing by forcing monomer molecules together [64]. Long macromolecule chains are then formed after the polymerization occurs inside interlayer spacing. Polymerized clay layers should be peeled off into the unstable state producing an exfoliated structure. Additive agglomeration remains at a minimum

Adv. App. of Micro and Nano Clay – Biopolymer-based Composites   Materials Research Forum LLC
Materials Research Foundations **125** (2022) 172-209   https://doi.org/10.21741/9781644901915-8

when nanocomposite production is done during in situ polymerization. So it is evident that the materials produced have a high degree of homogeneity [71, 72]. There are some types of polymerization techniques possible: ring-opening polymerizations, radical atom transfer polymerizations, living anionic/cationic polymerizations, and nitroxide-mediated polymerizations.

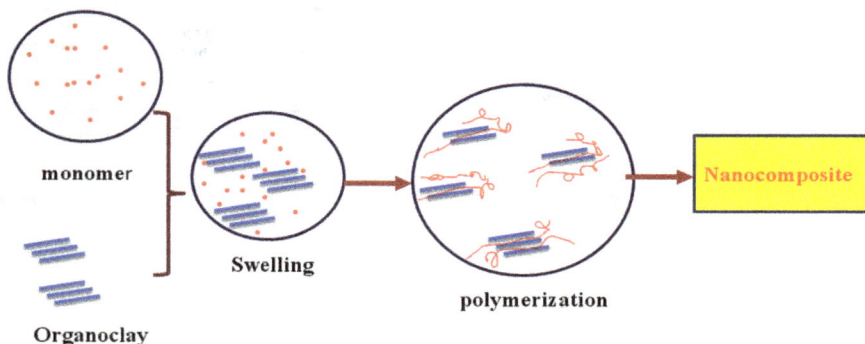

Figure 6. Steps involved in compiling nanocomposites through in situ polymerization Reprinted with permission from ref. [56], [Copyright © 2018 Elsevier]

4.2   Polymer solution intercalation

The intercalation of polymers in solvents including toluene, chloroform and water is what is required for this technique. The pre-polymer nanocomposites are soluble in an insolvent system and the nanoclay is dispersed to form the polymer nanocomposites. It involves the clay being bloated in solvent system in order to increase the layer of silicate [73, 64]. Then silicate solution is added to polymer, in which chains of polymer intercalate with the molecular solvent to form galleries in clay platelets. As a result of evaporation of the solvent, nanocomposites are formed under vacuum or by precipitation (Fig. 7). By using the thermodynamic concept, it is possible to explain the mechanisms which underlie the intercalation of macromolecular chains during solution. Efficacy of intercalation of polymer chains within clay layers is driven by the entropy enlarging due to desorption of solvent molecules [74, 75]. There may be health concerns, cost issues, and environmental concerns related to the solution intercalation method in industrial practice.

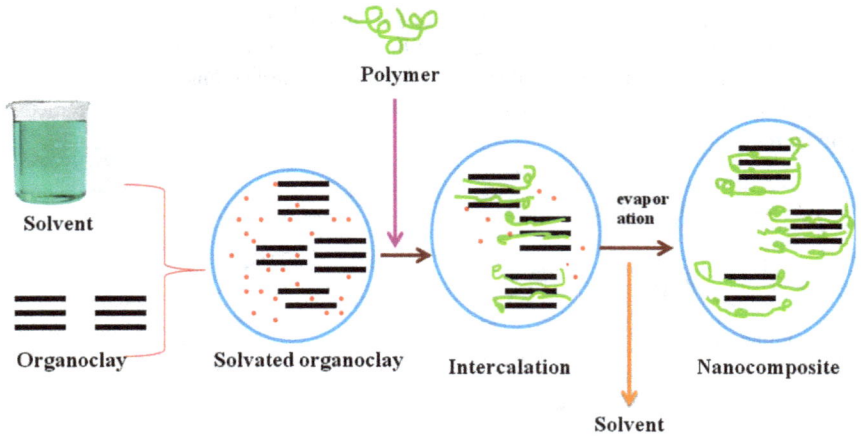

Figure 7. Nanocomposite synthesis through solution intercalation of polymers Reprinted with permission from ref. [56], [Copyright © 2018 Elsevier]

4.3   Intercalation of melt

Using the melt intercalation method, liquid thermoplastic polymer is blended with liquid organoclay by either extrusion or rotation into an internal mixer. The blended product is heated to a higher temperature than the polymer's glass transition temperature (Fig.8) and is then annealed to provide nanocomposites. In comparison to the previously discussed methods, melt intercalation provides the greatest versatility. This is due to the fact that, i) melting and intercalating are eco-friendly techniques as no solvents are used; ii) a polymer nanocomposite will be molded or extruded using this method; iii) melting can be used for polymers that cannot be polymerized in situ or through solution intercalation; iv) melting is simpler than in-situ polymerization or solution intercalation since no monomers or solvents are selected [76, 77]. There is a major concern for obtaining the delaminated morphology in various polymer matrix systems using melt intercalation, based on the extent of surface modification of nanomaterials. Melt compounding must consider the melt processing parameters in order to ensure maximum exfoliation. According to certain literature, has proposed that melt processing and mixing conditions impact the resulting polymer matrix due to the shear forces generated during melting and processing. Nevertheless, the properties of the polymer matrix like molecular weight, polarity can influence the two phases interaction, while the matrix types determines filler dispersion in polymer [78, 64].

Adv. App. of Micro and Nano Clay – Biopolymer-based Composites   Materials Research Forum LLC
Materials Research Foundations **125** (2022) 172-209   https://doi.org/10.21741/9781644901915-8

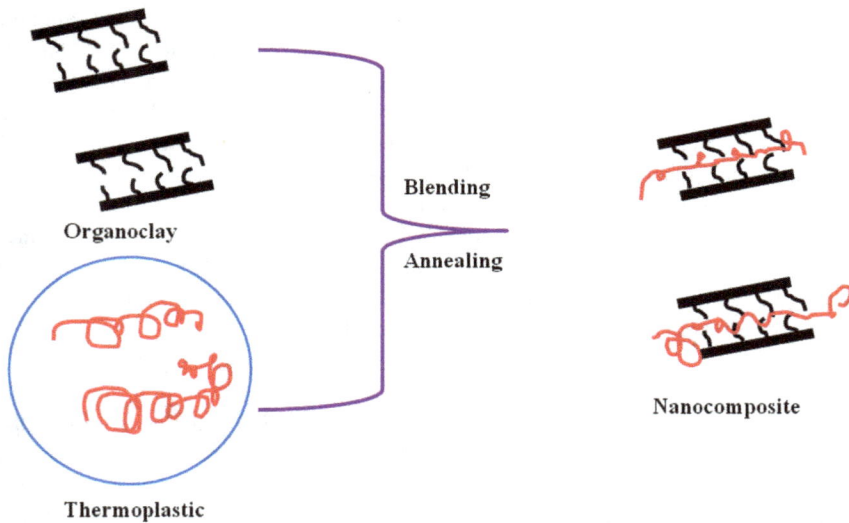

*Figure 8. The melting method used to form nanocomposite Reprinted with permission
from ref. [56] [Copyright © 2018 Elsevier]*

5.   Nanocomposites characterization techniques

A sample's degree of exfoliation must be determined by comparing it with other samples
when developing and optimizing nanocomposites. In the literature, numerous methods
have been proposed [79, 80]. Two common methods which are used to identify the
structure of nanocomposites are XRD or wide-angle X-ray diffraction (WAXD) analysis
as well as transmission electron micrography (TEM). A WAXD probe is typically
applicable to determine the structure and kinetics of polymer melt intercalation, and
sometimes used to construct nanocomposites. Those properties of nanocomposite
structures whose intercalated layers scatter silicate are determined by the shape, location,
and intensity of the basal reflections [81]. WAXD was used as an effective means of
discovering the interlayer spacing between the original silicate layers and intercalated
nanocomposites, but it was unable to determine the nature of silicate layer distribution [82].
Since certain layered silicates have highly diffuse basal reflections and broad peaks, it may
not be easy to understand the characteristics behind intensity decreases. It is therefore more
difficult to study the basal reflections of certain layer silicates after the broadening of the
peaks and the intensity decreases. As a result, the WAXD patterns are only tentative in
their description of nanocomposite structure and formation mechanisms. By examining

directly the defect structure and its spatial distribution by TEM, one gets a qualitative understanding of the defect structures. Nevertheless, it is important to ensure that the sample represents a representative cross-section [83]. The WAXD as well as TEM are both essential techniques for determining the nanocomposites structure [84]. The problem with TEM is that it is time-consuming and only provides qualitative information, while WAXD can quantify layer spacing changes through the measurement of low-angle peaks. Although the layers in intercalated nanocomposite exceed 6-7 nm in spacing and also the layers in exfoliated nanocomposite become reasonably disorganized, WAXD analysis becomes unusable. Consequently, recent studies combining small-angle X-ray scattering and wavelength-averaged X-ray diffractometer (WAVD) have identified nanostructure as well as crystallite structure in nanocomposites quantitatively [85].

5.1   X-ray diffraction studies

WAXD reveals how polymer chains can be constrained into MMT galleries, describing the extent of dispersion of MMT layers. A wide range of nanocomposite characterization methods rely on WAXD. By using WAXD, it is possible to identify the crystalline orientation and the interplaner spaces based on Bragg peak changes. Among the benefits of WAXD is its application to the assessment of the degree of exfoliation on clays and an interplanar expansion of polymer-layered nanocomposites following chemical surface treatment. Starch clay nanocomposites have also been successfully analyzed using this technique to analyze the degree of crystallinity and structure of starch/MMT materials [86,87]. As shown in Fig. 9, the peak of 1.01 to 1.25 and 2.08 nm is obtained from WAXD on starch/MMT nanocomposites.

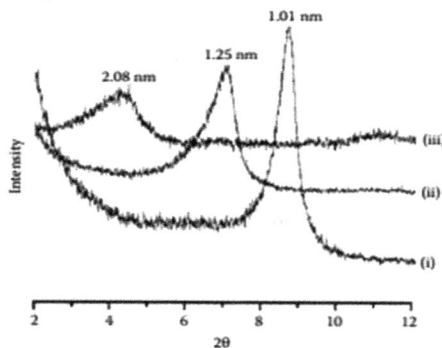

Figure 9. WAXD patterns of (i) untreated montmorillonite (MMT), (ii) activated montmorillonite (EMMT), and (iii) thermoplastic starch/activated montmorillonite nanocomposites. Reprinted with permission from ref. [78] [Copyright © 2005 Elsevier]

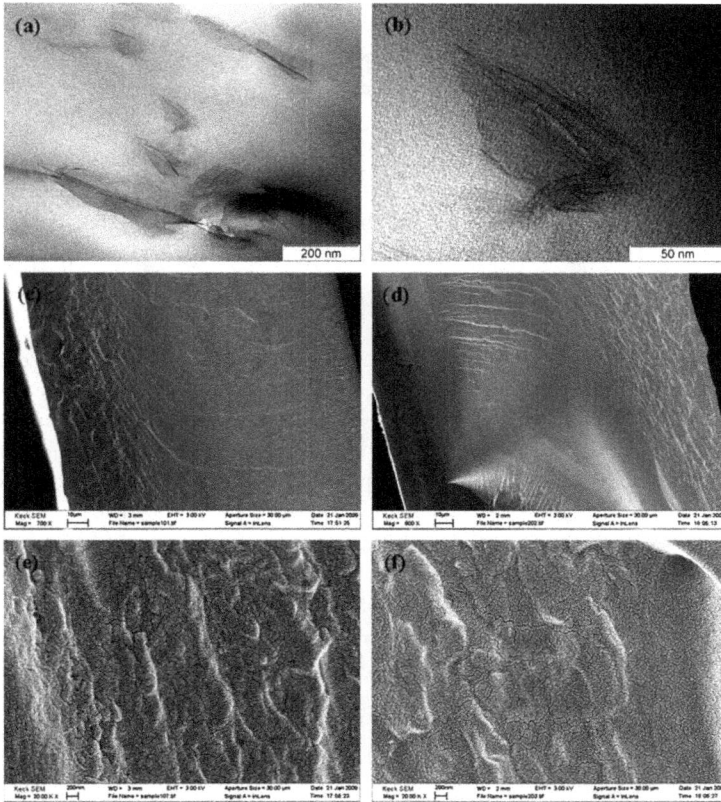

*Figure 10. (a and b) TEM images of starch–clay nanocomposites containing 5%
montmorillonite at different magnifications, and SEM micrographs of fractured surface:
(c and e) unfilled starch samples and (d and f) starch-5% montmorillonite
nanocomposites. Reprinted with permission from ref. [80] [Copyright © 2009 Elsevier]*

5.2 Transmission electron microscopy (TEM) and scanning electron microscopy (SEM)

An essential tool to characterize nanocomposites is transmission electron microscopy (TEM). In Fig.10a and 10b, TEM images of starch/MMT nanocomposites having 5% clay can be seen. A clay agglomerate coexists with exfoliated clay layers (Fig. 10a). While clay contains small tactodes with less than five layers. As can be seen in Fig. 10b, agglomerates of clay are often disordered and aligned in certain areas. It's distorted, shifted, and skewed in its relationship with other silicate platelets. Due to clay alignment during compression moulding, the clay shows some orientation. Furthermore, both the micro and macro scales need to be considered in dispersion of silicate layers.

Figs.10c-f show scans of fractured surfaces with SEM microscopy. Starch and starch nanocomposite exhibit homogeneous surfaces that reflect the ground up starch particles are completely broken down, and therefore, MMT clay is distributed throughout the polymer matrix. Higher magnifications do not reveal the presence of clay aggregates. Consequently, low amounts of clay with starches seem to be fairly compatible and also dispersible and result in well-dispersed nanocomposites [88, 89].

6.   Applications

As a result of their engineering properties, polymer/nanoclay nanocomposites possess superior mechanical properties like poor density, extreme damping, large specific strength and stiffness, and great fatigue endurance. These inherent properties make nanocomposite materials an attractive substance for many future applications [90]. There are so many companies invested billion dollars annually to improving nanocomposite materials, with 70 to 80 percent of them being used in the automotive, aeronautical, and packaging industries [91]. Here at the Applications section, we mainly focus on biological, rheological, food, and waste water applications.

6.1   Rheological control agent

In addition to being applied to various fields such as petroleum and pharmacy, rheology of nanocomposites is a powerful tool in the development of advanced products [92]. Although petroleum-based polymers bind to nanoclay, their surface energies are different and as a result various surfactants must be used to decrease the surface energy in clay layers [93]. Polymer-modified asphalt mixtures with nanocomposite additives have improved the storage stability and ageing resistance of the mixtures as well as the rutting and fatigue resistance [94]. The nanocomposites in this technology enhance the stability and formability of CO_2 foam, which in turn enhances the recovery of oil from porous media in a microfluidic device. These biopolymer-based films also feature enhanced properties from starch/MMT nanocomposites. A rheological analysis of a dilute solution explains the relationship between the structure and property of polymers and nanoparticles. Considering the two preparative methods of steady shear as well as dynamic examination, the concentration effect of MMT and the content of amylose on the rheological properties of nanocomposite forming solutions (NFS) is investigated. A component's addition order is a factor influencing the NFS level of organization. It is thought that interactions between starch components explain the reduced flow indices while amylose content rises. A higher

NFS can be observed in films soaked in MMT, as the storage modules grow larger, demonstrating a stronger matrix of starch. A second method can produce films with superior mechanical properties because of interactions in starch and MMT [95].

Therefore, the rheological analysis of solutions used in nanocomposite formation can be useful to understand the interaction of polymers and nanoparticles [96] and to optimize the casting and spreading processes for nanostructure film production [97].

6.2 Food packaging

Due to their environmental compatibility, readily availability, and cost effective, starch-clay nanocomposites are extensively studied for food packaging applications. Nanobio composites and other applied biotech technologies have become increasingly popular because they can reduce the dependency on fossil fuels and make materials more sustainable. Bio-nanocomposite materials can substitute petroleum-based conventional packing materials because they decompose more rapidly. However, starch and cellulose derivatives seem to be most commonly studied for packaging applications. Food packaging is highly focused to develop higher barrier strength against oxygen, carbon dioxide, flavor compounds and water vapor diffusion. In this way, nanostructures have the capability to develop functional as well as active properties in food packaging process, such as scavenging oxygen, immobilizing enzymes or indicating the degree to which a product becomes exposed to adverse factors such as inadequate temperature and levels of oxygen [98].

Biodegradable nanocomposites based on starch-clay have been investigated for many applications, including food packaging. As a outcome of the addition of MMT clay, the Young modulus and tensile strength of the material have been significantly boosted. Bottling and food packaging require barrier properties of the utmost importance. In general, nanocomposites have lower diffusion coefficients compared to starch alone. In this study, it was found that MMT reduces starch film water uptake possibly because of the tortuose structure formed by amorphous clay [99]. Many gases, including oxygen, carbon dioxide, water vapours and volatile compounds, cannot permeate through this type of bio-nanocomposites. The gas barrier properties along with its thermal and mechanical properties makes them very potential nanocomposites candidate for food packaging materials [100]. These innovative nanocomposites are also being developed into packaging materials that restrict the permeability of selected gases, like CO_2 and CH_4, thus increasing food shelf life during preservation [101].

Researchers recently found bio-based nanocomposites to have high barrier properties and eco-friendliness, which suggests that they could play an important role in developing stronger, eco-friendly and high barrier food packaging materials. Furthermore, the development of these materials will not only benefit the packaging industry, but would have a positive impact on other markets as well, including pharmaceutical and electronic packaging.

6.3   Biomedical applications of nanocomposites

People have been using polymeric materials for thousands of years, including wool, starch and cellulose, and proteins that are found in animal shells and horns. There are many industries that still utilize these polymeric materials. Natural polymeric substances and nanocomposites are produced from polymeric structures from plants or animals that break down either biologically or hydrologically in soil. Medical materials also have to be biocompatible. Applications must be proven to be safe for human health without causing side effects. Using nanocomposites from renewable sources for clinical use has recently attracted great interest due to their biocompatibility [102]. Nanocomposites, with their versatility and ease of adaptation in a variety of systems, may be used to develop many biomedical applications compared to synthetic materials. Medical applications are generally classified into four categories: drug delivery, gene therapy, antimicrobial activity, and tissue engineering.

A nanocomposite is prepared by adding inorganic components, such as MMT. In biomedical applications, the clay is mainly selected because of its large surface reactivity, swelling, intercalation properties and good rheological properties [103-107]. Clay MMT is an important component of the formation of a drug and has a significant impact on its pharmacokinetic characteristics such as half-life time, clearance, bioavailability and secure sustained drug delivery characteristics. Nanocomposites formed with MMT have improved in the mechanical properties (e.g., tensile strength), less diffusion rates and an increased capacity for adsorption, all due to the presence of MMT. Because the interactions between organic and inorganic materials occur with clay, it also increases mucoadhesion and solubility in acidic environments [108-111].

In the case of cell wall/polymer nanocomposites, drug encapsulation is carried out in two ways: (1) first, the drug is intercalated in the clay host, and then the clay is dispersed in the polymer matrix and (2) when the nanocomposite has been prepared, nanocomposite hydrogels are developed across drug adsorption (Fig. 11).

Three methods are used to build the clay drug complex or to intercalate drug compounds into the clay, namely ion exchange reactions, dual intercalation techniques, and dry methods. Incorporation of drugs in the interlayer space of clay is accomplished through ionic exchange (cationic/anionic). There are several example of drugs such as "cefazolin, fenbufen, chloramphenicol succinate sodium salt, amoxicillin, naproxen, flurbiprofen, ketoprofen, diclofenac, and acetylsalicylic acid" which are properly intercalated into the

Adv. App. of Micro and Nano Clay – Biopolymer-based Composites   Materials Research Forum LLC
Materials Research Foundations **125** (2022) 172-209   https://doi.org/10.21741/9781644901915-8

clay by ions exchange reaction [112-118]. Ionic exchange is based on the concept that clay rises in deionized water when dissolved. Its procedure similarly applies to hydrophilic drugs. Next, MMT suspension is prepared by adding dissolved drug dropwise to the suspension. Several times of deionized water washes are performed to remove non-intercalated drugs from the suspension after magnetic stirring for 24 hours. An ultrafine powder is produced by grinding the clay-drug complicated at a precise temperature [119, 120].

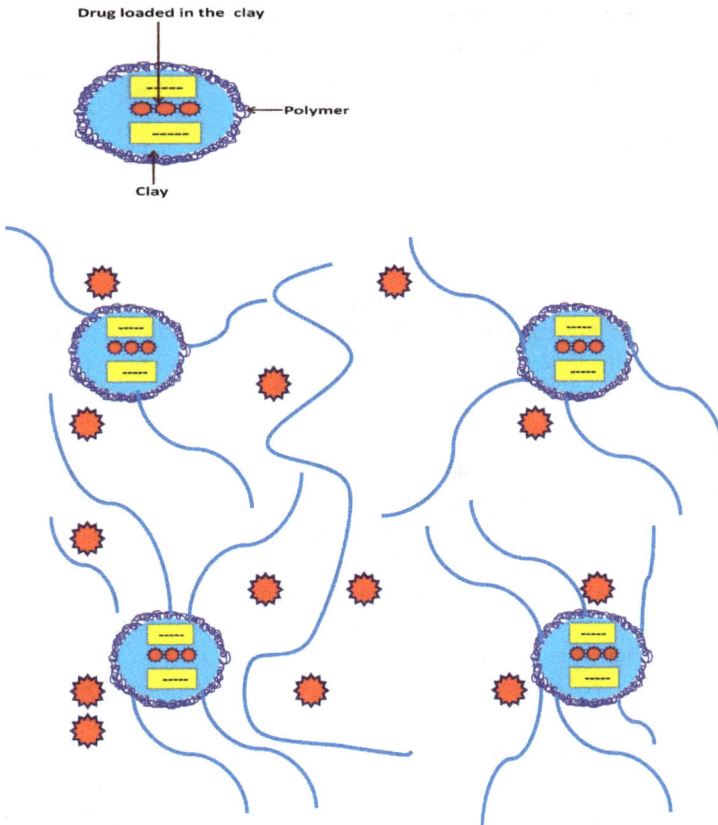

Figure 11. Possible routes for drug entrapment in the nanocomposites. Reprinted with permission from ref. [112] [Copyright © 2017 Elsevier]

Adv. App. of Micro and Nano Clay – Biopolymer-based Composites   Materials Research Forum LLC
Materials Research Foundations **125** (2022) 172-209   https://doi.org/10.21741/9781644901915-8

6.4   Water purification applications

As a result of heavy metals, aromatic molecules, dyes, and toxic substances in drinking water, water pollution is becoming a serious threat to the health of humans. As a result of these materials, many changes can be observed within the abiotic components physical, chemical, and mechanical properties, which results in numerous changes to biodiversity. Thus, the water resources are subjected to several contaminations from human activities including agriculture, production of chemicals, the use of various materials in manufacturing, electricity production, and mining [121]. According to the World Health Organization (WHO), nearly 3.4 million people each year die from water-associated diseases, and around 10% of all diseases can be prevented if water resources are distributed, managed, hygienized, and sanitized [122].

Various pollutants can be removed from water by using the adsorption technique. Adsorbents have been developed recently for a variety of purposes. Due to their easy processing, good exchange ability, extent surface area, comparatively low cost and less adsorption ability among the others, the starch/clay nanocomposites are thought to provide the best results in terms of waste water treatment or remediation [123]. While the adsorption results such as capacity, selectivity, and recycling ability are influenced by the existence of water molecules between the clay layers, the structure of the matrix, and the modification method. Researchers have shown through several current studies that synthesized nanocomposites act as potentially effective and powerful adsorbents for remediating pollution [124,125].

An in situ polymerization technique produces the starch-MMT nanocomposites. In batch mode, the synthesized composites are demonstrated to be efficient at removing Cr (VI) ions from aqueous solutions. In the synthesized nanocomposite, nanoclay is distributed uniformly throughout the polymer matrix and exhibits a dense, smooth surface. A larger adsorption capacity has been obtained using these nanocomposites than in others. Cr (VI) adsorbs best at a pH of 2 due to its low solubility. The Langmuir isotherm model very well fits the adsorption data and the adsorption capacity is measured at 208.33 mg g^{-1}. Adsorption is followed by a pseudo-second-order kinetic model. For the adsorption process and removal process, several thermodynamic parameters proved that it is a spontaneous and exothermic reaction. Using nanocomposites for three adsorption cycles revealed the reusability of nanocomposites, making this method almost cost-effective [126].

There are contaminations in water sources due to dyes and pigments. For this reason, wastewater dyes and pigments need to be removed. Chemical pollutants in water are contaminated by organic dyes, which are the most common. Various industries use organic dyes for various purposes including the manufacture of plastic bodies, tanning, food processing, paper production, textile manufacturing and cosmetic preparations. Industrial wastewater containing dyes is notoriously difficult to purify and these dyes negatively impact in human health as well as in the environment. Dye removal is extremely difficult because dyes are toxic and carcinogenic. Hence, the use of renewably sourced nanocomposites that have different structures and reinforcements is preferred when considering the potential impact on our environment and human health when removing organic dyes from dirty water [127]. Starch is intercalated into the space between the layers of MMT to form MMT-starch nanocomposites. Molecular starch increases the spacing between MMT interlayers, facilitating dye adsorption on nanocomposites of MMTs and starches. As compared to raw MMT, the nanocomposites exhibit a greater capacity for adsorption. Nanocomposites have a greater adsorption capacity for basic dyes due to their higher mass ratios. According to the Adsorption Isotherms studies, the Langmuir model closely matched the Basic Blue and Basic Yellow Adsorption Equilibrium on Nanocomposites [128]. Synthesis of MMT-starch/polyaniline nanocomposite is achieved through oxidative chemical polymerization. Reactive dyes can then be adsorbable to their surface using these nanocomposites. This batch experiment showed the dye was removed very quickly after contact time. In addition, electrostatic attraction between composite and dye molecules seems to be responsible for the removal mechanism. Nanocomposite has an adsorption capability of 91.74 mg g^{-1} for reactive dye. Nanocomposites have shown an ability to effectively remove organic dyes from textile effluents; they should prove useful as adsorbents [129]. Using a model dye, reactive blue is also removed from aqueous solutions using these types of nanocomposites. Accomplish to assess the adsorption capacity of a nanocomposite, response surface methodology is used. Across a variety of independent variables like original dye concentration, nanocomposite amount and pH of solution there is an empirical relationship relating adsorption capacity to a second-order model [130].

Conclusions and prospects

Recent times have seen an increase in interest in the preparation of biodegradable nanocomposite materials to preserve the environment. In light of the limited availability of fossil fuels and environmental concerns, we should develop new bio-nano composite-based materials. Furthermore, nanocomposite materials are primarily developed for their amalgamation of enhanced mechanical and thermal characteristic, their environment-friendliness, and their biodegradation and reusability properties. Natural nanomaterials are finding their way into a range of fields, including biomedical applications, sensors, energy storage, optical devices, automotive applications and flame-retardant construction. As a means of realizing these applications with nanocomposites efficaciously and more economically, further developments are needed.

Among the bio-based nanocomposites that have enhanced nanocomposites, the ones that are derived from renewable sources rank higher than those made from synthetic polymers. Several mechanical properties and tensile strengths have been achieved by starch/MMT nanocomposites. In the development of novel starch/MMT nanocomposites, physical properties have been found to be highly dependent on the synthesis approach and the nanoclay. The melt intercalation technique is both an economical and environmentally friendly synthesis procedure. While this method can successfully synthesize nanoclays, it generates variation in the synthesis process, the clay layer spacing and the distribution of various materials in the polymer matrix. Various synthesis techniques are discussed in this chapter, as are the effects they have on starch/MMT nanocomposites. There has been a significant increase in application of nanocomposites, as well as this emerging material has broad range of innovative applications.

In terms of using bio-based nanocomposites, there is some disadvantage. Microplastics (0.1 μm - 5 mm in size) that are produced in the biodegradation process are one of the major disadvantages. When exposed to oxygen and sunlight, the biopolymer nanocomposite materials break down quickly into microplastics. It is possible to accelerate the disappearance of nanocomposite materials compared to conventional plastics. However, the microplastics generated by this system are the same as those produced by petroleum-based plastics. Microplastic fragments also take a long time to completely disappear from natural environments. Therefore, biodegradable products should be used with caution as their biodegradation process is rapid. Creating biodegradable products is the simplest method to combat this problem. New systems are also needed to enhance the biodegradation process of these materials. Bio-based nanocomposites have the second drawback of being unable to achieve mechanical durability comparable to petroleum-based ones. There are several completely different strategies for overcoming this drawback. The nanoprocessing will specifically improve nanocomposites mechanical properties based on renewable energy resources. The biodegradability process will be enhanced significantly by this process.

Aside from its disadvantages, bio-nano composites made from renewable resources should be the main focus of looking for new uses and the unique properties of nanomaterials will provide green nanomaterials with tremendous opportunity in the next century. These nanocomposites will be used in various aspects of our lives in the future because of their biocompatibility, biodegradability and thermal as well as mechanical properties.

Acknowledgment

One of the authors M. K. Barman gratefully acknowledges the facilities provided by the Department of Chemistry, Sambhu Nath College, The University of Burdwan, Labpur, Birbhum, West-Bengal, India.   The authors thankfully acknowledge DS Kothari Postdoctoral Fellowship (No.F.42/ 2006 (BSR)/CH/16-17/ 0055), UGC, and India.

References

[1] V.K. Thakur, M.F. Lin, E.J. Tan, P.S. Lee, Green aqueous modification of fluoropolymers for energy storage applications, J. Mater. Chem. 22 (2012) 5951-5959. https://doi.org/10.1039/c2jm15665b

[2] V.K. Thakur, J. Yan, M.F. Lin, C. Zhi, D. Golberg, Y. Bando, R. Sim, P.S. Lee, Novel polymer nanocomposites from bioinspired green aqueous functionalization of BNNTs, Polym. Chem. 3 (2012) 962-969. https://doi.org/10.1039/c2py00612j

[3] V.K. Thakur, E.J. Tan, M.F. Lin, P.S. Lee, Polystyrene grafted polyvinylidenefluoride copolymers with high capacitive performance, Polym. Chem. 2 (2011) 2000-2009. https://doi.org/10.1039/c1py00225b

[4] M. Avella, J.J. De Vlieger, M.E. Errico, S. Fischer, P. Vacca, M.G. Volpe, Biodegradable starch/clay nanocomposite films for food packaging applications, Food Chem. 93 (2005) 467-474. https://doi.org/10.1016/j.foodchem.2004.10.024

[5] L. Avérous, P.J. Halley, Biocomposites based on plasticized starch, Biofuels, Bioprod. Biorefining. 3 (2009) 329-343. https://doi.org/10.1002/bbb.135

[6] J.A. Mbey, S. Hoppe, F. Thomas, Cassava starch-kaolinite composite film. Effect of clay content and clay modification on film properties, Carbohydr. Polym. 88 (2012) 213-222. https://doi.org/10.1016/j.carbpol.2011.11.091

[7] R. Zhao, P. Torley, P.J. Halley, Emerging biodegradable materials: Starch- and protein-based bio-nanocomposites, J. Mater. Sci. 43 (2008) 3058-3071. https://doi.org/10.1007/s10853-007-2434-8

[8] F. Zia, K.M. Zia, M. Zuber, S. Kamal, N. Aslam, Starch based polyurethanes: A critical review updating recent literature, Carbohydr. Polym. 134 (2015) 784-798. https://doi.org/10.1016/j.carbpol.2015.08.034

[9] L. Yu, K. Dean, L. Li, Polymer blends and composites from renewable resources, Prog. Polym. Sci. 31 (2006) 576-602. https://doi.org/10.1016/j.progpolymsci.2006.03.002

[10] D. Merino, T.J. Gutiérrez, A.Y. Mansilla, C.A. Casalongué, V.A. Alvarez, Critical Evaluation of Starch-Based Antibacterial Nanocomposites as Agricultural Mulch Films: Study on Their Interactions with Water and Light, ACS Sustain. Chem. Eng. 6 (2018) 15662-15672. https://doi.org/10.1021/acssuschemeng.8b04162

[11] Q. Wu, L. Zhang, Structure and properties of casting films blended with starch and waterborne polyurethane, J. Appl. Polym. Sci. 79 (2001) 2006-2013. https://doi.org/10.1002/1097-4628(20010314)79:11<2006::AID-APP1009>3.0.CO;2-F

[12] M. Bhattacharya, Stress relaxation of starch/synthetic polymer blends, J. Mater. Sci. 33 (1998) 4131-4139. https://doi.org/10.1023/A:1004449002240

[13] D.R. Coffin, M.L. Fishman, P.H. Cooke, Mechanical and microstructural properties of pectin/starch films, J. Appl. Polym. Sci. 57 (1995) 663-670. https://doi.org/10.1002/app.1995.070570602

[14] Y.X. Xu, K.M. Kim, M.A. Hanna, D. Nag, Chitosan-starch composite film: Preparation and characterization, Ind. Crops Prod. 21 (2005) 185-192. https://doi.org/10.1016/j.indcrop.2004.03.002

[15] P. Kampeerapappun, D. Aht-ong, D. Pentrakoon, K. Srikulkit, Preparation of cassava starch/montmorillonite composite film, Carbohydr. Polym. 67 (2007) 155-163. https://doi.org/10.1016/j.carbpol.2006.05.012

[16] H.M. Wilhelm, M.R. Sierakowski, G.P. Souza, F. Wypych, Starch films reinforced with mineral clay, Carbohydr. Polym. 52 (2003) 101-110. https://doi.org/10.1016/S0144-8617(02)00239-4

[17] I. Šimkovic, J.A. Laszlo, A.R. Thompson, Preparation of a weakly basic ion exchanger by crosslinking starch with epichlorohydrin in the presence of NH4OH1, Carbohydr. Polym. 30 (1996) 25-30. https://doi.org/10.1016/S0144-8617(96)00060-4

[18] G.E. Luckachan, C.K.S. Pillai, Biodegradable Polymers- A Review on Recent Trends and Emerging Perspectives, J. Polym. Environ. 19 (2011) 637-676 . https://doi.org/10.1007/s10924-011-0317-1

[19] J.C. Lin, Investigation of impact behavior of various silica-reinforced polymeric matrix nanocomposites, Compos. Struct. 84 (2008) 125-131. https://doi.org/10.1016/j.compstruct.2007.07.008

[20] M.W. Ho, C.K. Lam, K. tak Lau, D.H.L. Ng, D. Hui, Mechanical properties of epoxy-based composites using nanoclays, Compos. Struct. 75 (2006) 415-421. https://doi.org/10.1016/j.compstruct.2006.04.051

[21] D.R. Paul, L.M. Robeson, Polymer nanotechnology: Nanocomposites, Polymer (Guildf). 49 (2008) 3187-3204. https://doi.org/10.1016/j.polymer.2008.04.017

[22] K. Wilpiszewska, A.K. Antosik, T. Spychaj, Novel hydrophilic carboxymethyl starch/montmorillonite nanocomposite films, Carbohydr. Polym. 128 (2015) 82-89. https://doi.org/10.1016/j.carbpol.2015.04.023

[23] H.M. Park, W.K. Lee, C.Y. Park, W.J. Cho, C.S. Ha, Environmentally friendly polymer hybrids Part I mechanical, thermal, and barrier properties of thermoplastic starch/clay nanocomposites, J. Mater. Sci. 38 (2003) 909-915. https://doi.org/10.1023/A:1022308705231

[24] J.K. Pandey, R.P. Singh, Green nanocomposites from renewable resources: Effect of plasticizer on the structure and material properties of clay-filled starch, Starch/Staerke. 57 (2005) 8-15. https://doi.org/10.1002/star.200400313

[25] K. Bagdi, P. Müller, B. Pukánszky, Thermoplastic starch/layered silicate composites: Structure, interaction, properties, Compos. Interfaces. 13 (2006) 1-17. https://doi.org/10.1163/156855406774964364

[26] F. Chivrac, O. Gueguen, E. Pollet, S. Ahzi, A. Makradi, L. Averous, Micromechanical modeling and characterization of the effective properties in starch-based nano-biocomposites, Acta Biomater. 4 (2008) 1707-1714. https://doi.org/10.1016/j.actbio.2008.05.002

[27] J. Yang, K. Tang, G. Qin, Y. Chen, L. Peng, X. Wan, H. Xiao, Q. Xia, Hydrogen bonding energy determined by molecular dynamics simulation and correlation to properties of thermoplastic starch films, Carbohydr. Polym. 166 (2017) 256-263. https://doi.org/10.1016/j.carbpol.2017.03.001

[28] Y. Nakamura, S. Miyachi, Effect of temperature on starch degradation in chlorella vulgaris 11h cells, Plant Cell Physiol. 23 (1982) 333-341.

[29] M. Esmaeili, G. Pircheraghi, R. Bagheri, Optimizing the mechanical and physical properties of thermoplastic starch via tuning the molecular microstructure through co-plasticization by sorbitol and glycerol, Polym. Int. 66 (2017) 809-819. https://doi.org/10.1002/pi.5319

[30] N. Laohakunjit, A. Noomhorm, Effect of plasticizers on mechanical and barrier properties of rice starch film, Starch/Staerke. 56 (2004) 545-551. https://doi.org/10.1002/star.200300249

[31] X. Ma, J. Yu, The effects of plasticizers containing amide groups on the properties of thermoplastic starch, Starch/Staerke. 56 (2004) 2439-2448. https://doi.org/10.1002/star.200300256

[32] H. Li, M.A. Huneault, Comparison of sorbitol and glycerol as plasticizers for thermoplastic starch in TPS/PLA blends, J. Appl. Polym. Sci. 119 (2011) 2016-2026. https://doi.org/10.1002/app.32956

[33] H.D. Özeren, M. Guivier, R.T. Olsson, F. Nilsson, M.S. Hedenqvist, Ranking plasticizers for polymers with atomistic simulations: PVT, mechanical properties, and the role of hydrogen bonding in thermoplastic starch, ACS Appl. Polym. Mater. 2 (2020) 2405-2412. https://doi.org/10.1021/acsapm.0c00191

[34] H. Namazi, A. Dadkhah, Surface modification of starch nanocrystals through ring-opening polymerization of ε-caprolactone and investigation of their microstructures, J. Appl. Polym. Sci. 110 (2008) 281-293. https://doi.org/10.1002/app.28821

[35] J. Blazek, E.P. Gilbert, Application of small-angle X-ray and neutron scattering techniques to the characterisation of starch structure: A review, Carbohydr. Polym. 85 (2011) 1139-1153. https://doi.org/10.1016/j.carbpol.2011.02.041

[36] Y.I. Cornejo-ramírez, O. Martínez-cruz, C.L. Del, F.J. Wong-corral, J. Borboa-flores, J. Cinco-moroyoqui, Y. Isbeth, O. Martínez-cruz, C.L. Del, F.J. Wong-corral, J. Borboa-flores, F.J. Cinco-, Y.I. Cornejo-ramírez, O. Martínez-cruz, C.L. Del Toro-sánchez, F.J. Wong-corral, J. Borboa-flores, F.J. Cinco-moroyoqui, D. De Investigación, U. De Sonora, C.P. México, The structural characteristics of starches and their functional properties The structural characteristics of starches and their functional properties, CyTA - J. Food. 00 (2018) 1003-1017. https://doi.org/10.1080/19476337.2018.1518343

[37] S.G. You, M.S. Izydorczyk, Molecular characteristics of barley starches with variable amylose content, Carbohydr. Polym. 49 (2002) 33-42. https://doi.org/10.1016/S0144-8617(01)00300-9

[38] T. Verwimp, G.E. Vandeputte, K. Marrant, J.A. Delcour, Isolation and characterisation of rye starch, J. Cereal Sci. 39 (2004) 85-90. https://doi.org/10.1016/S0733-5210(03)00068-7

[39] F.H.G. Peroni, T.S. Rocha, C.M.L. Franco, Some structural and physicochemical characteristics of tuber and root starches, Food Sci. Technol. Int. 12 (2006) 505-513. https://doi.org/10.1177/1082013206073045

[40] Z. Ao, J. lin Jane, Characterization and modeling of the A- and B-granule starches of wheat, triticale, and barley, Carbohydr. Polym. 67 (2007) 46-55. https://doi.org/10.1016/j.carbpol.2006.04.013

[41] C.M. Brites, C.A.L. Dos Santos, A.S. Bagulho, M.L. Beirão-Da-Costa, Effect of wheat puroindoline alleles on functional properties of starch, Eur. Food Res. Technol. 226 (2008) 1205-1212 https://doi.org/10.1007/s00217-007-0711-z

[42] S. Sandeep, N. Singh, N. Isono, T. Noda, Relationship of granule size distribution and amylopectin structure with pasting, thermal, and retrogradation properties in wheat starch, J. Agric. Food Chem. 58 (2010) 1180-1188. https://doi.org/10.1021/jf902753f

[43] Y.I. Cornejo-Ramírez, F.J. Cinco-Moroyoqui, F. Ramírez-Reyes, E.C. Rosas-Burgos, P.S. Osuna-Amarillas, F.J. Wong-Corral, J. Borboa-Flores, A.G. Cota-Gastélum, Physicochemical characterization of starch from hexaploid triticale (X Triticosecale Wittmack) genotypes, CYTA - J. Food. 13 (2015). https://doi.org/10.1080/19476337.2014.994565

[44] E. Fuentes-Zaragoza, M.J. Riquelme-Navarrete, E. Sánchez-Zapata, J.A. Pérez-Álvarez, Resistant starch as functional ingredient: A review, Food Res. Int. 43 (2010) 931-942. https://doi.org/10.1016/j.foodres.2010.02.004

[45] C. Hernández-Jaimes, L.A. Bello-Pérez, E.J. Vernon-Carter, J. Alvarez-Ramirez, Plantain starch granules morphology, crystallinity, structure transition, and size evolution upon acid hydrolysis, Carbohydr. Polym. 95 (2013) 207-213. https://doi.org/10.1016/j.carbpol.2013.03.017

[46] D. Le Corre, J. Bras, A. Dufresne, Starch nanoparticles: A review, Biomacromolecules. 11 (2010) 1139-1153. https://doi.org/10.1021/bm901428y

[47] I. Diañez, I. Martínez, P. Partal, Synergistic effect of combined nanoparticles to elaborate exfoliated egg-white protein-based nanobiocomposites, Compos. Part B Eng. 88 (2016) 36-43. https://doi.org/10.1016/j.compositesb.2015.10.034

[48] C. Zhou, D. Tong, W. Yu, Smectite nanomaterials: Preparation, properties, and functional applications, in: Nanomater. from Clay Miner. A New Approach to Green Funct. Mater., 2019: pp. 335-364. https://doi.org/10.1016/B978-0-12-814533-3.00007-7

[49] C.H. Zhou, J. Keeling, Fundamental and applied research on clay minerals: From climate and environment to nanotechnology, Appl. Clay Sci. 74 (2013) 3-9. https://doi.org/10.1016/j.clay.2013.02.013

[50] T.T. Zhu, C.H. Zhou, F.B. Kabwe, Q.Q. Wu, C.S. Li, J.R. Zhang, Exfoliation of montmorillonite and related properties of clay/polymer nanocomposites, Appl. Clay Sci. 169 (2019) 48-66. https://doi.org/10.1016/j.clay.2018.12.006

[51] F. Jia, S. Song, Exfoliation and characterization of layered silicate minerals: A review, Surf. Rev. Lett. 21 (2014) 1-10. https://doi.org/10.1142/S0218625X14300019

[52] B. Ates, S. Koytepe, S. Balcioglu, A. Ulu, C. Gurses, Biomedical applications of hybrid polymer composite materials, in: Hybrid Polym. Compos. Mater. Appl., 2017: pp. 343-408. https://doi.org/10.1016/B978-0-08-100785-3.00012-7

[53] Y. Zare, Recent progress on preparation and properties of nanocomposites from recycled polymers: A review, Waste Manag. 33 (2013) 598-604. https://doi.org/10.1016/j.wasman.2012.07.031

[54] L. Yu, K. Dean, L. Li, Polymer blends and composites from renewable resources, Prog. Polym. Sci. 31 (2006) 576-602. https://doi.org/10.1016/j.progpolymsci.2006.03.002

[55] P.J. Jandas, S. Mohanty, S.K. Nayak, Green Nanocomposites from Renewable Resource-Based Biodegradable Polymers and Environmentally-Friendly Blends, in: Polym. Nanocomposites Based Inorg. Org. Nanomater., 2015: pp. 401-442. https://doi.org/10.1002/9781119179108.ch11

[56] H. Fischer, Polymer nanocomposites: From fundamental research to specific applications, Mater. Sci. Eng. C. 23 (2003) 763-772. https://doi.org/10.1016/j.msec.2003.09.148

[57] H.T. Liao, C.S. Wu, Synthesis and characterization of polyethylene-octene elastomer/clay/ biodegradable starch nanocomposites, J. Appl. Polym. Sci. 97 (2005) 397-404. https://doi.org/10.1002/app.21763

[58] H. Namazi, M. Mosadegh, A. Dadkhah, New intercalated layer silicate nanocomposites based on synthesized starch-g-PCL prepared via solution intercalation and in situ polymerization methods: As a comparative study, Carbohydr. Polym. 75 (2009) 665-669. https://doi.org/10.1016/j.carbpol.2008.09.006

[59] G. Madhumitha, J. Fowsiya, S. Mohana Roopan, V.K. Thakur, Recent advances in starch-clay nanocomposites, Int. J. Polym. Anal. Charact. 23 (2018) 331-345. https://doi.org/10.1080/1023666X.2018.1447260. https://doi.org/10.1080/1023666X.2018.1447260

[60] W. Wang, A. Wang, Recent progress in dispersion of palygorskite crystal bundles for nanocomposites, Appl. Clay Sci. 119 (2016) 18-30. https://doi.org/10.1016/j.clay.2015.06.030

[61] Q.X. Zhang, Z.Z. Yu, X.L. Xie, K. Naito, Y. Kagawa, preparation and crystalline morphology of biodegradable starch/clay nanocomposites, Polymer 48 (2007) 7193-7200. https://doi.org/10.1016/j.polymer.2007.09.051

[62] M.P. Guarás, V.A. Alvarez, L.N. Ludueña, Biodegradable nanocomposites based on starch/polycaprolactone/compatibilizer ternary blends reinforced with natural and organo-modified montmorillonite, J. Appl. Polym. Sci. 133 (2016) 6-11. https://doi.org/10.1002/app.44163

[63] A.S. Giroto, A. De Campos, E.I. Pereira, T.S. Ribeiro, J.M. Marconcini, C. Ribeiro, Photoprotective effect of starch/montmorillonite composites on ultraviolet-induced degradation of herbicides, React. Funct. Polym. 93 (2015) 156-162. https://doi.org/10.1016/j.reactfunctpolym.2015.06.013

[64] S.L. Bee, M.A.A. Abdullah, S.T. Bee, L.T. Sin, A.R. Rahmat, Polymer nanocomposites based on silylated-montmorillonite: A review, Prog. Polym. Sci. 85 (2018) 57-82. https://doi.org/10.1016/j.progpolymsci.2018.07.003

[65] A. A., F. K., N. K., PVA / Montmorillonite Nanocomposites: Development and Properties, in: Nanocomposites Polym. with Anal. Methods, 2011: pp. 29-50. https://doi.org/10.5772/18217

[66] J.M. Yeh, K.C. Chang, Polymer/layered silicate nanocomposite anticorrosive coatings, J. Ind. Eng. Chem. 14 (2008) 275-291. https://doi.org/10.1016/j.jiec.2008.01.011

[67] J. Ma, J. Xu, J.H. Ren, Z.Z. Yu, Y.W. Mai, A new approach to polymer/montmorillonite nanocomposites, Polymer (Guildf). 44 (2003) 4619-4624. https://doi.org/10.1016/S0032-3861(03)00362-8

[68] M. Panahi-Sarmad, M. Abrisham, M. Noroozi, A. Amirkiai, P. Dehghan, V. Goodarzi, B. Zahiri, Deep focusing on the role of microstructures in shape memory properties of polymer composites: A critical review, Eur. Polym. J. 117 (2019) 280-303. https://doi.org/10.1016/j.eurpolymj.2019.05.013

[69] R. Babu Valapa, S. Loganathan, G. Pugazhenthi, S. Thomas, T.O. Varghese, An Overview of Polymer-Clay Nanocomposites, in: Clay-Polymer Nanocomposites, 2017. https://doi.org/10.1016/B978-0-323-46153-5.00002-1. https://doi.org/10.1016/B978-0-323-46153-5.00002-1

[70] F. Gao, Clay/polymer composites: The story, Mater. Today. 7 (2004) 50-55. https://doi.org/10.1016/S1369-7021(04)00509-7

[71] J. Fawaz, V. Mittal, Synthesis of Polymer Nanocomposites : nanocomposite method, Synth. Tech. Polym. Nanocomposite. (2015) 1-30. https://doi.org/10.1002/9783527670307.ch1

[72] J.M. Yeh, K.C. Chang, Polymer/layered silicate nanocomposite anticorrosive coatings, J. Ind. Eng. Chem. 14 (2008) 275-291. https://doi.org/10.1016/j.jiec.2008.01.011

[73] Z. Shen, G.P. Simon, Y.B. Cheng, Comparison of solution intercalation and melt intercalation of polymer-clay nanocomposites, Polymer. 43 (2002) 4251-4260. https://doi.org/10.1016/S0032-3861(02)00230-6

[74] H. Namazi, M. Mosadegh, A. Dadkhah, New intercalated layer silicate nanocomposites based on synthesized starch-g-PCL prepared via solution intercalation and in situ polymerization methods: As a comparative study, Carbohydr. Polym. 75 (2009) 665-669. https://doi.org/10.1016/j.carbpol.2008.09.006

[75] B. Chen, J.R.G. Evans, Preferential intercalation in polymer-clay nanocomposites, J. Phys. Chem. B. 108 (2004) 14986-14990. https://doi.org/10.1021/jp040312e

[76] N.N. Bhiwankar, R.A. Weiss, Melt intercalation/exfoliation of polystyrene-sodium-montmorillonite nanocomposites using sulfonated polystyrene ionomer compatibilizers,Polymer. 47 (2006) 6684-6691. https://doi.org/10.1016/j.polymer.2006.07.017

[77] H.A. Stretz, D.R. Paul, R. Li, H. Keskkula, P.E. Cassidy, Intercalation and exfoliation relationships in melt-processed poly(styrene-co-acrylonitrile)/montmorillonite nanocomposites, Polymer. 46 (2005) 2621-2637. https://doi.org/10.1016/j.polymer.2005.01.063

[78] P. Motamedi, R. Bagheri, Investigation of the nanostructure and mechanical properties of polypropylene/polyamide 6/layered silicate ternary nanocomposites, Mater. Des. 31 (2010) 1776-1784. https://doi.org/10.1016/j.matdes.2009.11.013

[79] R.A. Vaia, K.D. Jandt, E.J. Kramer, E.P. Giannelis, Microstructural Evolution of Melt Intercalated Polymer-Organically Modified Layered Silicates Nanocomposites, Chem. Mater. 8 (1996) 2628-2635. https://doi.org/10.1021/cm960102h

[80] S. Sinha Ray, M. Okamoto, Polymer/layered silicate nanocomposites: A review from preparation to processing, Prog. Polym. Sci. 28 (2003) 1539-1641. https://doi.org/10.1016/j.progpolymsci.2003.08.002

[81] J.M. Yeh, K.C. Chang, Polymer/layered silicate nanocomposite anticorrosive coatings, J. Ind. Eng. Chem. 14 (2008) 275-291. https://doi.org/10.1016/j.jiec.2008.01.011

[82] M.H. Kim, O.O. Park, Fabrication of syndiotactic polystyrene nanocomposites with exfoliated clay and their properties, J. Appl. Polym. Sci. 125 (2012) 630-637. https://doi.org/10.1002/app.36289

[83] B. Yalcin, M. Cakmak, The role of plasticizer on the exfoliation and dispersion and fracture behavior of clay particles in PVC matrix: A comprehensive morphological study, Polymer. 45 (2004) 6623-6638. https://doi.org/10.1016/j.polymer.2004.06.061

[84] A.B. Morgan, J.W. Gilman, Characterization of polymer-layered silicate (clay) nanocomposites by transmission electron microscopy and X-ray diffraction: A comparative study, J. Appl. Polym. Sci. 87 (2002) 1329-1338. https://doi.org/10.1002/app.11884

[85] J.M. Mata-Padilla, C.A. Ávila-Orta, F.J. Medellín-Rodríguez, J.A. Valdéz-Garza, A. Torres-Martínez, Study of fracture behavior of polypropylene/MWCNT and polypropylene/m-MMT nanocomposites by small angle X-ray scattering (SAXS), in: Mater. Res. Soc. Symp. Proc., (2012): pp. 75-80. https://doi.org/10.1557/opl.2012.163

[86] M.F. Huang, J.G. Yu, X.F. Ma, P. Jin, High performance biodegradable thermoplastic starch - EMMT nanoplastics, Polymer 46 (2005) 3157-3162. https://doi.org/10.1016/j.polymer.2005.01.090

[87] K.M. Dean, M.D. Do, E. Petinakis, L. Yu, Key interactions in biodegradable thermoplastic starch/poly(vinyl alcohol)/montmorillonite micro- and nanocomposites, Compos. Sci. Technol. 68 (2008) 1453-1462. https://doi.org/10.1016/j.compscitech.2007.10.037

[88] Y.L. Chung, S. Ansari, L. Estevez, S. Hayrapetyan, E.P. Giannelis, H.M. Lai, Preparation and properties of biodegradable starch-clay nanocomposites, Carbohydr. Polym. 79 (2010) 391396. https://doi.org/10.1016/j.carbpol.2009.08.021

[89] W. Wang, P. Song, R. Wang, R. Zhang, Q. Guo, H. Hou, H. Dong, Effects of cationization of high amylose maize starch on the performance of starch/montmorillonite nano-biocomposites, Ind. Crops Prod. 117 (2018) 333-339. https://doi.org/10.1016/j.indcrop.2018.03.004

[90] A. Gürses, Introduction to polymer-clay nanocomposites, 2016. https://doi.org/10.1201/b18716

[91] S. Gul, A. Kausar, B. Muhammad, S. Jabeen, Research progress on properties and applications of polymer/clay nanocomposite, Polym. - Plast. Technol. Eng. 55 (2016) 684-703. https://doi.org/10.1080/03602559.2015.1098699

[92] A.Y. Malkin, A. Isayev, Rheology. Concepts, Methods, and Applications: Concepts, Methods, and Applications: 3rd Edition, 2017

[93] K. Majeed, M. Jawaid, A. Hassan, A. Abu Bakar, H.P.S. Abdul Khalil, A.A. Salema, I. Inuwa, Potential materials for food packaging from nanoclay/natural fibres filled hybrid composites, Mater. Des. 46 (2013) 391-410. https://doi.org/10.1016/j.matdes.2012.10.044

[94] J. Yang, S. Tighe, A Review of Advances of Nanotechnology in Asphalt Mixtures, Procedia - Soc. Behav. Sci. 96 (2013) 1269-1276. https://doi.org/10.1016/j.sbspro.2013.08.144

[95] C.A. Romero-Bastida, M. Chávez Gutiérrez, L.A. Bello-Pérez, E. Abarca-Ramírez, G. Velazquez, G. Mendez-Montealvo, Rheological properties of nanocomposite-forming solutions and film based on montmorillonite and corn starch with different amylose content, Carbohydr. Polym. 188 (2018) 121-127. https://doi.org/10.1016/j.carbpol.2018.01.089

[96] F. Xie, L. Yu, B. Su, P. Liu, J. Wang, H. Liu, L. Chen, Rheological properties of starches with different amylose/amylopectin ratios, J. Cereal Sci. 49 (2009) 371-377. https://doi.org/10.1016/j.jcs.2009.01.002

[97] M.F.C. Jorge, C.H. Caicedo Flaker, S.F. Nassar, I.C.F. Moraes, A.M.Q.B. Bittante, P.J. Do Amaral Sobral, Viscoelastic and rheological properties of nanocomposite-forming solutions based on gelatin and montmorillonite, J. Food Eng. 120 (2014) 81-87. https://doi.org/10.1016/j.jfoodeng.2013.07.007

[98] J.W. Rhim, H.M. Park, C.S. Ha, Bio-nanocomposites for food packaging applications, Prog. Polym. Sci. 38 (2013) 1629-1652. https://doi.org/10.1016/j.progpolymsci.2013.05.008

[99] A. Arora, G.W. Padua, Review: Nanocomposites in food packaging, J. Food Sci. 75 (2010) 43-49. https://doi.org/10.1111/j.1750-3841.2009.01456.x

[100] S. Fu, Z. Sun, P. Huang, Y. Li, N. Hu, Some basic aspects of polymer nanocomposites: A critical review, Nano Mater. Sci. 1 (2019) 2-30. https://doi.org/10.1016/j.nanoms.2019.02.006

[101] K.K. Kuorwel, M.J. Cran, J.D. Orbell, S. Buddhadasa, S.W. Bigger, Review of Mechanical Properties, Migration, and Potential Applications in Active Food

Packaging Systems Containing Nanoclays and Nanosilver, Compr. Rev. Food Sci. Food Saf. 14 (2015) 411-430. https://doi.org/10.1111/1541-4337.12139

[102] S. Thakur, R. V. Saini, P. Singh, P. Raizada, V.K. Thakur, A.K. Saini, Nanoparticles as an emerging tool to alter the gene expression: Preparation and conjugation methods, Mater. Today Chem. 17 (2020). https://doi.org/10.1016/j.mtchem.2020.100295

[103] M. Ghadiri, W. Chrzanowski, R. Rohanizadeh, Biomedical applications of cationic clay minerals, RSC Adv. 5 (2015) 29467-29481. https://doi.org/10.1039/C4RA16945J

[104] C.J. Ward, M. DeWitt, E.W. Davis, Halloysite nanoclay for controlled release applications, in: ACS Symp. Ser., 2012: pp. 209-238. https://doi.org/10.1021/bk-2012-1119.ch010

[105] S.A. Gârea, A.I. Mihai, A. Ghebaur, C. Nistor, A. Sârbu, Porous clay heterostructures: A new inorganic host for 5-fluorouracil encapsulation, Int. J. Pharm. 491 (2015) 299-309. https://doi.org/10.1016/j.ijpharm.2015.05.053

[106] F. Bazmi Zeynabad, R. Salehi, E. Alizadeh, H.S. Kafil, A.M. Hassanzadeh, M. Mahkam, PH-Controlled multiple-drug delivery by a novel antibacterial nanocomposite for combination therapy, RSC Adv. 5 (2015) 105678-105691. https://doi.org/10.1039/C5RA22784D

[107] G. V. Joshi, H.A. Patel, B.D. Kevadiya, H.C. Bajaj, Montmorillonite intercalated with vitamin B1 as drug carrier, Appl. Clay Sci. 45 (2009) 248-253. https://doi.org/10.1016/j.clay.2009.06.001

[108] J.H. Park, H.J. Shin, M.H. Kim, J.S. Kim, N. Kang, J.Y. Lee, K.T. Kim, J.I. Lee, D.D. Kim, Application of montmorillonite in bentonite as a pharmaceutical excipient in drug delivery systems, J. Pharm. Investig. 46 (2016) 363-375. https://doi.org/10.1007/s40005-016-0258-8

[109] M.L. Chan, K.T. Lau, T.T. Wong, M.P. Ho, D. Hui, Mechanism of reinforcement in a nanoclay/polymer composite, Compos. Part B Eng. 42 (2011) 1708-1712. https://doi.org/10.1016/j.compositesb.2011.03.011

[110] W.F. Lee, Y.C. Chen, Effect of bentonite on the physical properties and drug-release behavior of poly(AA-co-PEGMEA)/bentonite nanocomposite hydrogels for mucoadhesive, J. Appl. Polym. Sci. 91 (2004) 2934-2941. https://doi.org/10.1002/app.13499

[111] C.R.N. Jesus, E.F. Molina, S.H. Pulcinelli, C. V. Santilli, Highly Controlled Diffusion Drug Release from Ureasil-Poly(ethylene oxide)-Na+-Montmorillonite

Hybrid Hydrogel Nanocomposites, ACS Appl. Mater. Interfaces. 10 (2018) 19059-19068. https://doi.org/10.1021/acsami.8b04559

[112] M.S. San Román, M.J. Holgado, B. Salinas, V. Rives, Drug release from layered double hydroxides and from their polylactic acid (PLA) nanocomposites, Appl. Clay Sci. 71 (2013) 1-7. https://doi.org/10.1016/j.clay.2012.10.014

[113] E. Valarezo, L. Tammaro, S. González, O. Malagón, V. Vittoria, Fabrication and sustained release properties of poly(ε-caprolactone) electrospun fibers loaded with layered double hydroxide nanoparticles intercalated with amoxicillin, Appl. Clay Sci. 72 (2013) 104-109. https://doi.org/10.1016/j.clay.2012.12.006

[114] M. Del Arco, A. Fernández, C. Martín, V. Rives, Solubility and release of fenbufen intercalated in Mg, Al and Mg, Al, Fe layered double hydroxides (LDH): The effect of Eudragit® S 100 covering, J. Solid State Chem. 183 (2010) 3002-3009. https://doi.org/10.1016/j.jssc.2010.10.017

[115] L.N.M. Ribeiro, A.C.S. Alcântara, M. Darder, P. Aranda, P.S.P. Herrmann, F.M. Araújo-Moreira, M. García-Hernández, E. Ruiz-Hitzky, Bionanocomposites containing magnetic graphite as potential systems for drug delivery, Int. J. Pharm. 477 (2014) 553-563. https://doi.org/10.1016/j.ijpharm.2014.10.033

[116] Y.E. Miao, H. Zhu, D. Chen, R. Wang, W.W. Tjiu, T. Liu, Electrospun fibers of layered double hydroxide/biopolymer nanocomposites as effective drug delivery systems, Mater. Chem. Phys. 134 (2012) 623-630. https://doi.org/10.1016/j.matchemphys.2012.03.041

[117] F. Cao, Y. Wang, Q. Ping, Z. Liao, Zn-Al-NO3-layered double hydroxides with intercalated diclofenac for ocular delivery, Int. J. Pharm. 404 (2011) 250-256. https://doi.org/10.1016/j.ijpharm.2010.11.013

[118] R. Rojas, M.C. Palena, A.F. Jimenez-Kairuz, R.H. Manzo, C.E. Giacomelli, Modeling drug release from a layered double hydroxide-ibuprofen complex, Appl. Clay Sci. 62-63 (2012) 15-20. https://doi.org/10.1016/j.clay.2012.04.004

[119] S. Rajkumar, B.D. Kevadiya, H.C. Bajaj, Montmorillonite/Poly (L-Lactide) microcomposite spheres as reservoirs of antidepressant drugs and their controlled release property, Asian J. Pharm. Sci. 10 (2015) 452-558. https://doi.org/10.1016/j.ajps.2015.06.002

[120] S.A. Gârea, A.I. Voicu, H. Iovu, Clay-Polymer Nanocomposites for Controlled Drug Release, in: Clay-Polymer Nanocomposites, 2017. https://doi.org/10.1016/B978-0-323-46153-5.00014-8

[121] A. Verma, S. Thakur, G. Mamba, Prateek, R.K. Gupta, P. Thakur, V.K. Thakur, Graphite modified sodium alginate hydrogel composite for efficient removal of malachite green dye, Int. J. Biol. Macromol. 148 (2020) 1130-1139. https://doi.org/10.1016/j.ijbiomac.2020.01.142

[122] V. Bahadur, R. Gadi, S. Kalra, Clay based nanocomposites for removal of heavy metals from water : A review, J. Environ. Manage. 232 (2019) 803-817. https://doi.org/10.1016/j.jenvman.2018.11.120

[123] E.I. Unuabonah, A. Taubert, Clay-polymer nanocomposites (CPNs): Adsorbents of the future for water treatment, Appl. Clay Sci. 99 (2014) 83-92. https://doi.org/10.1016/j.clay.2014.06.016

[124] K.A.H. Hernández, Polymer-Clay Nanocomposites and Composites: Structures, Characteristics, and their Applications in the Removal of Organic Compounds of Environmental Interest, Med. Chem. (Los. Angeles). 6 (2016) 201-210. https://doi.org/10.4172/2161-0444.1000347

[125] A.M. Atta, H.A. Al-Lohedan, A.O. Ezzat, Z.A. Issa, A.B. Oumi, Synthesis and application of magnetite polyacrylamide amino-amidoxime nano-composites as adsorbents for water pollutants, J. Polym. Res. 23 (2016). https://doi.org/10.1007/s10965-016-0963-z

[126] A. Olad, M. Bastanian, H. Bakht Khosh Hagh, Thermodynamic and Kinetic Studies of Removal Process of Hexavalent Chromium Ions from Water by Using Bio-conducting Starch-Montmorillonite/Polyaniline Nanocomposite, J. Inorg. Organomet. Polym. Mater. 29 (2019) 1916-1926. https://doi.org/10.1007/s10904-019-01152-w

[127]] E. Forgacs, T. Cserháti, G. Oros, Removal of synthetic dyes from wastewaters: A review, Environ. Int. 30 (2004) 953−971. https://doi.org/10.1016/j.envint.2004.02.001

[128] S. Lawchoochaisakul, P. Monvisade, P. Siriphannon, Cationic starch intercalated montmorillonite nanocomposites as natural based adsorbent for dye removal, Carbohydr. Polym. 253 (2021) 117230. https://doi.org/10.1016/j.carbpol.2020.117230

[129] A. Olad, F.F. Azhar, Eco-friendly biopolymer/clay/conducting polymer nanocomposite: Characterization and its application in reactive dye removal, Fibers Polym. 15 (2014) 1321-1329. https://doi.org/10.1007/s12221-014-1321-6

[130] A. Olad, F.F. Azhar, M. Shargh, S. Jharfi, Application of response surface methodology for modeling of reactive dye removal from solution using starch-montmorillonite/polyaniline nanocomposite, Polym. Eng. Sci. 54 (2014) 1595-1607. https://doi.org/10.1002/pen.23697

Adv. App. of Micro and Nano Clay – Biopolymer-based Composites   Materials Research Forum LLC
Materials Research Foundations **125** (2022) 210-235   https://doi.org/10.21741/9781644901915-9

Chapter 9

Kaolinite-Starch based Nano-Composites and Applications

Preeti Gupta[1*], S.S. Das[1] and N.B. Singh[2*]

[1]Department of Chemistry, DDU Gorakhpur University, Gorakhpur, India

[2]Research Development Cell, Sharda University, Greater Noida, India

* preeti17_nov@yahoo.com and n.b.singh@sharda.ac.in

Abstract

Starch and kaolinite are low cost and easily available materials used in different domains. At the same time, they form biodegradable nanocomposites (NCs), attractive and important substitutes for non-biodegradable materials which are the major source of environmental pollution. Kaolinite-starch based NCs are promising materials because they possess good mechanical and thermal properties and water resistance capacity etc. Properties are dependent on the filler particles orientation, nature of kaolinite and starch interaction, filler–matrix interface and filler dosages. The properties also depend on the anisotropy of the particles of fillers, their surfaces and the filler dispersion. These NCs have enormous applications specially in food packaging, films, paper, etc. It is used in environmental remediation. In this chapter an overview of methods of preparation, characterization and applications of kaolinite-starch based nanocomposites have been discussed.

Keywords

Kaolinite, Starch, Clay, Composite, Food Packaging, Film

Contents

1.   Introduction

Petrochemical polymers are highly hazardous and attempts are being made to reduce their use [1]. Also such polymers produce wastes and are not easily degradable. They have lot of environmental problems [1]. On the other hand, plants contain polymers which are ecofriendly, renewable and easily degradable. Naturally derived biopolymers are available in plenty and inexpensive. In recent years' lot of research work has been done on nanocomposites made from renewable sources [1]. The naturally derived polymers used for synthesis of nanocomposites (NCs) provide considerable enhancement in mechanical and physical properties [1]. In recent years' natural polymer/clay NCs have got lot of importance because of eco friendliness, improved properties and applications in different areas [1].

Clay is a natural mineral, consisting of layers of linked silicate tetrahedral and octahedral units. About 30 types of nanoclays are known and have different mineralogical composition with different properties [2]. Classifications are mainly based on various combinations of sheets of octahedral and tetrahedral silicate and their stacking. Some of the clays are "(i) kaolinites, in which one tetrahedral sheet is attached with one octahedral in 1:1 ratio. Some examples are serpentine, kaolinite, halloysite); (ii) nonexpanding clays, as intercalation of one octahedral sheet between two tetrahedral sheets occurs in 2:1 ratio (e.g., mica and illite); (iii) The third type is limited expanding clays (2:1 ratio). Vermiculite is an example of this type; (iv) strongly expanding clays are found in the ratio of 2:1, for example Montmorillonite [Mt)]; (v) Pyrophyllite and talc are uncharged group in 2:1 ratio (e.g.,); (vi) The sixth type is 2:1:1 group having brucite octahedral sheets between a 2:1 mineral (e.g., chlorites) and (vii) silicates with fibrous structure (e.g., polygorskite and sepiolite)". Weak van der Waals forces are involved in holding the sheets forming an 'interlayer'. Negative and positive surface charges are generated due to the isomorphous and pH substitutions in the tetrahedral and octahedral sheets. The clay minerals classified above are nanoscale materials and are represented in Fig. 1 [2].

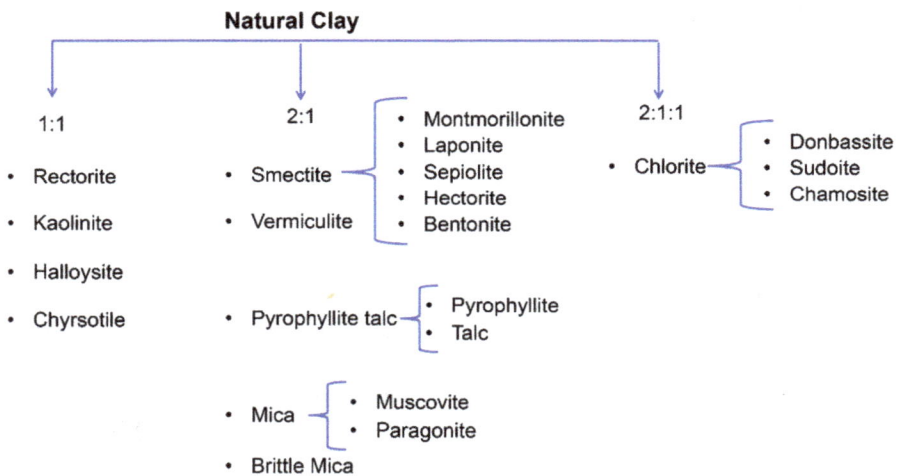

Fig. 1 Clay classification [2]

The kaolinite, "$Al_2[Si_2O_5](OH)_4$", a naturally occurring silicate clay mineral, has a layered structure (Fig.2) [3]. It consists of layers like siloxane and gibbsite and has many applications (Fig.3) [4]. Clays that are rich in Kaolinite mineral are known as Kaolin clay. In general kaolinite is always present in clay.

(a)   (b)

Si-centered tetrahedra

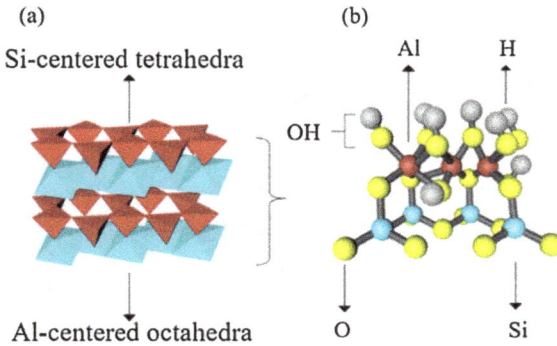

OH

Al

H

Al-centered octahedra   O   Si

Fig. 2 Structure of kaolinite [3]

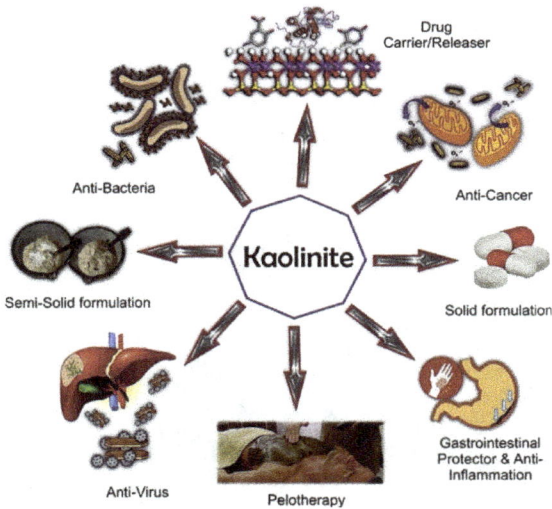

Drug
Carrier/Releaser

Anti-Bacteria

Anti-Cancer

Semi-Solid formulation

Kaolinite

Solid formulation

Anti-Virus   Pelotherapy

Gastrointestinal
Protector & Anti-
Inflammation

Fig. 3 Applications of kaoline [4]

Starch has polymeric structure consisting of long chains of glucose molecules. Amylase and amylopectin are the main constituents of starch (Fig.4) [5].

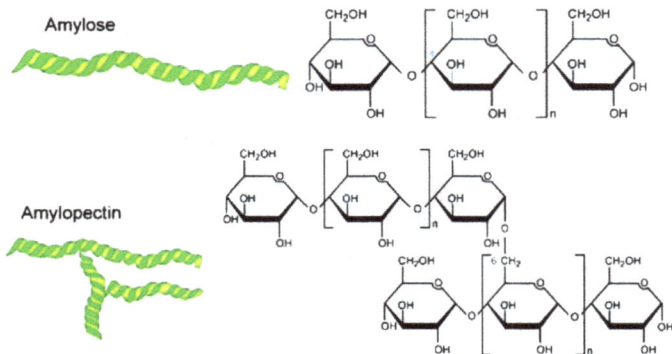

Fig. 4 Amylose and amylopectin units in starch. [5]

There are different sources for starch (Fig.5) and various aspects of starch are given in Fig.6[5].

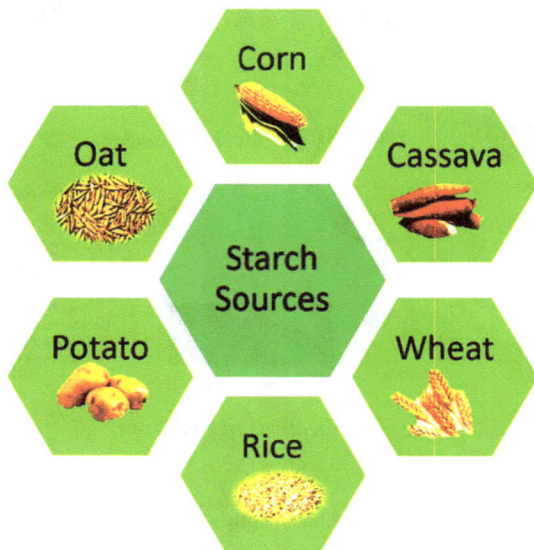

Fig. 5 Starch from different sources

Adv. App. of Micro and Nano Clay – Biopolymer-based Composites   Materials Research Forum LLC
Materials Research Foundations **125** (2022) 210-235   https://doi.org/10.21741/9781644901915-9

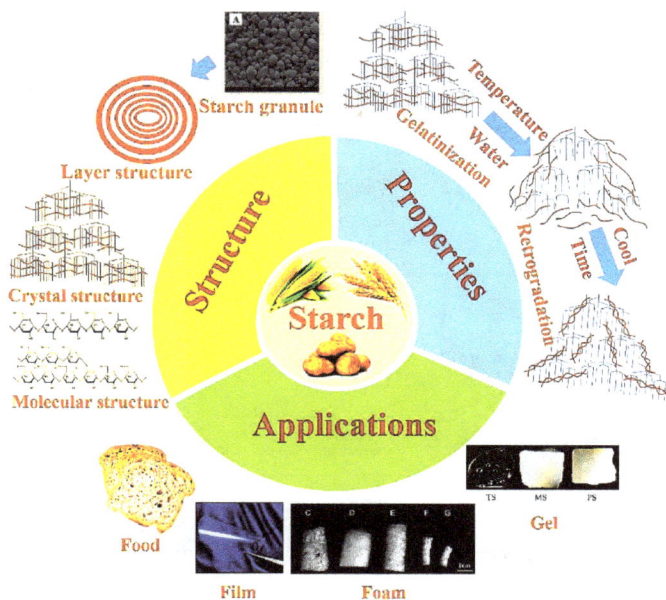

Fig. 6 Structure, properties and applications of starch [6].

Nowadays considerable focus is being made on polymer–clay NCs particularly on starch–kaolinite NC due to low cost, easy availability, biodegradability and better properties. Further, if suitable plasticizers are added, the properties are improved. In this chapter, preparation, characterization, properties and applications of Clay(kaolinite)-starch nanocomposite have been discussed in detail.

2. Synthesis of kaolinite-starch nano-composites

Based on the nature of interaction between starch matrix and kaolinite, it is possible to prepare three different types of NCs: "(a) NCs based on exfoliated structure, (b) NCs having intercalated structure and (c) nanocomposites having flocculated structure" [3]. Based on starting materials, three methods are used for the preparation of NCs-(a) Intercalative polymerization, (b) starch intercalation and (c) melt intercalation. Starch is a unique polysaccharide and is an important material due to its low cost. It is easily available and can be biodegradable. Various approaches such as solution blending, in situ polymerization and melt blending methods would be used for synthesizing starch kaolinite NC. The solution-blending method is often better in comparison to other techniques. However, the most important is uniform dispersion of one phase into the other phase. Different steps involved in all the three techniques are given in Figs.7-9[7-10]

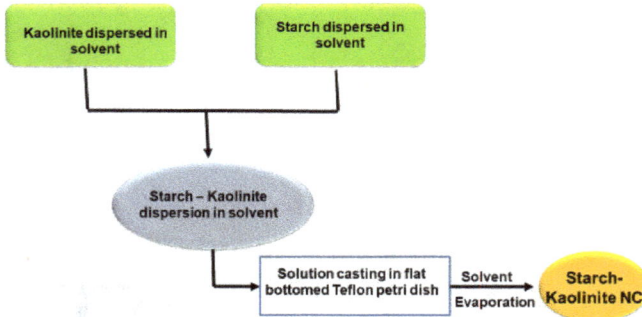

Fig. 7 Solution blending technique for preparation of starch –kaolinite NC

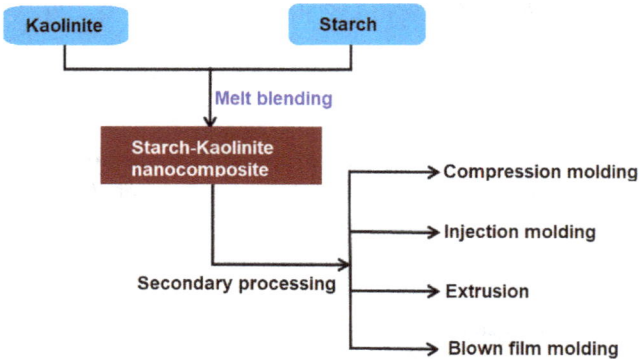

Fig. 8 Melt blending technique for preparation of starch – kaolinite NC

Adv. App. of Micro and Nano Clay – Biopolymer-based Composites   Materials Research Forum LLC
Materials Research Foundations **125** (2022) 210-235   https://doi.org/10.21741/9781644901915-9

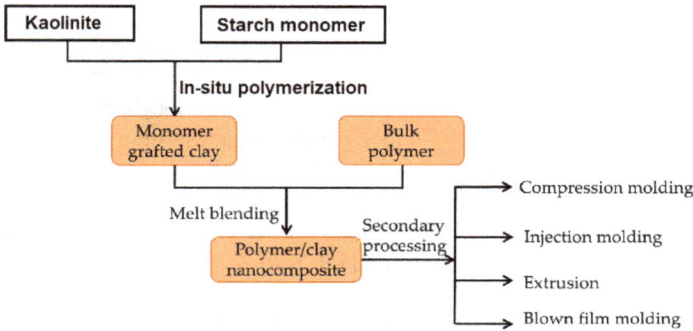

Fig. 9 In situ polymer for preparation of starch –kaolinite NC

Solution blending method is very common for synthesizing starch-kaolinite NC. In this technique as given in Fig.7, starch is mixed with water and heated at 100 °C. Aqueous kaolinite is dispersed into the starch solution. The mixture is stirred, where kaolinite-starch mixture gets precipitated on adding ethanol. This is kept overnight at 4 °C. After filtering, the precipitate is dried in vacuum at 50 °C. NC is blended with plasticizers and then made to swell the starch molecules at room temperature. Films are then obtained by hot pressing technique [6]. Three forms of NCs - phase separated structure, intercalated structure and exfoliated structures are formed (Fig.10) [2].

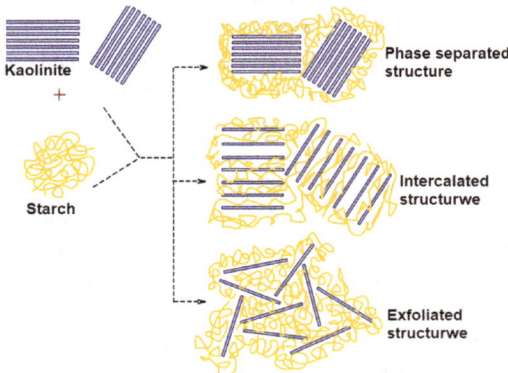

Fig. 10 Three forms of Kaolinite-starch NC [2]

Adv. App. of Micro and Nano Clay – Biopolymer-based Composites   Materials Research Forum LLC
Materials Research Foundations **125** (2022) 210-235   https://doi.org/10.21741/9781644901915-9

3.   Characterization and properties of kaolinite-starch NC

Starch is used for making biocomposites [11,12] due to its low cost, ready availability, easy extraction, chemical modification possibilities, solvent affinity, and versatile applications. On combination with plasticizing agent, starch becomes a tough material suitable for engineering purpose with increased thermoplastic characteristics [13]. Being water sensitive, large interacting nature, lack of flexibility and low mechanical strength, the starch-based composites have limited practical applications [14]. However, cellulosic fibers, when added into starch-based biocomposites, act as a good reinforcing material and improves the properties of the starch matrix [15]. Self-assembled NC films produced from kaolinite and starch in presence of plasticizers showed better properties and applications in different sectors. X-ray diffraction (XRD), TG/DSC, FTIR spectroscopic and other techniques have been used to characterize and study properties of kaolinite-starch NCs. Some of the techniques are discussed below.

3.1   X-ray diffraction studies

XRD patterns of thermoplastic starch film (TPS) with different additions of kaolinite (k) have been recorded and it is found that with the increase of kaolinite in the starch, crystalline character increases to some extent. [16]. When certain plasticizer like glycerin is added to NC, agglomeration of kaolinite is decreased and dispersion becomes much better. The XRD patterns of kaolinite, potato starch, film of starch-glycerin, kaolinite with starch-glycerin (10wt%) and starch-glycerin – kaolinite (2wt%) films have been shown in Fig.11 [12]. Starch shows a sharp peak at 17.05°due to crystalline nature and three broad peaks at "19.48, 22.14, and 24°" due to its amorphous character. Characteristic peaks in XRD pattern at 12.52 and 24.8° are due to crystalline character and other peaks are due to amorphous nature of kaolinite (Fig.11) [17]. The average crystallite size of kaolinite was found to be 25.9 nm. The peaks of kaolinite due to crystalline character are not seen in NC film of starch – glycerin–kaolinite (2 wt %), but with increasing kaolinite content, crystalline peaks are easily seen. It appears that when kaolinite content is high, distribution of particles of clay inside NC film networks occurs in an irregular way and as a result, separation of phases of kaolinite might take pace. Because of relative dispersion of kaolinite platelets, starch–kaolinite NCs of three types are formed. These are (i) "intercalated NC", (ii) "flocculated NC", and (iii) "exfoliated NC". Starch has good miscibility with kaolinite clays in water due to hydrophilic nature of starch. Because of this they can easily exfoliate/ intercalate into the interlayers.

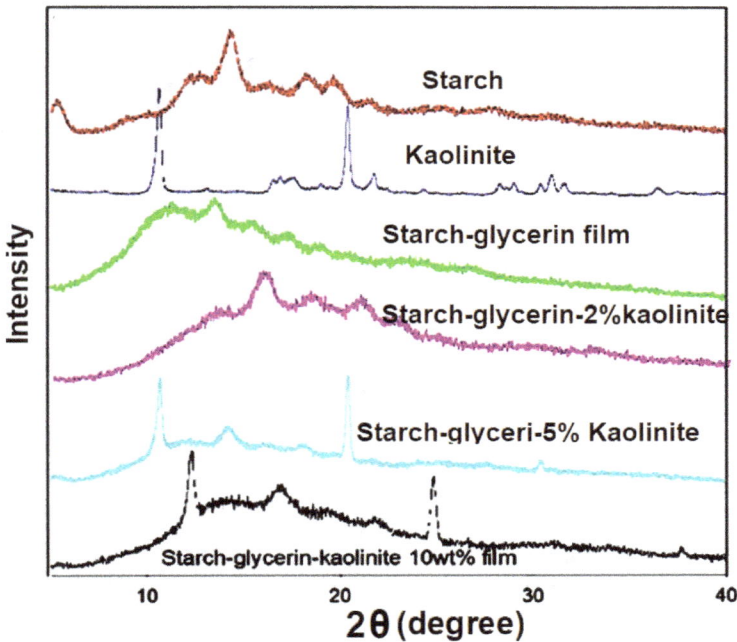

Fig. 11 XRD of different films [17]

3.2   Thermal property of kaolinite-starch film

TG/DTG/DTA studies of kaolinite - starch films containing different amounts of kaolinite have been studied [18]. The decomposition temperature in general increased with addition of kaolin and the percent decomposition decreased. The thermal decomposition of potato starch and its NC films follows three step processes. Between 50− 200 °C, the first step - loss of water and plasticizer molecules. Between 200−350 °C, i.e. the second step - starch degradation occurs, and in the third step - degradation of clay (400−900 °C) takes place (Fig. 12). From the Figure, it is seen that starch− glycerin film starts decomposing at 296 °C with 19% residue. However, kaolinite decomposes in two plateau regions; viz; 470 °C, and 790 °C. At 900 °C, the weight loss is 20%. The decomposition of starch−glycerin−kaolinite NC occurs at 311 °C yielding 13% ash residue. It appears that interactions in this system are more strong than those in the starch−glycerin system. This shows that NC films of starch- glycerin-kaolinite are thermally more stable.

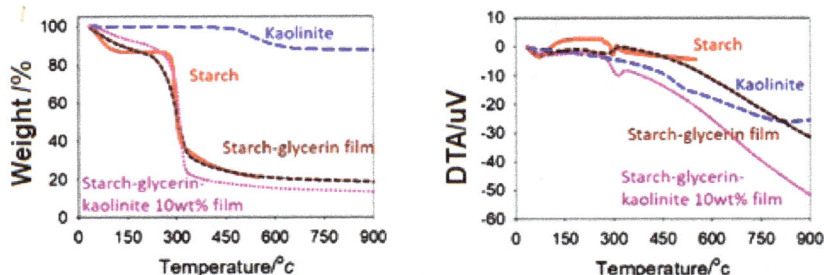

Fig. 12 TG-DTA curves for starch, kaolinite films

3.3   FT-IR spectral analysis of starch-kaolinite NC

FT-IR spectroscopic technique is a powerful method to understand the nature of interaction.   FT-IR spectra of starch, kaolinite, starch-glycerin and starch-glycerin-kaolinite (2wt%) are given in Fig.13 [17]. Due to stretching vibrations of -OH group, starch shows band at "2935 and 3450 cm^{-1}" respectively. CH$_2$ stretching vibrational frequency appears at "2924 and 1465 cm^{-1}", whereas bands at "1150, 1080 and 1040 cm^{-1}" are due to C−O stretching of C−O−C groups" in the starch molecule. Additional bands in starch−plasticizer films are also found. The band intensity on the lower side indicates strong interaction between plasticizer and the starch. It is found that when glycerin is incorporated in starch, two bands at "3450 and 1170 cm^{-1}" shift towards lower frequency. FT-IR spectra of kaolinite yield four bands at "3697, 3669, 3645 and 3620 cm^{-1}" respectively. In kaolinite, in addition to these bands, additional bands due to Si−O and Al−OH bonds are also found. All the characteristic peak frequencies of kaolinite are also observed in the starch-glycerin-kaolinite NC film. The results indicate that there are physical interactions.

When the plasticizer as a dispersing agent is changed, the properties of NCs get slightly changed and the vibrational frequencies are shifted. Within the composites, almost complete exfoliation of the kaolinite occurs due to DMSO treatment of the kaolinite [19]. The vibrational frequencies due to O-H and C-OH groups of the starch (S) at "995.1 cm^{-1}" and 1103.1 cm^{-1}" (Fig. 14a and b) are changed due to H-bonding in the presence of the kaolinite(k). Band at 2888.8 cm^{-1} due to stretching mode of the carbon skeleton of the starch chains is also affected (Fig. 14c), showing that interactions between chain−chain in the starch is weakened. The shift in frequencies in the presence of DMSO are more than those due to kaolinite alone. This indicates increased interaction between kaolinite−starch in presence of DMSO [19].

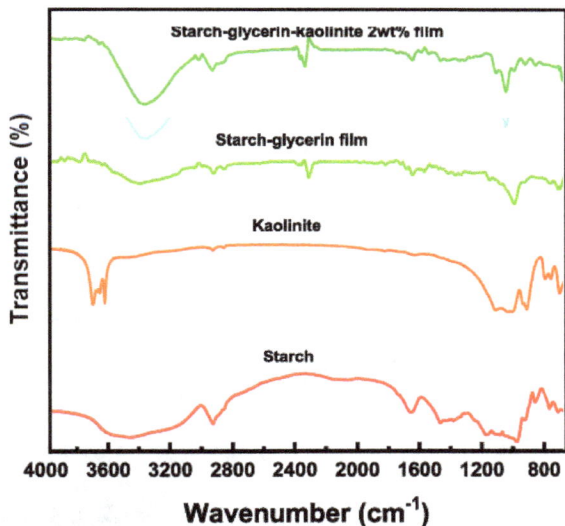

Fig. 13 FTIR spectra [17]

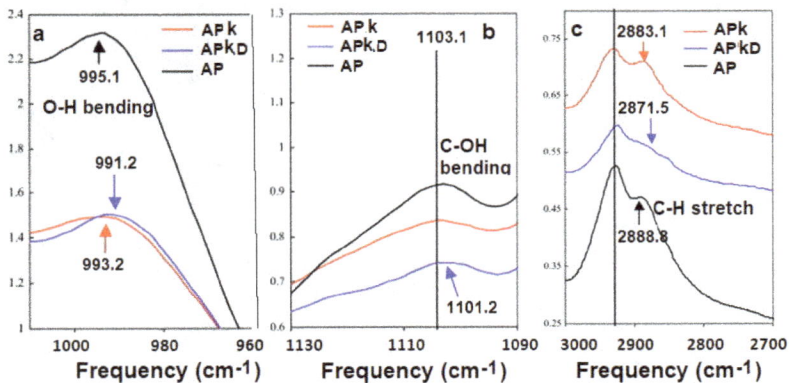

*Fig. 14 Vibrations (a) O-H (b) C-OH and (c) C-H in Aquous processed starch(AP), AP -
5% kaolinite(APk1) and APP1-5%DMSO(APkD1) in FTIR spectra [19]*

3.4 SEM studies of clay-starch nanocomposite

Thermoplastic starch (TPS)/clay NCs have been prepared by melt processing technique using modified clay (o-clay) and pristine clay (p-clay) and SEM pictures are given in Fig.15 [20]. Due to plasticizers, starch granules are swelled and the boundary between them after gelatinization, nearly disappears. Both TPS/p-clay and TPS/o-clay NCs exhibit continuous matrix texture. All the films are transparent and homogeneous. With the pristine clay. The color of the starch film becomes much deeper whereas colour becomes yellow in presence of o-clay. Homogeneous and transparent films are advantageous for packing materials [20].

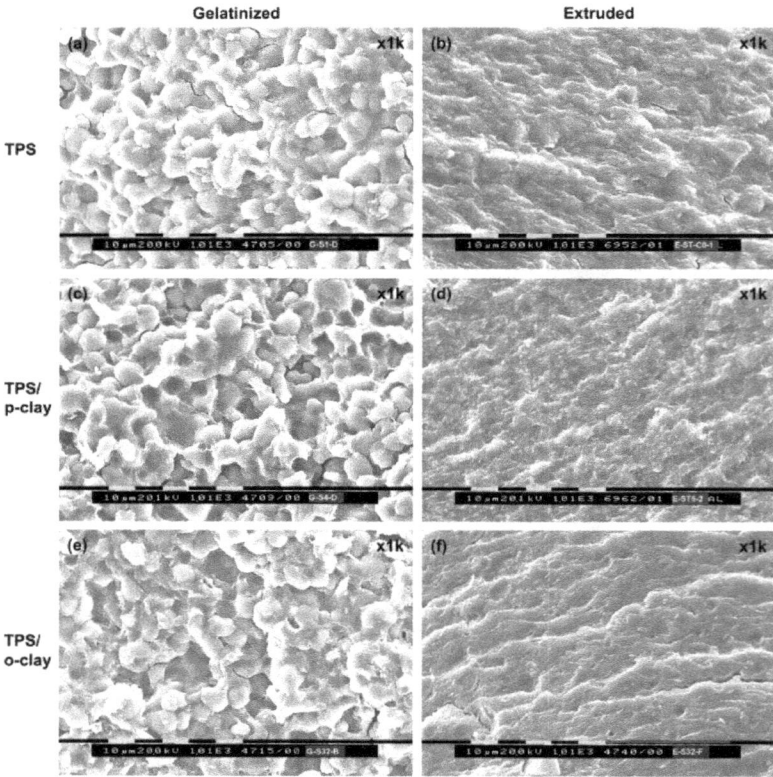

Fig. 15 SEM micrographs of different films [20]

Adv. App. of Micro and Nano Clay – Biopolymer-based Composites   Materials Research Forum LLC
Materials Research Foundations **125** (2022) 210-235   https://doi.org/10.21741/9781644901915-9

3.5   Water uptake

Hydrophobicity due to the presence of hydrophilic free hydroxyl groups, is an important property of starch films. These free hydroxyl groups of starch under higher relative humidity attract moisture and as a result, the NC film loses its strength and other barrier properties. On the other hand, when the humidity is low, the NC film becomes brittle. But, addition of clay into starch films in appropriate amounts address these issues [21]

The water uptake by kaolinite/starch film after 24 h is lower at 50% than 100% relative humidity and decreases with increasing kaolinite/starch ratio (Fig. 16). This is due to interaction [19]. Thus, increasing kaolinite content in NC film makes better for water barrier. It appears that interactions between clay-starch are responsible for low water uptake at 50% RH. In the presence of DMSO, Kaolinite the higher barrier effects on water uptake, may be due to clay particles of smaller size. This increases the starch–clay interactions, leaving limited number of free OH groups for water binding. The fine particles of clay probably well distributed well in starch matrix. This restricts the diffusion of water molecules inside the matrices of starch. Better dispersion of DMSO–kaolinite is confirmed by water uptake and as a result increases the barrier effect for water uptake.

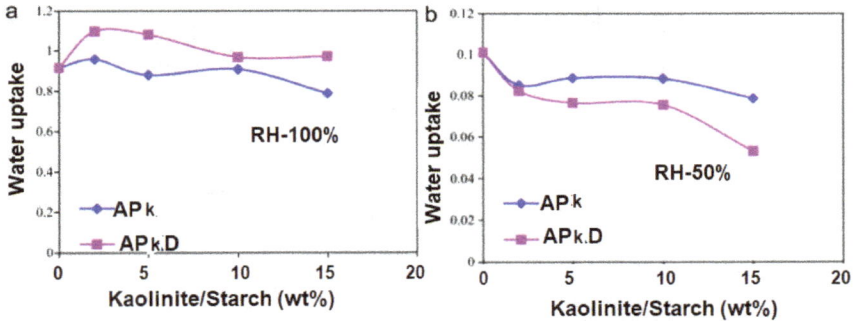

Fig.16 Water uptake after 24 h [19]

4.   Applications of kaolinite-starch based nano-composites

4.1   Water remediation using starch-kaolinite nanocomposite

Our environment is continuously degrading due to the presence of various types of toxic synthetic and natural-substances such as heavy metals, dyes, pesticides etc. These pollutants come from discharges coming out from different industries. Now this has become a big challenge particularly in developing countries. These toxic chemicals are to be removed in order to minimize pollution in water bodies. Out of different techniques, adsorption is found to be most economical and efficient technique. Number of adsorbents are being used for remediation of water [22]. In recent years, nanomaterials and nanocomposites have extensively been used as adsorbents.

Recently metakaolinite/CNT and kaolinite/starch NCs were used for removing Mn and Fe from water by using adsorption technique. Experiments have been done as a function of temperature, time, initial concentrations, pH and dose of adsorbent. Adsorption isotherm and kinetic models were tested. Thermodynamic parameters have also been calculated [23]. Effect of pH on the removal of Mn and Fe from aqueous solution by kaolinite-starch NC is shown in Fig. 17 [23]. As the pH increased, percent removal increased but above pH 6, the increase was very slow. It is also seen that Fe is removed faster as compared to Mn. This may be due to changes in ionic nature at higher pH.

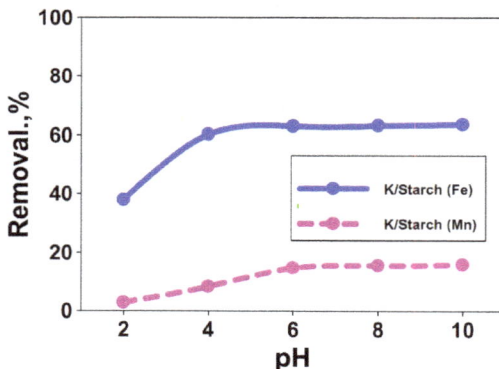

Fig.17 % removal of Fe/Mn from water using kaolinite-starch NC at different pH [23]

Mohamed et al. (2017) [24] studied in detail the adsorption process and verified different adsorption and kinetic models in removing Fe and Mn by using kaolinite-starch NC. Parameters (Tables 1-2) showed that for Mn removal, Freundlich adsorption isotherm model fits the data best, indicating multiple layer adsorption on the surface of NC. In the case of Fe removal, the data was fitted by Langmuir adsorption isotherm i.e. monolayer adsorption. Pseudo 2nd order kinetic model fitted the data for Fe removal, whereas Elovich model fitted the data for Mn removal.

Table 1 Parameters for adsorption isotherm models [23]

Model	Parameters	k/starch (Mn)	k/starch (Fe)
Langmuir	q_{max} (mg/g)	34.26	46.8
	b (L/mg)	0.225	1.361
	R^2	0.234	0.911
	RL	0.89-0.95	0.59-0.78
Freundlich	1/n	1.1	0.815
	k_F	27.509	4.601
	R^2	0.989	0.83
Temkin	B_T	7.188	14.62
	K_T	4.926	13.356
	R^2	0.918	0.902

Table 2 Kinetic parameters for adsorption [23]

Model	Parameters	K/starch (Fe)	K/starch (Mn)
Pseudo-second order	k_2 (g / mg min)	1.267×10^{-3}	3.12×10^{-4}
	q_e (Cal) (mg/g)	32.15	16.8
	R^2	0.9914	0.7571
Elovich	β (g/mg)	0.114	0.2795
	α (mg/ g min)	3.275	14.11
	R^2	0.9072	0.940

Effect of temperature on the removal of Mn and Fe using kaolinite-starch NC has also been studied and shown in Fig.18. The results show that % removal decreases with increase of temperature. Changes in thermodynamic functions such as Enthalpy (ΔH^0), Gibb's free energy (ΔG^0) and Entropy (ΔS^0) are listed in Table 3. [23]. The exothermic adsorption of Fe and Mn is revealed by thermodynamic parameters.

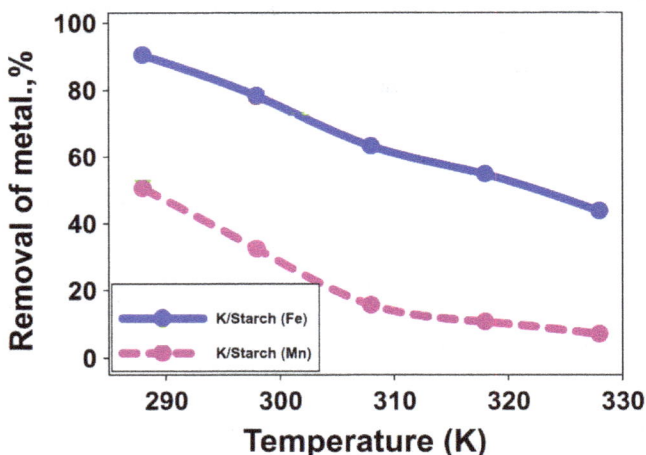

Fig. 18 Influence of Temperature on % removal of Mn and Fe from water by kaolinite-starch NC

Table 3 Values of ΔG^0, ΔH^0, ΔS^0 for Mn and Fe adsorption on kaolinite-starch surface [23]

Thermodynamic Parameter		k/starch (Mn)	k /starch (Fe)
ΔG^0 (J mol^{-1})	288 K	-9.4	-14.73
	298 K	-7.85	-12.85
	308 K	-5.67	-11.4
	318 K	-4.7	-10.84
	328 K	-3.59	-9.96
ΔH^0 (kJ mol^{-1})		-52.6	-48.65
ΔS^0 (J K^{-1} mol)		-0.15	-0.12

Kaolinite-starch NC film appears to be an effective adsorbent for removing metal ions from a solution. It seems that a strong H-bond interaction takes place between the two components of NC and make its surface homogeneous and smooth and also modifies the microstructure of NC. A model structure for metakaolinite-starch interaction is shown in Fig. 19 [24].

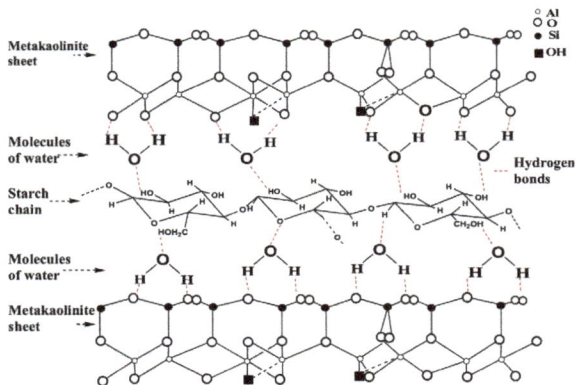

Fig. 19 Model for interaction between metakaolinite and starch [24]

4.2 Films

When solution containing starch, kaolinite, glycerol and water is mixed thoroughly, casted on a glass surface and covered by a Teflon ribbon, formation of thin film takes place. This on atmospheric exposer, evaporates solvent from the solution and a film is formed. As a result, when solvent evaporates, the content of starch increases in the film and hence the interaction within starch matrix also increases. This film is formed as a result of hydrogen bonding. These films are transparent and can be used for food packaging [Fig.20]. This type of films may prove to be an excellent food packaging films [18].

Fig. 20 Kaolinite-starch film covering fruits

4.3   Paper making

For lowering the cost, improving printability, optical properties, and the water removal rate during papermaking, suitable fillers are used. But due to increased filler amounts, the paper stiffness and strength are decreased. Starch is the conventional material for making papers. Clay-starch composites have been used to make papers by using this composite of different aggregates sizes and different starch to clay ratios. However, paper strength increased when modified clay is used instead of raw clay in starch [25].

4.4   Food packaging

Food packaging is required for maintaining the food products quality during storage and transportation and also protect from microorganisms, moisture, light, external force, chemical contaminants, oxygen, etc. For these purposes, number of suitable physicochemical properties in the materials, from which food packaging substances are made, are needed. Some of the properties essential for food packaging materials are given in Fig.21 [26].

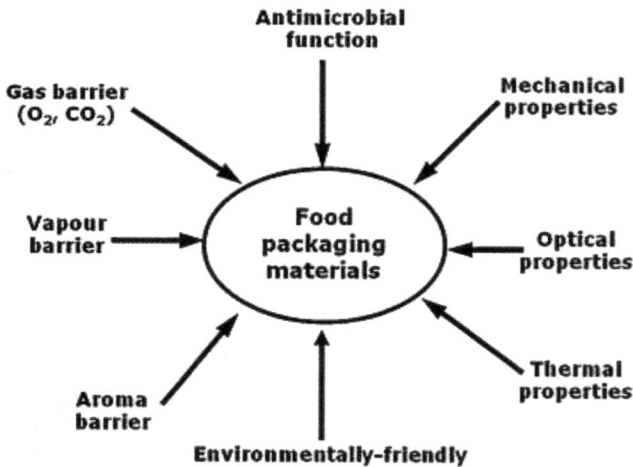

Fig. 21 Qualities of food packaging materials [26]

With the increasing ready-to-eat foods, packaging has become very important. Out of different materials, plastic packaging is used maximum. However, these plastic packaging after use, pose lot of environmental problems. Therefore, attempts are being made to use biodegradable plastics. In general bioplastics containing some fillers are being used (Fig.22) [27].

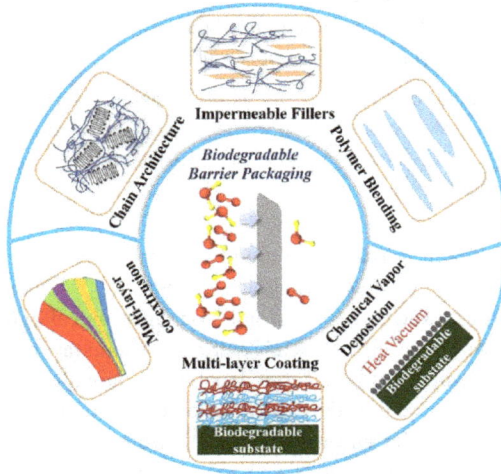

Fig. 22 Different biodegradable packaging materials [27]

So far, for suitable packaging, NCs of polycaprolactone (PCL), starch and cellulose derivatives, poly(butylene succinate) (PBS), polylactic acid (PLA), and polyhydroxybutyrate (PHB) with nanofillers such as nanoclays - montmorillonite and kaolinite, layered silicate, etc. are being used. In food packaging applications, importance is given the development of high barrier properties in order to protect from O_2, CO_2 diffusions, flavor compounds, and water vapors. Apart from these, their mechanical and surface properties are also an important criterion. For this purpose, extensive studies on starch- clay NCs have been done [28,29]

Ruamcharoen et al. [30] have discussed the mechanical behaviour and improved barrier properties of sago starch-kaolinite composite in detail. Water vapour transmission of starch, and starch-kaolin composite with the intercalation of DMSO has been studied. It is found that with increasing kaolinite clay, water transmission rate is decreased [31,32]. These results were explained due to the formation of H-bonding [33, 34].

Mechanical strength including tensile and modulus of starch-kaolinite composite film with different wt % of kaolinite content and starch-kaolinite composite with DMSO has been investigated. It is found that tensile strength and modulus values were maximum for 4% of kaolinite intercalated with DMSO. The elongation at break decreases when the content of kaolinite was increased. This can be described as the incorporation of inorganic filler which maximize the brittleness [32]. It can be said that starch-kaolinite NC is one of the suitable packaging materials as it is natural and biodegradable.

Considering environmental problems, synthetic polymers as packaging materials must be changed. However, cost weighs over environmental concerns. Therefore, research efforts are needed to develop low cost biopolymer-based eco-friendly packaging materials [35].

4.5   Paper packaging

Because of the lightweight, biodegradable and recyclable nature of paper made from starch-clay composite, it is used as packaging materials. For achieving improvement in water barrier properties, water vapor transmission, mechanical property etc., surface coating is a significant tool [36-40]. Starch is extremely useful raw material because it has high molecular weight which easily undergo depolymerization, hydrophilic in nature and its hydroxyl group can be used in variety of oxidation or substitution reactions which may adjust its rheological properties and its retrogradation could be eliminated. However, it showed undesirable physical properties and mechanical strength [41-43]. Clay when incorporated into starch-based composites films increased its water barrier behavior and improved its mechanical properties [42-49].

Kaovilu et al. [50] have studied Starch-kaolinite coating as a barrier layer on the paper board to increase the properties of fibre-containing packaging material to protect against mineral oil. The coating has been performed by dispersion coating method.

Conclusions

Kaolinite, an important clay material when mixed with starch obtained from different sources, kaolinite-starch NC is formed. There are three different methods for the synthesis of nanocomposite. In order to have uniform distribution, a plasticizer is mixed. NC formed has been characterized by number of techniques. Results have indicated the compactness of structure due to H-bonding. NC can be used in water purification, film formation and packaging materials. Seeing its importance, further research is needed.

References

[1] G. Madhumitha, J. Fowsiya , S. Mohana Roopan1 , V.K. Thakur, Recent advances in starch-clay nanocomposites. Inter. J Polym. Anal. Charact. 23 (2018) 331-345. https://doi.org/10.1080/1023666X.2018.1447260

[2] M.M. Orta, J. Martin, J.L. Santosb, I. Apariciob, S.M. Carrascoc, E. Alonsob, Biopolymer-clay nanocomposites as novel and eco-friendly adsorbents for environmental remediation, Appl. Clay Sci. 198 (2020) 105838. https://doi.org/10.1016/j.clay.2020.105838

[3] C. Zhou, W. Qizhao, C. Hongfei, Recent advances in kaolinite-based material for photocatalysts, Chin. Chem. Let. 32 (2021) 2617-2628 https://doi.org/10.1016/j.cclet.2021.01.009

[4] M.E. Awada, A.L. Galindo, M. Setti, M.M. El-Rahmany, C.V. Iborr, Kaolinite in pharmaceutics and biomedicine, Inter. J. Pharma. 533 (2017) 34-48. https://doi.org/10.1016/j.ijpharm.2017.09.056

[5] H. Cheng, L. Chen, D. J. McClements, T. Yang, Z. Zhang, F. Ren, M. Miao, Y. Tian, Z. Jin, Starch-based biodegradable packaging materials: A review of their preparation, characterization and diverse applications in the food industry, Trends Food Sci. Tech. 114 (2021)70-82. https://doi.org/10.1016/j.tifs.2021.05.017

[6] F. Zia, K. M. Zia, M. Zuber, S. Kamal, N. Aslam, Starch based polyurethanes: A critical review updating recent literature, Carbohydr. Polym. 134 (2015) 784 -798. https://doi.org/10.1016/j.carbpol.2015.08.034

[7] F. Guo, S. Aryana, Y. Han, Y. Jiao, A Review of the Synthesis and Applications of Polymer-Nanoclay Composites, Appl. Sci. 8 (2018) 1696. https://doi.org/10.3390/app8091696

[8] M. Alexandre, P. Dubois, Polymer-layered silicate nanocomposites: Preparation, properties and uses of a new class of materials, Mater. Sci. Eng. R Rep. 28 (2000) 1-63. https://doi.org/10.1016/S0927-796X(00)00012-7

[9] N. Yarahmadi, I. Jakubowicz, T. Hjertberg, Development of poly(vinyl chloride)/montmorillonite nanocomposites using chelating agents, Polym. Degrad. Stab. 95 (2010) 132-137. https://doi.org/10.1016/j.polymdegradstab.2009.11.043

[10] M. Asensio, M. Herrero, K. Núñez, R. Gallego, J. C. Merino, J.M. Pastor, In situ polymerization of isotactic polypropylene sepiolite nanocomposites and its copolymers by metallocene catalysis, Eur. Polym. J. 100 (2018) 278-289. https://doi.org/10.1016/j.eurpolymj.2018.01.034

[11] K.K. Yang, X.L. Wang, Y.Z. Wang, Progress in nanocomposite of biodegradable polymer, J. Ind. Eng. Chem. 13 (2007) 485-500.

[12] J.K. Pandey, A.P. Kumar, M. Misra, A. K. Mohanty, L. T. Drzal, R.P. Singh, Recent advances in biodegradable nanocomposites, J. Nanosci. Nanotechnol. 5 (2005) 497-526. https://doi.org/10.1166/jnn.2005.111

[13] A. Ujcic, M. Nevoralove, J. Dybal, A. Zhigunov, J. Kredatusova, S. Krejcikova, I. Fortelny, M. Slouf, Thermoplastic starch composites filled with isometric and elongated TiO2 -based nanoparticles, Front. Mater. 6 (2019) 1-13. https://doi.org/10.3389/fmats.2019.00284

[14] M.R. Amin, M.A. Chowdhury, M.A. Kowser, Characterization and performance analysis of composite bioplastics synthesized using titanium dioxide nanoparticles with corn starch, Heliyon 5 (2019) 1-12. https://doi.org/10.1016/j.heliyon.2019.e02009

[15] I. Spiridon, C. A. Teaca, R. Bodirlau, M. Bercea, Behavior of cellulose reinforced cross-linked starch composite films made with tartaric acid modified starch microparticles. J. Polym. Environ. 21 (2013) 431-440. https://doi.org/10.1007/s10924-012-0498-2

[16] A. Kwaśniewska, D. Chocyk, G. Gładyszewski, J. Borc , M. Świetlicki, B. Gładyszewska, The influence of kaolin clay on the mechanical properties and structure of thermoplastic starch films, Polymers 12 (2020) 73. https://doi.org/10.3390/polym12010073

[17] H. Banna, M.Z. Islam, Md. A.B.H. Susan, B. Imran, Effects of plasticizers and clays on the physical, chemical, mechanical, thermal, and morphological properties of potato starch-based nanocomposite films, ACS Omega 5 (2020) 17543−17552. https://doi.org/10.1021/acsomega.0c02012

[18] Md. Ashaduzzaman, D. Saha, M. M.Rashid, Mechanical and thermal properties of self-assembled kaolin-doped starch-based environment-friendly nanocomposite films, J. Compos. Sci. 4 (2020) 38. https://doi.org/10.3390/jcs4020038

[19] J.A. Mbey, S. Hoppe, F. Thomas, Cassava starch-kaolinite composite film. Effect of clay content and clay modification on film properties, Carbohy. Polym. 8 (2012) 213-222. https://doi.org/10.1016/j.carbpol.2011.11.091

[20] Q.X. Zhang, Z.Z. Yu, X.L. Xie , K. Naito , Y. Kagawa, Preparation and crystalline morphology of biodegradable starch/clay nanocomposites, Polym. 48 (2007) 7193-7200. https://doi.org/10.1016/j.polymer.2007.09.051

[21] S. Agarwal, Major factors affecting the characteristics of starch based biopolymer films, Europ. Polym. J. 160 (2021) 110788. https://doi.org/10.1016/j.eurpolymj.2021.110788

[22] N.B.Singh, Garima Nagpal, Sonal Agrawal and Rachna, Water purification by using Adsorbents: A Review, Journal of Environmental Technology & Innovation 11 (2018)187-240 https://doi.org/10.1016/j.eti.2018.05.006

[23] M. Shaban, M.E.M. Hassouna, F.M. Nasief, M.R. Abu Khadra, Adsorption properties of kaolinite-based nanocomposites for Fe and Mn pollutants from aqueous solutions and raw ground water: kinetics and equilibrium studies, Environ. Sci. Pollut. Res. 24 (2017) 22954-22966. https://doi.org/10.1007/s11356-017-9942-0

[24] N. Méité, L. K. Konan, M. T. Tognonvi, B. I. H. Goure, Doubi, M. Gomina, S. Oyetola, Properties of hydric and biodegradability of cassava starch-based bioplastics reinforced with thermally modified kaolin, Carbohy Polym. 254 (2021) 117322. https://doi.org/10.1016/j.carbpol.2020.117322

[25] S.Y. Yoon, Y. Deng, Clay-starch composites and their application in papermaking, J. Appl. Polym. Sci. 100 (2006) 1032-1038. https://doi.org/10.1002/app.23007

[26] J.W. Rhim, H.M. Park, C.S. Ha, Bio-nanocomposites for food packaging applications, Prog. Polym. Sci. 38 (2013) 1629-1652. https://doi.org/10.1016/j.progpolymsci.2013.05.008

[27] F. Wua, M. Misraa, A.K. Mohanty, Challenges and new opportunities on barrier performance of biodegradable polymers for sustainable packaging, Prog. Polym. Sci. 117 (2021) 101395. https://doi.org/10.1016/j.progpolymsci.2021.101395

[28] K. Kaewtatip, V. Tanrattanakul, W. Phetrat, Preparation and characterization of Kaolin/starch foam, Appl. Clay Sci. 80-81 (2013) 413-416. https://doi.org/10.1016/j.clay.2013.07.011

[29] A. Kwaśniewska, M.Świetlicki, A. Prószyński, G.Gładyszewski, The quantitative nanomechanical mapping of starch/kaolin film surfaces by peak force AFM, Polym. 13 (2021) 244. https://doi.org/10.3390/polym13020244

[30] J. Ruamcharoen, R. Munlee, P. Ruamcharoen, Improvement of water vapor barrier and mechanical properties of sago starch-kaolinite nanocomposite, Polym. Compos. 41 (2019) 201-209. https://doi.org/10.1002/pc.25360

[31] S. Wang, C. Li, L. Copeland, Q. Niu, S. Wang, Starch retrogradation: A comprehensive review, Comp. Rev. Food Sci. Food Saf. 14 (2015) 568-585. https://doi.org/10.1111/1541-4337.12143

[32] P. Müller, E. Kapin, E. Fekete, Effects of preparation methods on the structure and mechanical properties of wet conditioned starch/montmorillonite nanocomposite films, Carbohyd. Polym. 113 (2014) 569-576. https://doi.org/10.1016/j.carbpol.2014.07.054

[33] G. Harikrishnan, T.U. Patro, D.V. Khakhar, Polyurethane foam−clay nanocomposites: nanoclays as cell openers, Ind. Eng. Chem. Res. 45 (2006) 7126-7134. https://doi.org/10.1021/ie0600994

[34] M.K.S. Monteiro, V.R.L. Oliveira, F.K.G. Santos, E.L. Barros Neto, R.H.L. Leite, E.M.M. Aroucha, R.R. Silva, K.N.O. Silva, Incorporation of bentonite clay in cassava starch films for the reduction of water vapor permeability, Food Res. Int. 105 (2018) 637-644. https://doi.org/10.1016/j.foodres.2017.11.030

[35] S.P. Bangar, W.S. Whiteside, A.O. Ashogbon, M. Kumar, Recent advances in thermoplastic starches for food packaging: A review, Food Packag. Shelf Life 30 (2021) 100743. https://doi.org/10.1016/j.fpsl.2021.100743

[36] J. Shen, P. Fatehi, Y. Ni, Biopolymers for surface engineering of paper-based products, Cellulose 21 (2014) 3145-3160. https://doi.org/10.1007/s10570-014-0380-6

[37] K. Khwaldia, E. Arab-Tehrany, S. Desobry, Biopolymer coatings on paper packaging materials, Compr. Rev. Food Sci. Food Saf. 9 (2010) 82-91. https://doi.org/10.1111/j.1541-4337.2009.00095.x

[38] V. Rastogi, P. Samyn, Bio-based coatings for paper applications, Coatings 5 (2015) 887-930. https://doi.org/10.3390/coatings5040887

[39] C. Andersson, New ways to enhance the functionality of paperboard by surface treatment - a review, Packag. Technol. Sci. 21 (2008) 339-373. https://doi.org/10.1002/pts.823

[40] R. Bollström, R. Nyqvist, J. Preston, P. Salminen, M. Toivakka, Barrierproperties created by dispersion coating, Tappi J. 12 (2013) 45-51. https://doi.org/10.32964/TJ12.4.45

[41] F.A. Aouada, L.H.C. Mattoso, E. Longo, New strategies in the preparation of exfoliated thermoplastic starch-montmorillonite nanocomposites, Indus. Crops Products 34 (2011) 1502-1508. https://doi.org/10.1016/j.indcrop.2011.05.003

[42] K.M. Ardakani, A.H. Navarchian, F. Sadeghi, Optimization of mechanical properties of thermoplastic starch/clay nanocomposites, Carbohy. Polym. 79 (2010) 547-554. https://doi.org/10.1016/j.carbpol.2009.09.001

[43] C. Zeppa, F. Gouanve, E. Espuche, Effect of a plasticizer on the structure of biodegradable starch/clay nanocomposites: thermal, water-sorption, and oxygen-barrier properties, J. Appl. Polymer Sci. 112 (2009) 2044-2056. https://doi.org/10.1002/app.29588

[44] B. Chen, J.R.G. Evans, Thermoplastic starch-clay nanocomposites and their characteristics. Carbohydr. Polym. 61 (2005) 455-463. https://doi.org/10.1016/j.carbpol.2005.06.020

[45] D. Schlemmer, R. Angelica, M. J. A. Sales, Morphological and thermomechanical characterization of thermoplastic starch/montmorillonite nanocomposites, Comp. Struct. 92 (2010) 2066-2070. https://doi.org/10.1016/j.compstruct.2009.10.034

[46] H. Liu, D. Chaudhary, S. I. Yusa, M. O. Tade, Glycerol/starch/Na+-montmorillonite nanocomposites: A XRD, FTIR, DSC and 1H NMR study, Carbohyd. Polym. 83 (2011) 1591-1597. https://doi.org/10.1016/j.carbpol.2010.10.018

[47] C.M.O. Müller, J.B. Laurindo, F. Yamashita, Composites of thermoplastic starch and nanoclays produced by extrusion and thermopressing, Carbohy. Polym. 89 (2012) 504-510. https://doi.org/10.1016/j.carbpol.2012.03.035

[48] A.C. Souza, R. Benze, E.S. Ferrão, C. Ditchfield, A.C.V. Coelho, C.C. Tadini, Cassava starch biodegradable films: Influence of glycerol and clay nanoparticles content on tensile and barrier properties and glass transition temperature, LWT - Food Sci. Technol. 46 (2012) 110-117. https://doi.org/10.1016/j.lwt.2011.10.018

[49] P. Müller, E. Kapin, E. Fekete, Effects of preparation methods on the structure and mechanical properties of wet conditioned starch/montmorillonite nanocomposite films, Carbohyd. Polym. 113 (2014) 569-576. https://doi.org/10.1016/j.carbpol.2014.07.054

[50] H.M. Koivula, L. Jalkanen, E. Saukkonenb, S. S. Ovaska, J. Lahti, H. Christophliemk, K. S. Mikkonen, Machine-coated starch-based dispersion coatings prevent mineral oil migration from paperboard, Prog. Org. Coatings 99 (2016) 173-181. https://doi.org/10.1016/j.porgcoat.2016.05.017

Adv. App. of Micro and Nano Clay – Biopolymer-based Composites   Materials Research Forum LLC
Materials Research Foundations **125** (2022) 236-253   https://doi.org/10.21741/9781644901915-10

Chapter 10

Cellulose based Nano-Composites and Applications

Phumlani Tetyana[1]*, Nikiwe Mhlanga[1]

[1]DSI/Mintek Nanotechnology Innovation Centre, Advanced Materials Division, Mintek, 200 Malibongwe Drive, Hans Strydom, Randburg, South Africa

Phumlani.tetyana@gmail.com (indicate corresponding authors with a *)

Abstract

Cellulose is an abundant, naturally occurring and bio-degradable material that has been examined as a possible replacement for conventional materials such as plastics which are known to be toxic to the environment. Cellulose nanomaterials, which can be produced directly from cellulose, offer unique properties and structures that have proven useful for a myriad of applications globally. Currently, cellulose nanomaterials have found widespread use in numerous fields including the biomedical, pharmaceutical, packaging, and the food technology industries. Thus, this chapter reports on the properties of cellulose nanomaterials and cellulose nanocomposites and their use in various fields or industries.

Keywords

Cellulose, Nanomaterials, Nanocrystals, Nanofibers, Morphology

Contents

1.   Introduction

Literature has recorded an increase in the need for green, renewable and sustainable materials that can be used in the fabrication of superior products and industrial processes, with a low environmental impact [1]. These are intended to offer suitable alternatives to traditional petroleum-based plastics and related products across a myriad of applications. Cellulose is amongst the most widely researched biodegradable materials or polymers which have been earmarked as a possible alternative to the use of unpleasant materials [2,3,4].

Cellulose is a biopolymer that is largely found in plants (including wood and hemp) and microorganisms. In plants, cellulose forms the major structural component of the cell wall where it maintains stability, strength and rigidity. Various microorganisms are known to produce cellulose and include bacteria, algae and tunicates. Microorganisms, especially bacteria, produce cellulose for several reasons including to facilitate flocculation, to maintain suitable environmental conditions (aerobic conditions) and also to facilitate attachment to plants and other surfaces [5,6,7]. Plant based cellulose and cellulose produced in microorganisms are structurally similar but are different in terms of their composition or purity. Both plant cellulose and bacterial cellulose are in the form of polysaccharide chains which are arranged into microfibrils which are bundled into ribbons. Bacterial cellulose is regarded as the purest form of cellulose as it does not contain lignin which is mostly found in plant cellulose. Moreover, bacterial cellulose has been reported to have superior water retention properties compared to plant cellulose, thus allowing bacterial cellulose to have crystallinity [8,9,10].

Cellulose is a linear macromolecule composed of glucose monomers ($> 10\ 000$) adjoined by β-(1-4)-glycosidic bonds. This biopolymer is popular due to its superior properties such as biocompatibility, hydrophilicity and its non-toxic nature. Owing to these properties, cellulose has been used extensively, with an estimated production of approximately 1.5×10^{12} tons each year [11,12]. Similarly, the production of cellulose nanomaterials (nanocellulose) has increased immensely, with cellulose fibers or nanocrystals applied in

Adv. App. of Micro and Nano Clay – Biopolymer-based Composites   Materials Research Forum LLC
Materials Research Foundations **125** (2022) 236-253   https://doi.org/10.21741/9781644901915-10

composite materials. Nanocellulose does not only retain the excellent properties of cellulose but also amplifies some of the properties such as mechanical stiffness and specific surface area [13, 14, 15]. In this chapter, we explore cellulose nanomaterials that have been used in nanocomposites and their subsequent applications in various fields.

2.   Chemistry of cellulose

Anselme Payen, in 1838, first reported on an organic, fibrous, and tough, water-soluble polymer. The polymer known as cellulose constitutes the cell wall of plants and hence maintains the structure of plant cell walls. Plants, therefore, are the main source of cellulose and other living species such as algae, fungi, bacteria, invertebrates, amoeba, and sea animals e.g., tunicates [16, 17]. Table 1 summarizes the known sources of cellulose and examples of each category.

The structure of cellulose is characterized by linear β-1,4-glucopyranose units connected covalently by β-1,4-glycosidic bonds [18, 19]. The cellulose units are packed together into microfibrils that are bound by intramolecular H-bonds and intramolecular van der Waals forces. Each unit of the β-1,4-glucopyranose is corkscrewed 180 ° to its adjacent units [16, 17]. The cellobiose repeating units are glucose dimers. The glucopyranose unit has three highly reactive hydroxyl groups which proffer the cellulose's hydrophilicity, chirality, and biodegradability properties. Furthermore, the hydroxyl groups are enablers of the strong hydrogen bonds within the microfibrils and hence impact the crystallinity and highly cohesiveness nature of cellulose [16, 17]. Fig. 1 adapted from ref [16] depicts the structure of cellulose with non-reducing and reducing ends on the periphery of the cellobiose chain.

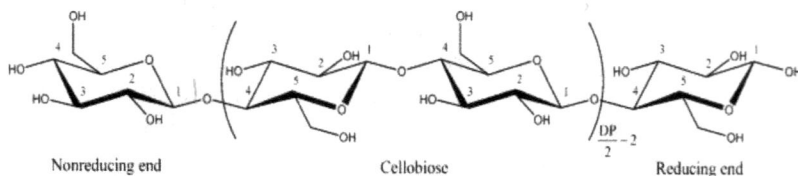

Figure 1. Chemical structure of cellulose. Figure reprinted from ref [16].

Adv. App. of Micro and Nano Clay – Biopolymer-based Composites   Materials Research Forum LLC
Materials Research Foundations **125** (2022) 236-253   https://doi.org/10.21741/9781644901915-10

Table 1. Sources of cellulose.

Source	Examples
Plants	Wood pulp, cotton fibers, jute, ramie, sisal, flax, hemp, water plants, grasses, leaves, stem, fruits, agricultural waste e.g., wheat, rice, straw, sugarcane bagasse, sawdust, and cotton stables.
Algae	Red, green, and yellow algae. However, yellow algae dominate – cladophorales (cladophora, chaetomorpha, Rhizoclonium, and Microdyction), Siphonocladiales (Valonia, Dictyosphaeria, Siphonocladus, boergesenia)
Bacteria	Komagataeibacterxylnus, Acetobacterxylinum [5] Rhizobium, agrobacterium, Psedomonas, and Salmonella
Tunicates	Ascidians, Larvaceans, Salps

3.   Synthesis and extraction of cellulose

Extraction of cellulose fibres is done in biosynthesis platforms, where cellulose individual molecules are converted into cellulose fibers. Fig. 2A pictorially summarises the process. Individual cellulose chains form an an-assemble of basic fibrils characterized by a diameter of 3.5 nm. This diameter can range from 2 - 20 nm depending on the source of the cellulose. Also, the biosynthesis procedure results in different packaging arrangements. The fibrils amalgamate to form microfibrils and this is catalyzed by the Van der Waal's forces and intra- and molecular hydrogen bond formation. The amalgamation results in the formation of highly crystalline and stable microfibrils and some less-ordered chains forming the amorphous regions (Fig. 2B) [17]. The highly ordered isolated regions are generated and isolated as nanocellulose [21]. Nanocellulose is classified into nanostructured materials and nanofibers. Microcrystalline and cellulose microfibrils are classified as nanostructured materials, while nanocrystals, nanofibrils, and bacterial cellulose (Fig. 2c) are nanofibers. Nanocellulose has garnered popularity in numerous applications because it has been reported to contain intrinsic properties such as a high aspect ratio, renewability, biocompatibility, high strength, low density, stability, and surface-functionalization [21].

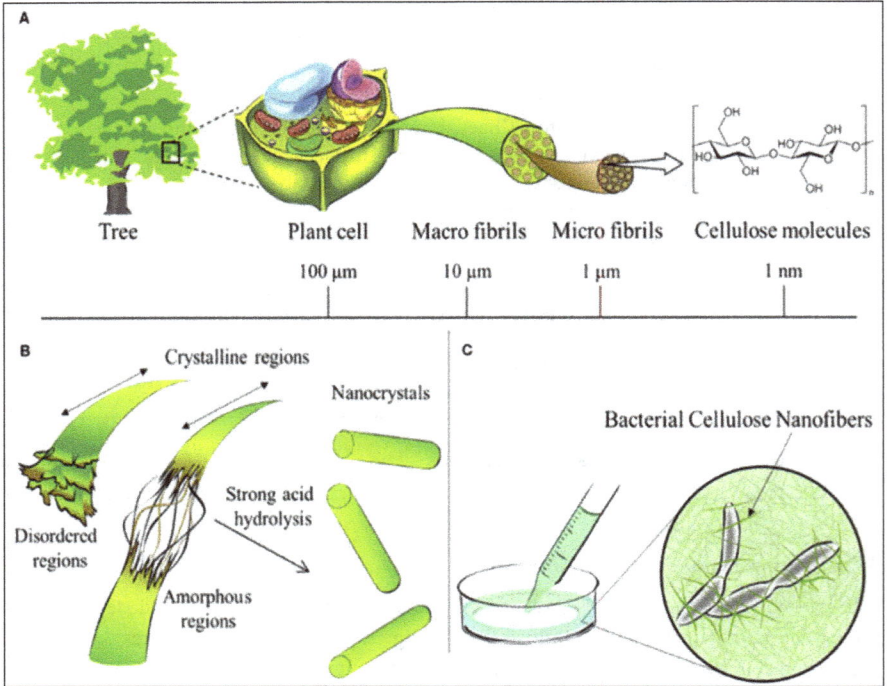

Figure 2. Adapted from ref [21] shows A: the preparation of cellulose molecules from a plant source. B: different regions of the microfibrils. C: bacterial cellulose nanofibers.

Cellulose exist in different polymorphs. The different polymorphs include: cellulose I, II, III_I, III_{II}, IV_I, IV_{II}. Cellulose I is the native polymorph and its also subdivided into Iβ characterized by a monoclinic $P2_1$ structure and Iα with a triclinic P1 structure [17, 18], based on hydrogen bonding patterns. The Iβ monoclinic is made up of two chains that enhance its stability while Iα has one chain [18]. The cellulose I polymorphs exist in different ratios depending on the source of the cellulose. i.e., algae and bacteria are dominated by Iα and higher plans have both polymorphs [16]. The second most studied of the cellulose polymorphs is cellulose II. Cellulose II is made from cellulose I via chemical regeneration and mercerization processes.

During the conversion, the hydroxyl group of the cellulose I rotate from a t_g to g_t conformation and hence cellulose II is a monoclinic $P2_1$ structure [16]. Cellulose II and I are precursors for cellulose III; their exposure to gaseous/liquid ammonia or various amines produces cellulose III. Cellulose I produce cellulose III$_I$while cellulose II affords cellulose III$_{II}$. The III$_I$ inherits a monoclinic P2$_1$structure and a $g_r$conformation of the hydromethyl group. The structure of Cellulose III$_{II}$is still missing and it's known as a disordered phase of cellulose. Cellulose III$_{I/II}$ is heated in glycerol at a temperature of 260°C to yield the IV$_I$ and IV$_{II}$cellulose polymorphs. However, cellulose IV has also been discovered in the primary cell walls of several plants [16].

4.   Cellulose based nanomaterials and nanocomposites

Cellulose has been identified as promising candidate in the development of nanocomposites due to its excellent properties. Cellulose nanomaterials or nanocellulose has emerged as the material choice in these nanocomposites. Nanocellulose is defined as cellulose that has a diameter in the nanometer range, less than 100 nm [22]. Nanocellulose can be grouped in three categories based on their nature, dimensions and the method of extraction used, as depicted in table 2. Cellulose nonofibers (CNFs) or nanofibrillated cellulose, cellulose nanocrystals (CNCs) (sometimes referred to as nanowhiskers) or cellulose nanoparticles, and bacterial cellulose (BC). Furthermore, the properties of these nanomaterials are highly dependent on the source of cellulose used and conditions employed during processing [23,24]

Table 2. Different types of cellulose nanomaterials and their average diameters.

Type	Dimensions	Synthesis Method & Starting Material
CNC	Diameter: 3 - 15 nm	Alkali treatment and acid hydrolysis; Wood
CNF	Diameter: 5 - 50 nm	Mechanical treatment; Wood & Tunicates
BC	Diameter: 10 - 75 nm	Biological treatment

4.1   Cellulose nanofibers

Cellulose nanofibers represent long, flexible and cross-linked form of nanocellulose which is less 100 nm in size. These materials have come out as promising candidates for reinforcements in polymers, as they have been observed to contain numerous properties that are divergent to those of normal cellulose. These include their reduced cost, outstanding mechanical properties, high aspect ratio, large specific surface area and low coefficient of thermal expansion [23,25]. Moreover, cellulose nanofibers are increasingly becoming popular due to their biocompatibility, biodegradability, ability to be recycled and also their environmentally friendly nature [26]. Cellulose nanofibers can be produced directly from cellulose. However, this is not a simple process as the fibers comprising cellulose are tightly bound together via hydrogen bonds. Various separation methods have been used to achieve this and include mechanical, chemical and chemo-mechanical methods [27,28].

Although cellulose nanofibers are known to be insoluble in aqueous solvents, they are dispersible in such solvents thus making it easier for them to be used in nanocomposite applications. These nanofibers appear translucent at smaller quantities and have a gel-like morphology. Cellulose nanofibers show great promise as biodegradable polymeric materials but have also been reported to have drawbacks that have prevented their industrial production up until now. One of the major drawbacks that have deterred the use of CNFs in many other applications is the need for use of excessive energy during their production. CNFs are produced via mechanical disintegration of cellulose microfibrils to form nanofibers [29,30]. However, this can now be circumvented through the use of pretreatment processes such as mechanical refining, enzymatic hydrolysis and catalytic oxidation to name a few. Following the pre-treatment process, cellulose microfibrils are then subjected to the mechanical disintegration process and many other treatments thereafter. To date, CNFs can be successfully produced using a combination of chemical and mechanical processes [31, 32].

4.2   Cellulose nanocrystals

Cellulose nanocrystals, also known as cellulose nanoparticles or nanowhiskers, are elongated crystalline needle-like nanometric particles ranging from 5 – 20 nm in width and 5 – 30 nm in length [33]. Cellulose nanocrystals are reported to be highly crystalline (up to 80%) and are composed mainly of 1β fraction structure. Raw cellulose contains amorphous regions which contain a low order of microfibrils and crystalline regions which have a high order of microfibrils [34]. Cellulose nanocrystals are produced through the splitting or separation of amorphous and crystalline domains of cellulose nanofibrils, with the crystalline region forming cellulose nanocrystals. This is achieved through a process known as acid hydrolysis, whereby the amorphous regions of native cellulose, which are most susceptible to acid attack, are hydrolyzed into solution whereas the crystalline region still remains in a solid form [35]

Adv. App. of Micro and Nano Clay – Biopolymer-based Composites   Materials Research Forum LLC
Materials Research Foundations **125** (2022) 236-253   https://doi.org/10.21741/9781644901915-10

Initially, the feed stock (cellulosic material) or raw cellulose undergoes purification and bleaching to remove all unwanted constituents such as lignin and hemicelluloses. Thereafter, cellulose is treated with a strong acid such as hydrochloric acid or sulfuric acid under carefully controlled conditions. Cellulose nanocrystals produced using sulfuric acid during hydrolysis have been reported to contain a net negative charge that has been reported to improve the phase durability of the resulting nanoparticles in solution. The degree of crystallinity and sizes of cellulose nanocrystals is greatly dependent upon the source of the cellulose material together with the conditions (concentration, temperature, agitation, and time) under which the preparation was conducted. One of the main sources of cellulose is cotton. This cellulosic material is known to produce nanocrystals with a degree of crystallinity ranging from 55 – 90%. One significant property of CNCs is the aspect ratio. This parameter is reported to be approximately 10 for nanocrystals produced from cotton. This is determined by the ratio of the nanocrystal length to its width and is an important indicator of the formation of the anisotropic phase and reinforcing properties [36,37,38].

4.3 Nanocomposite materials

A nanocomposite is defined as a material with enhanced properties resulting from the combination of two or more materials with differing physical and chemical properties, with one of the constituents having dimensions less than 100 nm. A nanocomposite differs from a normal composite in that they have distinct properties that emanate from their high surface area to volume proportion. Thus, the nanocomposite material results in properties that surpass that of individual components from which it is made. Also, nanocomposites are reported to involve a nanosized filler (usually nanoparticles), a matrix (to disperse the filler) and an interfacial region [39,40].

Cellulose nanomaterials are reported to have improved mechanical, thermal and structural attributes brought about by their nanometer size range. Cellulose and its nanomaterial counterparts can be used in nanocomposites in one of various ways. They can either be used as fillers in a polymer matrix, or as a matrix reinforced by various fillers, or can be used as both a matrix and a reinforcing filler at the same time. Typically, the various phases that make up a nanocomposite differ with regards to their chemical composition and properties and are separated by an interphase. Thus, the properties of the resulting composite material are a combination of the characteristics or properties of the constituent phases [26,41,42].

As indicated earlier, nanocellulose has a high aspect ratio together with better stiffness and strength which has been seen to give rise to enhanced mechanical properties when used as a reinforcing material with a polymer matrix. Use of cellulosic nanomaterials in polymer nanocomposites was first demonstrated by Favier [43] and colleagues in 1995. This gave rise to many subsequent studies which have increased enormously over the years [44]. As such, various nanocomposite materials which with cellulose nanomaterials been developed and many of them are currently undergoing commercialization [45].

5. Applications

Cellulose is amongst the most ubiquitous and abundant materials on earth. The application of cellulose materials across various disciplines has increased significantly, thus replacing the use of synthetic materials. Some of the applications that are using cellulose nanomaterials include the aerospace, automotive, textile packaging, printed electronics and the pharmaceutical industries to mention but a few.

5.1. Biofuel

About 71% of global anthropogenic greenhouse gas (GHG) emissions is from energy production which includes transportation. Fossil fuel causes adverse effect on the environment and carbon emissions and is depleting due to the increasing demands of the population [46]. To offset these issues, alternative renewable energy sources are envisioned. Biofuel is an attractive, sustainable, clean (green) and cost-effective energy source. Cellulose-derived biofuel offers less carbon emission in addition to the aforementioned biofuel advantages [47]. Cellulosic biofuel, the second generation of biofuel, derives ethanol-based cellulosic biofuel or biogas from cellulosic waste such as straw, forest residues, and urban litter. The cellulosic waste ethanol surpasses corn ethanol because it reduces GHG by about 55 %. Despite the urgency of the need for biofuel, the adoption of cellulosic and algae biofuel is much slower. Their poor economic viability undermines the large-scale and long-term sustainability. The cost of the cellulosic biofuel product negates the process. Cellulosic biomass has a low content of fermentable hydrocarbons and this affects the yield for ethanol production. Also, a costly pre-treatment of cellulose is required to break the lignin seal to access the hydrocarbons [48]. Uncertainties of the biomass-fuelled by climate changes and investment risk have also delayed the cellulosic biofuel sector [49]. Governments must mitigate these aforementioned challenges into a sustainable green world.

5.2 Biomedical and pharmaceutical applications

Several biomedical applications: tissue regeneration, tissue repair/implants, drug delivery, biosensing, hemodialysis membranes, absorbable hemostats, biocatalyst, and anti-bacterial benefit from cellulose [21]. Cellulose is applied in its pristine format or engineered with other chemical groups to prepare composites with enhanced properties suitable for biomedical applications. Drug delivery systems are characterized by improved solubility, sustained drug release, targeting or specificity, reduced clearance, drug stability, and therapeutic effect. Nanocelluloce, especially cellulose nanocrystals (CNCs) are candidates for this application owing to their properties [21].

Bacterial nanocellulose (BNC) has unique properties; high purity, large surface area, high porosity, high water absorption capacity, and high mechanical strength. As a result, it is classified as a type of hydrogel owing to its high water content. The BNC hydrogels are converted into aerogels via freeze-drying. Both gels are exceptional candidates for biological scaffolds used as wound dressers, tissue engineering, scaffolds, drug delivery, and artificial blood vessel repair [50].

The skin fundamentally controls body temperature, maintains water and electrolytes balance, prevents infections, allows gas exchange, and senses feelings. Injury/cut on the skin hampers its functionality and wound dressers are used to cover and promote healing of the skin. Biomedical dressers promote moisture in the environment, block infections, insulate, allow gas movements and should have low adhesion to the wound. BNC qualified by its non-toxicity, porosity, liquid adsorption, flexibility, biocompatibility is a suitable wound dresser material [50].

5.3. Food technology

Food technology also benefits from the abundant polysaccharide cellulose. Cellulose and its derivatives CNCs, nanofibers, and BNC due to their properties are useful in the food technology sectors: food packaging, functional foods, and food stabilizers. Food packaging remains essential for protection, preservation of nutrition capacity, and food quality [51]. CNCs are the best material for food packaging and their crystallinity and network structure proffer an excellent oxygen barrier.

The world has embraced processed foods. However, they suffer from instability with prolonged storage, e.g., fluids will normally thin and separate (water phase and the other constituents). Stabilizers mitigate the instability of the processed foods and several of them have been produced e.g. xanthan gum. Nonetheless, cellulose is the cheaper alternative to the stabilizers because they naturally emulsify and stabilizes. Cellulose composites such as hydroxyethyl are used in ice cream, frozen milk, and milk drinks as a stabilizer to extend shelf-life.

Although processed foods have saturated the world, enriched food for healthy processes referred to as functional foods are gaining ground. Functional foods prevent diseases, i.e., dietary fiber foods that resolve digestive issues. Cellulose advantaged by its low calorie is added to food to improve its fiber content [47]. Obesity is one of the challenges of our time and low calorie/energy foods are gaining attention to solve the obesity issue [51].

Although CNCs have a vast potential in food and technology, a few concerns still require attention. Their unknown properties at the nanoscale are a concern especially with human and environmental exposure; pending regulatory issues; establishment of their health benefits; improvement of their yield and investigation of their human gut fermentation process [51].

5.4   Sensors

This section explores the usability of cellulose in the preparation of sensors, particularly humidity sensors. An ultimate humidity sensor is characterized by the subsequent attributes: a wide detection range, high sensitivity, rapidity, stability, and accuracy. Humidity sensors are of paramount importance in industrial manufacturing, food and drug storage/packaging, medical health, climatology, agriculture, daily life, and aviation [52, 53]. Cellulose qualified by its humidity responsiveness is ideal in the manufacturing of green-humidity sensors [52]. Cellulose and its derivatives are endowed with –OH functional groups which proffer good water retention and swelling performance [52]. This induces changes in its chiral nematic pitch and reflected wavelength. The technique borrows from the nature of the *Tmesisternus isabellae* insect [54]. High-performance humidity sensors can be fabricated from composites of cellulose with polymers/carbon materials. Sun et al [53] coupled CNCs with hydrazone groups modified poly(N-isopropyl acrylamide) (PNIPAM) copolymers to prepare a dual, humidity, and heat sensor. Wang et al [54] fabricated a capacitivewood-derived cellulose nanopapers humidity sensor. The humidity sensors exhibited ultrahigh sensitivity, fast response, small hysteresis, and more importantly, a wide working range of relative humidity. The cellulose-derived humidity sensors achieve sensing using different mechanisms; resistance, gravity, optical fiber, microwave, colorimetry, capacitance, and shape deformation. Table 3 summarizes the sensing mechanism of humidity sensors.

Table 3. Various sensing mechanisms of humidity sensors.

Humidity Sensor	Mechanism Principle	Advantages	Examples of Studies
Resistance sensor	Based on the changes in the material's resistance with exposure to external force, heat, and humidity.	Simple architecture, high precision, convenient reading,	Han et al [55]

Adv. App. of Micro and Nano Clay – Biopolymer-based Composites   Materials Research Forum LLC
Materials Research Foundations **125** (2022) 236-253   https://doi.org/10.21741/9781644901915-10

Capacitance sensor	Convert humidity changes of material into capacitance variations.	Reduced energy usage and hysteresis, increased linear response, affordable	Ducéré et al [56]
Colorimetry sensor	Based on Colour change.	Convenient, rapidity, real-time detection, high sensitivity, simple, field operation capacity.	Kumar et al [57]
Gravity sensors	Established according to the association of frequency shift and mass change. The changes of a material's mass with the absorption of water molecules are transformed into frequency changes.	Very high sensitivity, affordable, excellent stability, rapid response.	Dai et al [58]
Shape deformation sensors	Based on the response of some plants like ice plants and pinecones to humidity.		Wang et al [59]
Optical sensors	Changes of light optical properties	Enhanced sensitivity, and selectivity, remoter sensing, real-time detection, corrosive resistance, etc	Hsieh et al [60]
Microwave sensors	Dielectric changes with humidity variations	The packaging doesn't interfere with reading, therefore most suitable for crops.	Eyebe et al [61]

References

[1] K. Dhali, M. Ghasemlou, F. Daver, P. Cass, B. Adhikari, A review of nanocellulose as a new material towards environmental sustainability, Science of the Total Environment 775 (2021) 145871. https://doi.org/10.1016/j.scitotenv.2021.145871

[2] J.H. Song, R.J. Murphy, R. Narayan, G.B.H. Davies, Biodegradable and compostable alternatives to conventional plastics, Philos Trans R Soc Lond B Biol Sci. 364 (2009) 2127-2139. https://doi.org/10.1098/rstb.2008.0289

[3] A. Kołodziejczyk, Bacterial cellulose: Multipurpose biodegradable robust nanomaterial, In: A. Sand, S. Banga (Eds.), Cellulose Science and Derivatives, IntechOpen, 2021 https://doi.org/10.5772/intechopen.98880

[4] C.L. Reichert, E. Bugnicourt, M-B. Coltelli, P. Cinelli, A. Lazzeri, I. Canesi, F. Braca, B.M. Martínez, R. Alonso, L. Agostinis, S. Verstichel, L. Six, S. De Mets, E.C. Gómez, C. Ißbrücker, R. Geerinck, D.F. Nettleton, I. Campo, E. Sauter, P. Pieczyk, M. Schmid, Bio-based packaging: materials, modifications, industrial applications and sustainability, Polymers, 12 (2020) 1558-1593. https://doi.org/10.3390/polym12071558

[5] S. Rongpipi, D. Ye, E.D. Gomez, E.W. Gomez, Progress and opportunities in the characterization of cellulose - an important regulator of cell wall growth and mechanics, Front. Plant. Sci. 9 (2019) 1894-1922. https://doi.org/10.3389/fpls.2018.01894

[6] R. Zhong, Z.H. Ye, Secondary cell walls: biosynthesis, patterned deposition and transcriptional regulation, Plant Cell Physiol. 56 (2015) 195-214. https://doi.org/10.1093/pcp/pcu140

[7] C. Brigham, Biopolymers: Biodegradable Alternatives to Traditional Plastics, in: B. Torok, T. Dransfield (Eds.), Green Chemistry: An Inclusive Approach, Elsevier Inc., New York, 2018, pp. 753-770 https://doi.org/10.1016/B978-0-12-809270-5.00027-3

[8] R. Naomi, R.B.H. Idrus, M.B. Fauzi, Plant- vs. bacterial-derived cellulose for wound healing: a review, Int. J. Environ. Res. Public Health 17 (2020) 6803 https://doi.org/10.3390/ijerph17186803

[9] L.R. Lynd, P.J. Weimer, W.H. van Zyl, I.S. Pretorius, Microbial cellulose utilization: fundamentals and biotechnology, Microbiol Mol Biol Rev. 66 (2002) 506-577. https://doi.org/10.1128/MMBR.66.3.506-577.2002

[10] H. Seddiqi, E. Oliaei, H. Honarkar, J. Jin, L.C. Geonzon, R.G. Bacabac, J. Klein-Nulend, Cellulose and its derivatives: towards biomedical applications, Cellulose 28 (2021) 1893-1931. https://doi.org/10.1007/s10570-020-03674-w

[11] S.P. Gautam, P.S. Bundela, A.K. Pandey, Jamaluddin, M.K. Awasthi, S. Sarsaiya, A review on systematic study of cellulose, J. Appl. & Nat. Sci. 2 (2010) 330-343. https://doi.org/10.31018/jans.v2i2.143

[12] S. Kalia, A. Dufresne, B.M. Cherian, B.S. Kaith, L. Averous, J. Njuguna, E. Nassiopoulos, Cellulose-based bio- and nanocomposites: a review, Int. J. Polym. Sci. 2011 (2011) 1-35 https://doi.org/10.1155/2011/837875

[13] A. Dufresne, Nanocellulose, From nature to high performance tailored materials, Walter de Gruyter GmbH, Berlin/Boston (2012). https://doi.org/10.1515/9783110254600

[14] D. Trache, A.F. Tarchoun, M. Derradji, T.S. Hamidon, N. Masruchin, N. Brosse, M.H. Hussin, Nanocellulose: from fundamentals to advanced applications, Front. Chem. 8 (2020) 392-425. https://doi.org/10.3389/fchem.2020.00392

[15] A. Isogai, Development of completely dispersed cellulose nanofibers, Proc. Jpn. Acad., Ser. B 94 (2018) 161-179. https://doi.org/10.2183/pjab.94.012

[16] Y. Habibi, L.A. Lucia, O.J. Rojas, Cellulose nanocrystals: chemistry, self-assembly, and applications, Chem. Rev, 110 (2010) 3479-3500. https://doi.org/10.1021/cr900339w

[17] J. George, S. Sabapathi, Cellulose nanocrystals: synthesis, functional properties, and applications. Nanotechnol. Sci. Appl, 8 (2015) 45-54 https://doi.org/10.2147/NSA.S64386

[18] F-W. Bai, S. Yang, and N.W. Ho, Fuel ethanol production from lignocellulosic biomass, In: M. Moo-Young (Ed.), Comprehensive biotechnology, Elsevier, 2019, Pages 49-65. https://doi.org/10.1016/B978-0-444-64046-8.00150-6

[19] C. Brigham, Biopolymers: biodegradable alternatives to traditional plastics, In: B. Torok, T. Dransfield (Eds.), Green Chemistry, Elsevier, 2018, Pages 753-770. https://doi.org/10.1016/B978-0-12-809270-5.00027-3

[20] I.M. Saxena, R. Brown Jr, Biosynthesis of cellulose, In: N. Morohoshi, A. Komamine (Eds.), Progress in Biotechnology, Elsevier, 2001, Pages 69-76. https://doi.org/10.1016/S0921-0423(01)80057-5

[21] D. Trache, A.F. Tarchoun, M. Derradji, T.S. Hamidon, N. Masruchin, N. Brosse, M.H. Hussin, Nanocellulose: from fundamentals to advanced applications. Front. Chem., 8 (2020) 392-425. https://doi.org/10.3389/fchem.2020.00392

[22] P. Phanthong, P. Reubroycharoen, X. Hao, G. Xu, A. Abudula, G. Guan, Nanocellulose: extraction and application, Carbon Resour. Convers. 1 (2018) 32-43 https://doi.org/10.1016/j.crcon.2018.05.004

[23] K. Zhang, A. Barhoum, C. Xiaoqing, H. Li, P. Samyn, Cellulose nanofibers: fabrication and surface functionalization techniques, in: A. Barhoum, H. Li, P. Samyn (Eds.), Handbook of nanofibers, Psringer Nature, Switzerland, 2019, pp. 409-449. https://doi.org/10.1007/978-3-319-53655-2_58

[24] K.J. Nagarajan, N.R. Ramanujam, M.R. Sanjay, S. Siengchin, B.S. Rajan, K.S. Basha, P. Madhu, G.R. Raghav, A comprehensive review on cellulose nanocrystals and cellulose nanofibers: Pretreatment, preparation, and characterization, Polym. Compos. 42 (2021) 1588-1630. https://doi.org/10.1002/pc.25929

[25] A. Sharma, M. Thakur, M. Bhattacharya, T. Mandal, S. Goswami, Commercial application of cellulose nano-composites - A review, Biotechnol. Rep. (2018). https://doi.org/10.1016/j.btre.2019.e00316

[26] M.M. Khattab, N.A. Abdel-Hady, Y. Dahman, Cellulose nanocomposites: Opportunities, challenges, and applications, In: M. Jawaid, A.H.P.S. Khalil, S. Boufi (Eds.), Cellulose-Reinforced Nanofibre Composites: Production, Properties and Applications, Woodhead Publishing, 2017, Pages 483-516. https://doi.org/10.1016/B978-0-08-100957-4.00021-8

[27] I.P. Mahendra, B. Wirjosentono, Tamrin, H. Ismail, J.A. Mendez, Thermal and morphology properties of cellulose nanofiber from TEMPO-oxidized lower part of empty fruit bunches (LEFB), Open Chem. 17 (2019) 526-536. https://doi.org/10.1515/chem-2019-0063

[28] S.J. Eichhorn, A. Dufresne, M. Aranguren, N.E. Marcovich, J R. Capadona, S.J. Rowan, C. Weder, W. Thielemans, M. Roman, S. Renneckar, W. Gindl, S. Veigel, J. Keckes, H. Yano, K. Abe, M. Nogi, A.N. Nakagaito, A. Mangalam, J. Simonsen, A.S. Benight, A. Bismarck, L.A. Berglund, T. Peijs, Review: current international research into cellulose nanofibres and nanocomposites, J Mater Sci. 45 (2010) 1-33. https://doi.org/10.1007/s10853-009-3874-0

[29] J.A. Sirviö, M. Lakovaara, A fast dissolution pretreatment to produce strong regenerated cellulose nanofibers via mechanical disintegration, Biomacromolecules. 22 (2021) 3366–3376. https://doi.org/10.1021/acs.biomac.1c00466

[30] S. Jonasson, A. Bunder, T. Niittyla, K. Oksman, Isolation and characterization of cellulose nanofibers from aspen wood using derivatizing and non-derivatizing

pretreatments, Cellulose, 27 (2020) 185-203. https://doi.org/10.1007/s10570-019-02754-w

[31] E. Espinosa, F. Rol, J. Bras, A. Rodríguez, Production of lignocellulose nanofibers from wheat straw by different fibrillation methods. Comparison of its viability in cardboard recycling process, J. Clean. Prod., 239 (2019) 118083-118091. https://doi.org/10.1016/j.jclepro.2019.118083

[32] J. Zeng, Z. Zeng, Z. Cheng, Y. Wang, X. Wang, B. Wang, W. Gao, Cellulose nanofbrils manufactured by various methods with application as paper strength additives, Scientifc Reports 11 (2021) 11918. https://doi.org/10.1038/s41598-021-91420-y

[33] T.C. Mokhena, M.J. John, Cellulose nanomaterials: new generation materials for solving global issues, Cellulose. 27 (2020) 1149-1194. https://doi.org/10.1007/s10570-019-02889-w

[34] H. Zhao, J.H. Kwak, Z.C. Zhang, H.M. Brown, B.W. Arey, J.E. Holladay. Studying cellulose fiber structure by SEM, XRD, NMR and acid hydrolysis, Carbohydr. Polym. 2 (2007) 235-241. https://doi.org/10.1016/j.carbpol.2006.12.013

[35] D. Zielinska, K. Szentner, A. Waskiewicz, S. Borysiak, Production of Nanocellulose by Enzymatic Treatment for application in polymer composites, Materials. 14 (2021) 2124-2150. https://doi.org/10.3390/ma14092124

[36] E.C. Ramires, A. Dufresne, Cellulose nanoparticles as reinforcement in polymer nanocomposites, In: F. Gao (Eds.), Advances in polymer nanocomposites. Types and applications, Woodhead Publishing Limited, 2012, pp. 131-163. https://doi.org/10.1533/9780857096241.1.131

[37] J. Rojas, M. Bedoya, Y. Ciro, Current Trends in the Production of Cellulose Nanoparticles and Nanocomposites for Biomedical Applications, In: M. Poletto (Eds.), Intechopen, 2015. https://doi.org/10.5772/61334

[38] R.J. Moon, A. Martin, J. Nairn, J. Simonsen, J. Youngblood, Cellulose nanomaterials review: structure, properties and nanocomposites, Chem. Soc. Rev. 40 (2011) 3941-3994 https://doi.org/10.1039/c0cs00108b

[39] I. Siro, D. Plackett, Microfibrillated cellulose and new nanocomposite materials: a review, Cellulose 17 (2010) 459-494. https://doi.org/10.1007/s10570-010-9405-y

[40] M.A. Hubbe, O.J. Rojas, L.A. Lucia, M. Sain, Cellulosic nanocomposites: a review, BioResources. 3(2008) 929-980. https://doi.org/10.15376/biores.3.3.929-980

[41] E. Roduner, Size matters: why nanomaterials are different, Chem. Soc. Rev.35 (2006) 583-592. https://doi.org/10.1039/b502142c

[42] J. Blessy, V.K. Sagarika, S. Chinnu, K. Nandakumar, T. Sabu, Cellulose nanocomposites: Fabrication and biomedical applications, Journal of Bioresources and Bioproducts 5 (2020) 223-237. https://doi.org/10.1016/j.jobab.2020.10.001

[43] V. Favier, G.R. Canova, J.Y. Cavaillé, H. Chanzy, A. Dufresne, C. Gauthier, Nanocomposite materials from latex and cellulose whiskers, Polymers for Advanced Technologies. 6 (1995) 351-355 https://doi.org/10.1002/pat.1995.220060514

[44] K-Y. Lee, Y. Aitomäki, L.A. Berglund, K. Oksman, A. Bismarck, On the use of nanocellulose as reinforcement in polymer matrix composites, Composite Science and Technology. 105 (2014) 15-27. https://doi.org/10.1016/j.compscitech.2014.08.032

[45] D. Feldman, Cellulose Nanocomposites, Journal of Macromolecular Science, Part A: Pure and Applied Chemistry 52 (2015) 322-329. https://doi.org/10.1080/10601325.2015.1007279

[46] N.E. Mahbub, E. Gemechu, H. Zhang, A. Kumar, The life cycle greenhouse gas emission benefits from alternative uses of biofuel co-products, Sustain. Energy Technol. Assess., 34 (2019) 173-186. https://doi.org/10.1016/j.seta.2019.05.001

[47] P.K. Gupta, S.S. Raghunath, D.V. Prasanna, P. Venkat, V. Shree, C. Chithananthan, S. Choudhary, K. Surender, K. Geetha, An update on overview of cellulose, its structure, and applications, In: A.R. Pascual, M.E.E. Martin (Eds.), Cellulose, Intechopen, 2019, Pages 846-1297.

[48] L. Li, Y. Ge, M. Xiao, Towards biofuel generation III+: A sustainable industrial symbiosis design of co-producing algal and cellulosic biofuels, J. Clean. Prod., 306 (2021) 127144-12758. https://doi.org/10.1016/j.jclepro.2021.127144

[49] B.P. Sharma, T.E. Yu, B.C. English, C.N. Boyer, J.A. Larson, Impact of government subsidies on a cellulosic biofuel sector with diverse risk preferences toward feedstock uncertainty, Energy Policy, 146 (2020) 111737-111747. https://doi.org/10.1016/j.enpol.2020.111737

[50] W. Liu, H. Du, M. Zhang, K. Liu, H. Liu, H. Xie, X. Zhang, C. Si, Bacterial cellulose-based composite scaffolds for biomedical applications: a review, ACS Sustain. Chem. Eng., 8 (2020) 7536-7562. https://doi.org/10.1021/acssuschemeng.0c00125

[51] R. Mu, X. Hong, Y. Ni, Y. Li, J. Pang, Q. Wang, J. Xiao, Y. Zheng, Recent trends and applications of cellulose nanocrystals in food industry, Trends Food Sci Technol., 93 (2019) 136-144. https://doi.org/10.1016/j.tifs.2019.09.013

[52] Z. Li, J. Wang, Y. Xu, M. Shen, C. Duan, L. Dai, Y. Ni, Green and sustainable cellulose-derived humidity sensors: A review, Carbohydr. Polym., (2021) 118385. https://doi.org/10.1016/j.carbpol.2021.118385

[53] C. Sun, D. Zhu, H. Jia, K. Lei, Z. Zheng, X. Wang, Humidity and heat dual response cellulose nanocrystals/poly (N-isopropylacrylamide) composite films with cyclic performance, ACS Appl. Mater. Interfaces, 11 (2019) 39192-39200. https://doi.org/10.1021/acsami.9b14201

[54] Y. Wang, S. Hou, T. Li, S. Jin, Y. Shao, H. Yang, D. Wu, S. Dai, Y. Lu, S. Chen, Flexible capacitive humidity sensors based on ionic conductive wood-derived cellulose nanopapers, ACS Appl. Mater. Interfaces, 12 (2020) 41896-41904. https://doi.org/10.1021/acsami.0c12868

[55] J.W. Han, B. Kim, J. Li, M. Meyyappan, Carbon nanotube-based humidity sensor on cellulose paper, J. Phys. Chem. C., 116 (2012) 22094-22097. https://doi.org/10.1021/jp3080223

[56] V. Ducéré, A. Bernès, C. Lacabanne, A capacitive humidity sensor using cross-linked cellulose acetate butyrate, Sens. Actuators B Chem, 106 (2005) 331-334. https://doi.org/10.1016/j.snb.2004.08.028

[57] P. Kumar, A. Ghosh, D.A. Jose, A simple colorimetric sensor for the detection of moisture in organic solvents and building materials: applications in rewritable paper and fingerprint imaging, Analyst, 144 (2019) 594-601. https://doi.org/10.1039/C8AN01042K

[58] L. Dai, Y. Wang, X. Zou, Z. Chen, H. Liu, Y. Ni, Ultrasensitive physical, Bio, and chemical sensors derived from 1-, 2-, and 3-D nanocellulosic materials, Small, 16 (2020) 1906567. https://doi.org/10.1002/smll.201906567

[59] M. Wang, X. Tian, R.H. Ras, O. Ikkala, Sensitive Humidity-Driven Reversible and Bidirectional Bending of Nanocellulose Thin Films as Bio-Inspired Actuation, Adv. Mater. Interfaces, 2 (2015) 1500080-1500087. https://doi.org/10.1002/admi.201500080

[60] C.Y. Hsieh, C.D. Liao, W.C. Wang, Evanescent wave sensor using cellulose nanocrystals composite fiber coating for humidity measurement, Proc. SPIE, 11233 (2020) 1123302. https://doi.org/10.1117/12.2545580

[61] G.F.V.A. Eyebe, B. Bideau, É. Loranger, F. Domingue, TEMPO-oxidized cellulose nanofibre (TOCN) films and composites with PVOH as sensitive dielectrics for microwave humidity sensing, Sens. Actuators B Chem., 291 (2019) 385-393. https://doi.org/10.1016/j.snb.2019.04.070

Chapter 11

HNT-Cellulose based Nano-Composite and Applications

Anirudh Pratap Singh Raman[1], Prashant Singh[2], Pallavi Jain[1]*

[1]Department of Chemistry, SRM Institute of Science and Technology, Delhi-NCR Campus,Modinagar, Ghaziabad, 201204, India

[2]Department of Chemistry, Atma Ram Sanatan Dharma College, New Delhi-110021, India

*palli24@gmail.com

Abstract

Halloysite, a naturally occurring nanoclay has unique characteristics like nanometric size range, tubular structure, high biocompatibility, low cost and opposite charges on its surface is driving attraction as a versatile and important component of biomaterials. The incorporation of functionalized Halloysite nanotubes (HNT) with the cellulose biopolymer is to obtain enhanced and desirable properties like mechanical strength, and barrier properties and to have controlled pore size. The enhanced properties of bio-nano composites are employed in wide areas including food packaging, the automobile industry, drug delivery and many more. In this chapter, extended work on various HNT-cellulose-based bio- nanocomposites is provided along with their structural, and functional properties including their applications.

Keywords

Halloysite, Cellulose Bio-Nanocomposite, Food Packaging, Nanoclay, HNT

Contents

1.   Introduction

In recent years, the adverse effect of environmental issues accompanying the huge disposal of plastics. The other various non-biodegradable wastes are pushing researchers to look into biodegradable and renewable materials. Another concern is meeting the requirement of fresh food for theconsumer is a concern to be investigated. The quality of food is not only judged by the "expiry date" on the packet, although it might be affected by various parameters like temperature, mishandling, etc. Bio-nanotechnology is the interdisciplinary branch that deals with the development of nanomaterials with the help of material science, nanotechnology and biotechnology to find solutions to current problems. Recently, these nanomaterials have received great attention from researchers in diverse areas. Bio-nanocomposites are the material formed with the combination of biopolymers and inorganic solids at the nano scale. Among the biopolymers, cellulose-based biomaterials are the promising alternative as they are abundant, non-toxic and edible. The structure of cellulose is semicrystalline providing mechanical strength and flexibility.

Mostly, these cellulosic biomaterials consist of waste materials containing more than 50% cellulose content. Cellulose has been used for the manufacturing of packaging films, drug delivery, tissue engineering, gene therapy, automobile industry, etc. Incorporating reinforcement agents into the matrix is the most prominent way to enhance the physical properties of nanocomposite films. Clays are natural, biocompatible, non- toxic and biodegradable materials. Halloysite nanotubes (HNT) are the naturally available aluminum silicate nanotubes having a similar structure as kaolin. The extensive use of HNT in the production of ceramics, crucible products, or thin-walled porcelain is due to its high-temperature resistant properties and high L/D ratio. Due to their unique properties such as economical, nanoscale lumens, researchers for their application in wide areas. HNT has dominant morphology of hollow tubular structure and nanoscale dimensions, resembling carbon nanotubes (CNT). HNT shows high dispersion in polymer matrix due to the surface having siloxane moieties and some hydroxy (-OH) group anticipating the forming hydrogen bond. Moreover, the higher stiffness and high length/diameter ratio (L/D) aspect ratio of tubes favors excellent nano reinforcement of the matrix of the polymer. The study on HNT-based nanocomposite has been done for several decades due to the features like hydrophobicity, ion exchange and nanotubular structure. The application of these materials in drug delivery has been done by various scientists who observed the chemical stability, cytotoxicity, loading and releasing of drugs. This chapter is focused on the HNT-cellulose-based nanocomposites and their applications.

2.   Bio-nanocomposites

Bio-nanotechnology is the interdisciplinary branch that deals with the development of nanomaterials with the help of material science, nanotechnology and biotechnology. Bio-nanocomposites are the material formed with the combination of biopolymers and inorganic solids at a nano scale. These hybrids are extremely versatile as they are formed from the different biopolymers (polysaccharides, polypeptides, proteins and nucleic acids, etc). These bio-nano composites are biodegradable, biocompatible compounds that make them highly useful in advanced bio-medical applications such as tissue engineering, gene therapy or artificial bones. Other possible applications are dependent on the properties like thermal, mechanical and barrier making this class attractive for potential use in a controlled amount delivery of drugs and pesticides, water purification, food packing and oxygen barrier films. Additionally, these nanocomposites are designed for Pickering emulsion to meet the present demand for eco-friendly materials. Moreover, these nanocomposites are designed for Pickering emulsion that is to meet the current demand for eco-friendly materials using particles [1].

2.1 Structural enhancement

Biopolymers have flexible and semi-flexible primary, second and tertiary structures. The intra and intermolecular bonding, as well as folding in the polymer structure, have boosted the mechanical and functional performance, allowing it to be used in various applications. For instance, silk II has high strengthdue to intermolecular bonding which directs the self-polymerization of poly-(Gly-Ala) into β- nanosheet. These β-nanosheets are responsible for making spider silk fiber with higher tensile strength and expansibility when these are covalently combined with the silk I polymorph domains. Similarly, a nano cellulose structure is stabilized by the intermolecular van der Waal forces and hydrogen bonding. The degree of crystallinity and elastic moduli is very high. For example, nano-cellulose is extensively used for increasing the mechanical strength and used for embedding a wide selection of nanomaterials like carbon matrices, and synthetic polymers. Graphene papers have low mechanical strength due to weak intermolecular interaction within the laminated structure. The worth mentioning disadvantage inthe processing of nanocomposites is that they face critical challenges in their uncontrolled aggregation and sometimes even deteriorate mechanical performance. On contrary, to common synthetic nanomaterials, biomaterials have a variety of functional groups that allowed multiscale hydrogen bonding, π-π interaction, hydrophobic-hydrophilic interactions and van derWaals interaction. Further, we have discussed the most popular cellulose bio-polymer andtheir interaction with HNT which are recently explored for the fabrication of nanocomposites, including metallic and mineral nanocomposites [2].

2.2 Environmental benefits

The bio nanocomposites can be easily degraded into their constituent building blocks. Therefore, the perceptive is biodegradable, renewable and characteristics of green synthesis can remarkably reduce the emission of carbon dioxide and hazardous waste. To improve the properties and green benefits, various nanocomponents possess complementary and strong interactions with biopolymers that are incorporated into the biopolymer matrix procuring desired functionalities and mechanical strength [3]. Composting can degrade biodegradable waste into soil improvement material. For instance, the decomposition of HNT-based cellulose nanocomposite completely degraded into the soil within a few monthsof composting and exhibit high water adsorption and enhanced properties by the filler. Although the composting requires basic components like carbon, nitrogen, water and oxygen. The rate of degradation increases with the increase in the molecular weight of nanocomposites[4].

Adv. App. of Micro and Nano Clay – Biopolymer-based Composites   Materials Research Forum LLC
Materials Research Foundations **125** (2022) 254-274   https://doi.org/10.21741/9781644901915-11

3.   Halloysite nanotubes

In the present study, halloysite nanotubes are of immense significance to polymers. Halloysite is a foremost member of the kaolin group of nano-clay having a structural composition of $Al_2Si_2O_5(OH)4n.H_2O$ where 'n' varies according to the group. According to the hydration state, HNT is classified into two categories, Hydrated form has the crystalline structure of 10 Å with d_{001} spacing and 7 Å d_{001} spacing in dehydrated form. The interlayer of water in the HNT acts as a special feature that distinguishes it from the kaolinite class. In the dehydration, the d_{001} spacing goes from 10 Å to 7 Å with the irreversible process. The high use of HNT is in the production of ceramics, crucible products, or thin-walled porcelain due to its high-temperature resist property and high L/D ratio. HNT has dominant morphology of hollow tubular structure and nanoscale dimensions, resembling carbon nanotubes (CNT). HNT shows high dispersion in polymer matrix due to the surface having siloxane moieties and some hydroxy group an anticipating formation of the hydrogen bond. Moreover, the higher stiffness and high length/diameter aspect ratio of tubes favor excellent nano- reinforcement in the matrix of the polymer. The study-based nanocomposites have been done for several decades due to the features like hydrophobicity, ion exchange and nanotubular structure [5]. The applications of these materials in drug delivery have been done by many scientists who observed the chemical stability, cytotoxicity, loading and releasing of drugs [6-9]. It is also used in catalytic, optical, magnetic and electrical studies, and energy storage [10]. In many cases, HNT is incorporated into the matrix of polymer without any chemical alteration. The smectic and kaolin have varying compositions and structural morphology which lead to their applications in different areas. Many researchers published work on the properties of nanocomposites. They summed up organically modified nanocomposites with HNT and worked on the passive and impassive properties of PHBV nanocomposites derived from the melting process. Researchers have worked on the heat and flame-retardant property of polymeric material with HNT. The combining process of HNT with other biomaterials works on the properties like heat sink barrier effect and flame retardant in the condensed phase. The next step of this process during combustion involves the development of an inorganic layer on the surface, which prevents the material from exposure to heat and delay the mass transfer taking place during the transformation of the condensed and gas phase. For developing nanoclay composites, they preferred PES due to its high T_g and low thermal stability of surfactant. However, due to the environmental difficulties associated with the excessive use of organic solvents, scaling up to industrial manufacturing is not feasible. [11]. Therefore, it is requisite to find an eco-friendlier alternative way to make temperature-resistant PES nanocomposites. For producing a cost-effective and highly compatible PES composite system by using conventional melt processing technique [12, 13].

4.   Halloysite biocompatibility

The increasing attentiveness to HNT nanomaterial and its application on large scale results in accumulation in the environment, in consequence, leads to damage to human and vegetation health. The phototoxicity study of halloysite was experimentally investigated through a hazardous assessment model for concluding the impact of HNT on the *Raphanus sativus L* plant [14].

Several works of the literature concluded that halloysite being biocompatible material, the toxicity and uptake of halloysite have been explored on human cells including hepatic cancer cells, thyroid cancer cells, breast cancer and adenocarcinoma cells. It comes up with that it is non-toxic and less harmful than sodium chloride. The activity of HNT can be enhanced by the introduction of functional moieties such as thiazolium salt anti-proliferative activity. For example, chitosan doped with halloysite formed scaffold for tissue engineering shows no adverse side effects on fibroblast [15].

5.   HNT-cellulose based nanocomposites

5.1   Carboxymethyl cellulose-based HNT nanocomposite

L.F. Wanga and J.W. Rhim in their work did acidification of halloysite to develop carboxymethyl cellulose-based nanocomposites films. The preparation of HNT-A was conducted by absorbing it into the saturated solution of different metal salts including copper acetate, silver nitrate and zinc nitrate. The fraction of metal ions linked and their performance completely depends on their metal ions. The functionalized HNT-A showed high antibacterial activity against food-borne pathogenic bacteria. Moreover, it is a promising material to be utilized as a reinforcing filler for the manufacturing of nanocomposite films, which can be used in food packing or biomedical applications. CMC nanocomposites integrated with functionalized nanocomposites showed an increase in mechanical, thermal stability and barrier and anti-bacterial activity. The change in the morphology of functionalized HNT-A obtained was an enlargement in the diameter of the lumen because of the transformation in the structure into porous nanorods due to the partial loss in the tubular morphology after the removal of the alumina group. The functionalized HNT has highzeta potential with metal ions as compared to simple halloysite due to the subjection of the surface with silica groups. All the CMC- based films prepared were transparent, flexible, and free-standing. The CMC/ HNT-A film exhibit less agglomeration which indicates thefilm was highly compatible due to an increase in the surface charge density that helps in the dispersion in the matrix as observed with zeta potential reports. Liu et. al. observed thatthe hydroxy group of biopolymeric interacts with the Si-O bond of HNT by forming hydrogenbonding. Since the acidification of HNT enhanced the number and strength of hydrogen bonding with uniform distribution [16].

5.2 Carrageenan based HNT nanocomposites

A considerable amount of plastic waste is added to marine through rivers. And this adequate amount of plastic accumulates in the marine ecosystem, degrading after several years or millennia in the water. So, this topic has become a seriousconcern of studies and the utmost topic for the abundance of microplastic particles. In thisregard, many researchers work on developing various bio-based polymers as biodegradable alternatives. Hence carrageenan comes as an attractive polysaccharide for producingbiodegradable packaging materials. Carrageenan being biocompatible, renewable andhaving good film-forming property attract great interest in forming nanofilms with nanofillers [17] HNT is the prominent nanofiller but the drawback is that it lacks functionality.There are various methods of functionalization of the surface of HNT. Researchers have worked on the functionalization of silver-exchanged zeolite (NaY) using 3-amino propyl trimethoxy silane (APTES) with amine that could improve the activity against bacteria against *S. aureus* and *E. coli* [18]. Silvers show the strong antibacterial activity against various bacteria easily incorporated into the surface of HNT. Saedi et. al. workedon the preparation of silver nanoparticles by reducing silver nitrate using ascorbic acid and adsorbing on the surface of HNT and then treating with sodium dodecyl sulfate (SDS). The nanofiller used HNT-AgNP for preparing carrageenan-based nanocomposite. This incorporation increases the mechanical strength, transparency, UV barrier property and antibacterial activity. The unique physical and functional properties make the use of thesenanocomposite films in food packing [19].

5.3 Chitosan based HNT polymer

Pullulan is a material used in packing, biomedical, coating and cosmetic emulsions as it is non-toxic, biodegradable, and biocompatible. Despite these advantages, it is soluble in water which leads to its limitation in the application of water treatment and also it has high cost and low mechanical strength. Hence, this pullulan grafted with chitosan forms a semipermeable polymer network whichovercomes its limitation of solubility in water and enhances mechanical strength. Chitosanis a natural derivative of polysaccharides having β-(1-4)-linked D-glucosamine and N- acetyl-D glucosamine which is formed by the deacetylation of chitin. It attracts great attention in the preparation of film and food packing areas due to its excellent film formability,non-toxicity, transparency and antibacterial properties. In the reported work the synthesis of nanocomposite film by integrated

Adv. App. of Micro and Nano Clay – Biopolymer-based Composites   Materials Research Forum LLC
Materials Research Foundations **125** (2022) 254-274   https://doi.org/10.21741/9781644901915-11

polymerization of pullulan and chitosan in the 50:50ratios [20][21]. The addition of a reinforcing agent during the preparation of chitosan-basednanocomposite film enhances the thermal, mechanical, water resistance, and gas barrier properties. Liu et. al. reported high transmittance in the visible light chitosan-basednanocomposite film with HNT. Similarly, interfacial interactions were developed betweenchitosan and HNT resulting in CHT-added films [22]. On the contrary, rutin films formedwere yellowish due to the strong absorption in the UV region and low transmittance. The UVlight absorption function is due to the presence of the aromatic -OH group, which has raised the excitation of an electron from n- π^* absorbance at 200-400nm [23]. From the reports by T280 and T660 respectively, it is understandable that rutin plays a significant role in blocking the visible and UV rays through absorption.

5.4   Functionalized HNT with sodium alginate

Sodium alginate is a naturally occurring polysaccharide compound comprising α-L-guluronic acid (M) and β-D-mannuronic acid (G). This is derived from the cell wall of brown algae. Sodium alginate has attractive properties such as less toxicity, low-cost polymer and biocompatible. Sodium alginate can be functionalized using a different approach for application in biomedical like drug delivery, wound repair, cell proliferation, tissue regeneration, cell transplantation and restoring tissue [24]. Poly(vinyl alcohol) incorporated HNT is a biodegradable nanocomposite having promising properties such as hydrophilic, water-soluble, biocompatible and non-toxic which makes it useful for tissue engineering applications [25]. Scientists have worked on the synthesis of bio-nanocomposite films using HNT, PVA and alginate. Additionally, functionalized HNT is reinforced in the matrix of PVA as a dopant that makes it more dispersive. The evaluation of physicochemical properties of prepared bio-composite films was done. Moreover, the evaluation of cell proliferation and cell adhesion properties was tested through cell line study (NH3T3) and their hemocompatibility using erythrocytes cells. Literature review proved that there was no work published on the functionalization of HNT with sodium alginateapplicable in tissue engineering that arouses to do the study. Bio-nanocomposites are an enormous application in cell proliferation, gene therapy and tissue engineering [26].Sodium alginate does not possess cell-binding nature and thus it is important to modify the structure by crosslinking with other suitable biomaterials.

5.5   β-galactosidase-halloysite

Immobilization makes it easy to recover, and reuse enzymes and also increased stability that allowing prolonged use. B-galactosidase immobilized into the matrix of HNT with the simple adsorption process at 55 °C temperature. At first, bacterial cellulose nanocrystals (BCNC) were prepared by acid hydrolysis and mechanical treatment of bacterial cellulose.

BCNC is utilized as a stabilizer for food, cosmetics, and pharmaceuticals due to its promising properties such as non-toxic, good wettability, porous structure with a large surface area, and hydrophilic. The BCNC properties can be enhanced with the entrapment of HNT-BGAL which reduces the leakage of the enzyme during storage and improves reusability. The novel combination of halloysite nanotubes and cellulose nanocrystals has interesting advantages and the immobilization process is rapid, easy and non-toxic [27].

5.6   PET-halloysite nanotubes

The increase in consumption of plastic in the packaging material affects health and is environmentally hazardous. With the fluctuating oil prices and greater consumption, there is an increasing demand for bio-plastic for packaging. Recently bio-poly terephthalic acid (PET) was manufactured using ethylene glycol is using as the packaging material. It is made up of 30% monoethylene glycol obtained from the sugar molasses and the left amount made from terephthalic acid. The motive of the researcher is to substitute with bio-based PET. The research on PET is limited due to the difficulties in the preparation of PET nanocomposites by melt compounding method that leads to easy degradation of polymer chain at high temperature. The sustainable method for melt mixing has been done with high-energy ball milling (HEBM)at ambient temperature. The preparation of PET/HNTs nanocomposites involves mechanical energy to promote the mixing of two or more species at normal temperatures. Thefunctionalization of HNT was done using sodium benzoate and PET nanocomposite incorporated into halloysite nanofiller. The evaluation of the thermal, morphology, mechanical and crystallization properties of HNT effect on PET. The fundamental analysisof antimicrobial activity revealed the efficiency of the PET/HNTs-Benzoate system as a promising bio-material in active-packaging applications [28].

6.   Applications

HNT-Cellulose-based nanocomposites have wide applications in different areas. Some of the applications of HNT-Cellulose-based nanocomposites are demonstrated in figure 1.

6.1   Food: packing and edible coating

Although the demand for fresh food is high yet it's not available to all the people who are far from the source of production due to their short life. Therefore, it is highly recommended to invent new technologies to prevent stale and increase the life of fresh food. The early deterioration of the food results in consumer health, economic loss and impact on the environment. The quality of fresh food is maintained with various methods such as water activity, cold chain transportation systems, use of preservatives and pH. Food packing plays a vital role in maintaining the quality of food and safety [29]. Active food packing is

Adv. App. of Micro and Nano Clay – Biopolymer-based Composites   Materials Research Forum LLC
Materials Research Foundations **125** (2022) 254-274   https://doi.org/10.21741/9781644901915-11

the promising key to the improvement of fresh food life and the continuous release of antimicrobial agents proposed by the Food and Agriculture Organisation of the United Nation. The combination of HNT and cellulose-based nanocomposite is favorable in remodelingboth the mechanical and barrier properties of films. The hydroxy group of biopolymeric interacts with the Silica oxide bond of HNT by forming hydrogen bonding. Since the acidification of HNT enhanced the number and strength of hydrogen bonding with uniformdistribution. Jaewon Jong et. al. developed ethylene nanocomposite film by the integration of functionalized HNT into an LDPE matrix. The functionalization of HNT was done using potassium permanganate. The properties of P-HNT/LDPE nanocomposite like thermal stability, morphology, and mechanical were studied [30]. In addition, the impact of these nano-composites films on the difference in the quality of cherry tomatoes was analyzed. The HNT-AgNP was used as a nanofiller for preparing carrageenan-based nanocomposite prepared by Saedi et al. 2020. This incorporation increased the mechanical strength, transparency, UV barrier property and antibacterial activity. The unique physical and functional properties make the use of these nanocomposite films in food packing. PET/HNT-based nanocomposite was developed as a promising type of active packing. The PET is incorporated into functionalized HNT benzoate system with the help of a high-energy ball milling process at ambient temperature without solvent. The efficacy against antibacterial activity was analyzed and the results suggested that the PET/HNT benzoate system shows antimicrobial activity and can be used in active packaging applications.

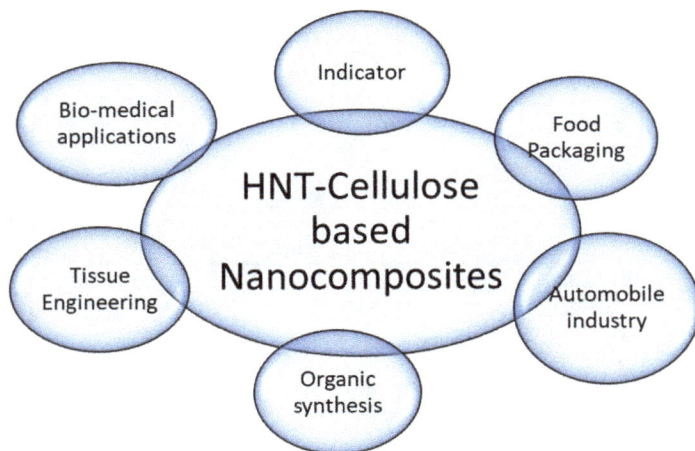

Fig. 1 Diagrammatic representation of various applications

Table 1. Application of HNT as nanofiller in food packaging

HNT origin	Matrix	Barrier Properties	Mechanical Properties	Ref.
USA	Potato Starch	Water and gas permeability reduced Contact angle increased.	TS increase and EB decreased	[1]
France	Starch	Enhance thermal stability	TB increased by 30.75% and EB decreased by 4.02%	[3]
New Zealand and Australia	Pectin	Even dispersion	TS increased	[3]
USA	Corn starch	WVP decreased	TS increase and EB decreased	[3]
India	Starch and Chitosan	WVTR values increased	TS increased b y 7.12% and EB decreased by 0.15%	[3]
China	Potato Starch	N/A	TS raised 55.5%	[9]
China	Chitosan	VWP and thermal stability increased	TS and EB increased by 53.41%	[9]
USA	Chitosan	WVP decreased by 13%	EB decreased 50% by 30% HNT	[11]
USA	Carrageenan and gelatin	Thermal stability	TS enhance with 17.75%	[20]
China	Chitosan	Light transmittance reduced	EB and TS increased by 7.5% MPa and 3% MPa, respectively	[22]
USA	Alginate	WVP decreased	TS increased with ZnO	[24]
England	PVA	Enhanced water barrier	TS and EB increased	[25]
China	Chitosan	N/A	TS increased 135%	[49]
Malaysia	Polylactic acid	Low water and oxygen properties	TS increased and EB decreased	[50]
New Zealand	PA and Starch	WVP decreased by 735.9%	N/A	[51]

6.2   Coating

The coating is the active process in food packaging used to increase the time limit and the efficiency of food substances due to the direct contact. The process of dipping is well known to be the universal method, as the direct dipping of food into the solution favors the complete rinsing of the substance [31]. The process helps in preserving the quality of food and increased the shell life by lowering the rate of ripening. The solution composing Eos impregnated within cyclodextrin and HNTs and coated on the packaging boxes was synthesized. The results observed that the treatment of packaging material increased the quality and life of tomatoes which are inobedience with physiochemical quality data [32]. Fibers can be manufactured by electrospinning in the presence of an electric field. Electrospinning is the method of producing continuous nanoscale fibrous films. It is a low-cost, effective and versatile technique utilized for the fashioning of packaging material with antimicrobial and antioxidant properties. One of the literature concluded that adding 15% of the weight of HNT enhances the tensile strength of nano-composite from 2.33 MPa to 7.38MPa of composite fiber. Another example of nano-fiber is chitosan with HNT and Fe_3O_4 nanoparticles, these nanofibers were widely used as adsorbents. Additionally, the compositemembrane exhibit efficient antibacterial activity against some food pathogenic bacteria such as *E. coli* and *S. aureus [33]*.

6.3   Drug delivery

In the biomedical area, nanotechnology has advanced to the point that it can currently detect and treat a wide range of diseases. . Few examples are carbon, dextran, phospholipids, polyethylene glycol, and chitosan is incorporated with other materials.The dissolved drug is entrapped and then assimilated into the matrix of nanoparticles as the size range responsible for the biodistribution of the particles and enhanced the activityof the drug, therefore, known as a good drug carrier. The drug is loaded in the hydrophobicpart of the matrix while the hydrophilic part occluded opsonization that helps the movement making easier. The application of nanoparticles in the branch of oncology has been boomed. They are directly concentrated on the infectious sites. Recent reports saidthat HNT has been proven biocompatible, non-toxic and has effective encapsulation capacity on cell culture [34]. In particular, studies, where the cytotoxicity of HNT was investigatedagainst breast cells and fibroblasts after incubation for 48 hours. The reports revealed that it is less harmful than sodium chloride salt. Ganguly et. al. worked on the alginate-based hydrogels for drug delivery systems [35]. The surface of the HNT was functionalized withself-polymerized dopamine and incorporated into alginate hydrogels. PDA coated HNT filler acts as both chelating agent and loading of diltiazem as calcium channel broker by insitu method. Recently the study of the preparation of HNT porous based nanocomposite with the sodium tripolyphosphate (STPP) and modified HNT as a cross-linking compound. The suspension of HNT or modified HNT was mixed with oxidized starch for preparing chitosan starch nanocomposite hydrogels. These hydrogels had the potential to serve as a drug delivery system. One of the studies is done using HNT with cellulose urea solution to synthesize hydrogels using epichlorohydrin crosslinking at high temperatures. The properties of HNT composite hydrogels such as mechanical, microstructure, cytocompatibility, swelling properties and behavior of delivery of drug were investigated [36]. The addition of HNT into the solution raised the viscosity. The mechanical properties of resulting nanocomposite hydrogels are high as compared to cellulose hydrogels. The composite hydrogels have mechanical strength with HNT 66.7% is 128KPa while pure cellulose hydrogels have 29.8 KPa. The morphological structure was investigated by X-ray diffraction and FTIR which showed there is no chemical change in the structure of HNT and cellulose fiber in the nanocomposite. The cytotoxicity of these hydrogels was investigated against MC3T3-E1 cells and MCF-7 cells. The SEM results revealed that composite hydrogels have a permeable structure having nano-pores of the size of 400 μm. Further, hydrogels are loaded with curcumin through physical adsorption [37]. The cytotoxicity of curcumin-loaded hydrogels is explored on cancer cells and has found prominent applications such as drug delivery in cancer and anti-inflammatory wound dressing [38].

6.4   Tissue engineering

Tissue engineering is an interdisciplinary area that involves combining cells, engineering and physiochemical methods to revamp and maintain living tissues. The area involves looking into waste materials and optimizing those for the application area. Polysaccharides materials and their nanocomposites like gums xanthan emerged as they are biocompatible, as well as have excellent tunable rheological and mechanical behavior [39]. These materials are favorable for the formation of hydrogels scaffold. Xanthan and Gellan-based hydrogels have a 3D microstructure that allows the attachment with the cells, and their proliferation and differentiation for proper regeneration, or revamping of organs. The gellan and xanthan-based nanocomposite biomaterials are helpful in tissue engineering including skin, neural, retinal, cartilage, bone, intervertebral disc, or other tissue [40]. The major challenge we are facing at present time is bone regeneration. Many researchers approach artificial substitutes, autografts and xenograft methods for restoring the function and structure of damaged bone. The application of bacterial cellulose was investigated for bone repair or regeneration involving tissue engineering techniques. Dorozhkin [41] reported that the material needed for the bone graft should be of varying sizes, shapes, porosity and composition as the human skeletal system differs in size and shape depending on the body function and location. The various techniques for preparing polymer-based

Adv. App. of Micro and Nano Clay – Biopolymer-based Composites   Materials Research Forum LLC
Materials Research Foundations **125** (2022) 254-274   https://doi.org/10.21741/9781644901915-11

scaffolds involve self-assembly, electrospinning, phase separation and chemical polymerization of aniline reported. Abshar Hasan [42] worked on the development of a highly efficient polymeric nanocomposite with governed pore size, composition and mechanical strength. They worked on the carboxylated CNWs decorated with silver nanoparticles providing antibacterial activity and mechanical strength. These scaffolds comprising chitosan and carboxylated cellulose with different percentages were prepared by the freeze-drying method. The morphology study was done using a different technique such as XRD, and FESEM which revealed nanocomposite has a high crystalline structure and interaction between CCNW and AgNPs are confirmed by FTIR. The nanocomposites added during preparation give desirable pore structure and improved mechanical strength and the range of these carboxylated CNWs is similar to the mechanical strength of cancellous bone. The decrease in the swelling capacity and scaffold degradation was adapted to vascularization and angiogenesis. Along with antimicrobial activity, it also showed MG63 cell adhesion and proliferation as well as improved biomineralization for bone growth. All the literature confirmed the excellent fabricated scaffold to overcome bone problems [43].

6.5   Indicator

Indicators are substance that indicates the change in pH in reaction or food quality. Alizarin is one of the indicators obtained from the roots of the madder plant. It has been used as an indicator in many varieties of industries, colorant, food industries, the reagent in tissue chemistry, or inorganic chemistry as an analytical agent. The transfer of proton between the hydroxyl group and carbonyl group through hydrogen bonding results in the color change. Alizarin has been utilized to synthesize the pH indicator films with various biopolymers including chitosan, starch, and cellulose nanofibers [44]. Alizarin exhibits excellent properties like antioxidant activity, definite pH change detection and UV blocking, hydrophobicity and thermal stability when incorporated into chitosan polymer. Although having many advantages, the negative point is that it can't show reversible color changeability. When the acid-base reaction occurs, the pH repeatedly changed, their color change was irreversible and did not obtain the same color. Hence, there is a need to enhance the reproducibility of color change. The CMC and CNF films were produced having distinguishable colors. The CMC films were transparent while CNF films were translucent. However, these films when impregnated with alizarin produced uniformly distributed orange-brown color under the formation of hydrogen bonding between the -OH group and a phenolic group of cellulosic material and alizarin. The quality of food is not only judged by the "expiry date" on the packet, although it might be affected by various parameters like temperature, mishandling, etc. Therefore, there is a need for working on pH-sensitive nanofilms [45]. Intelligent packaging and preservatives can be used to maintain the quality of food till they reach the consumer. Somework has been done on the film developed by combining PVA, starch, and anthocyanin. They prepared PSPE or RCE films of different concentrations having distinguishable colorand pH buffer. Remarkably, PSPE films exhibit better mechanical properties, color changeand low transmittance. Anthocyanin is abundant content and has excellent properties, the PS-PSPE film was picked and combined to form films and monitored against shrimp freshness and revealed the change of color during the spoilage of shrimp spoilage. HNT- based cellulose nanocomposites are highly preferential for forming the pH indicator filmsthat help in determining the quality of food majorly utilized in the food industries in the packaging of food [46].

6.6   Automotive application

Growing research and awareness in the area of nanocomposites have been reviewed about the development of these materials with polymer composites. Nanotechnology plays an important in enhancing the qualities of existing materials, which have been in application in the automobile industry for 25 years due to the attractive properties and performance such as strength, thermal stability, dispersion, tensile strength and stiffness. Many scientists have been investigating the enhancement of distinctive properties such as heat resistance, physical properties and mechanical properties with recyclable and biodegradable polymers like oil palm, and bamboo. Buchholz used polymer nanocomposites in the General Motors Safari as "step assists" as these composites have hardness, strength and high scratch resistance, reduced weight and rustproof that increase the life span of vehicles [47]. The recent work on the material has been used in the interior of the cars. In this process, the fiber was proposed to the pressed mats and then for giving shape it was processed into carding machine, and then impregnated with polylactic acid resin then the hot-pressing process turns into the final product. The results satisfied the requirement of automotive manufacturers as the material has high mechanical properties and low volatile organic emissions. Besides, it is also used in the manufacturing of blades of turbines. The PLA-based nanocomposites are known for the highest flexural strength and rigidity, thermal stability, impact strength and ductility. However, the previous study reveals there is a compromise between high rigidity/and ductility. The properties of PLA-based nanocomposite are similar to the PP used in automotive. However, they both are different and allow the comparison of evaluation of the mechanical performance of bio-based nanocomposites [48]

Conclusion and future perceptive

In this chapter, we focused on the recent research on the bio-based polymers their structural design after being impregnated with HNT, fabrication and applications of nanocomposites derived from natural biopolymers mostly focusing on the cellulose due totheir abundance, low toxicity, high mechanical strength and can produce complementary interfaces with other materials with functionalization. The combination of functional halloysite nanotubes with cellulose materials results in nanocomposites bearing excellent characteristics such as structural enhancement, high-performance multi functionalities, etc.[49]. Furthermore, the functionalization of the surface of the HNT through the incorporation of metallic nanoparticles increased its chemical and physical properties suchas catalytic ability, and electrical conductivity for prospective applications in biomedicine, sensing, energy harvesting and structural nanomaterials. Thinking about future perceptivethis chapter gives the solution to many problems of our time, enabling the preservation of food,enhancing drug delivery at a specific site, and tissue engineering. Many other various bio- nanocomposites play a vital role in conveying the challenges in near time as actuators andsensors, self-healing bones, and drug delivery [50].

References

[1]   F. Sadegh-Hassani, A. Mohammadi Nafchi, Preparation and characterization of nanocomposite films based on potato starch/halloysite nanoclay, Int. J. Biol. Macromol. 67 (2014) 458-462. https://doi.org/10.1016/j.ijbiomac.2014.04.009

[2]   V. Vergara, E. Abdullayev, Y.M. Lvov, A. Zeitoun, R. Cingolani, R. Rinaldi, S. Leporatti, Cytocompatibility and uptake of halloysite clay nanotubes, Biomacromolecules. 11 (2010) 820-826. https://doi.org/10.1021/bm9014446

[3]   Z.W. Abdullah, Y. Dong, Biodegradable and water resistant poly(vinyl) alcohol (PVA)/starch (ST)/glycerol (GL)/halloysite nanotube (HNT) nanocomposite films for sustainable food packaging, Front. Mater. 6 (2019). https://doi.org/10.3389/fmats.2019.00058

[4]   Z. Rozynek, T. Zacher, M. Janek, M. Čaplovičová, J.O. Fossum, Electric-field-induced structuring and rheological properties of kaolinite and halloysite, Appl. ClaySci. 77-78 (2013) 1-9. https://doi.org/10.1016/j.clay.2013.03.014

[5]   D.A. Dean, T. Ramanathan, D. Machado, R. Sundararajan, Electrical impedance spectroscopy study of biological tissues, J. Electrostat. 66 (2008) 165-177. https://doi.org/10.1016/j.elstat.2007.11.005

[6]   J.H. Kim, H. Eguchi, M. Umemura, I. Sato, S. Yamada, Y. Hoshino, T. Masuda, I. Aoki, K. Sakurai, M. Yamamoto, Y. Ishikawa, Magnetic metal-complex-conducting copolymer core-shell nanoassemblies for a single-drug anticancer platform, NPG Asia Mater. 9 (2017). https://doi.org/10.1038/am.2017.29

[7]   Y. Zhang, H. Yang, Halloysite nanotubes coated with magnetic nanoparticles, Appl. Clay Sci. 56 (2012) 97-102. https://doi.org/10.1016/j.clay.2011.11.028

[8]   D.G. Shchukin, G.B. Sukhorukov, R.R. Price, Y.M. Lvov, Halloysite nanotubes as biomimetic nanoreactors, Small. 1 (2005) 510-513. https://doi.org/10.1002/smll.200400120

[9]   F. Xie, J. Bao, L. Zhuo, Y. Zhao, W. Dang, L. Si, C. Yao, M. Zhang, Z. Lu, Towardhigh-performance nanofibrillated cellulose/aramid fibrid paper-based composites viapolyethyleneimine-assisted decoration of silica nanoparticle onto aramid fibrid, Carbohydr. Polym. 245 (2020). https://doi.org/10.1016/j.carbpol.2020.116610

[10]   K. Yoo, A. Deshpande, S. Banerjee, P. Dutta, Electrochemical Model for Ionic Liquid Electrolytes in Lithium Batteries, Electrochim. Acta. 176 (2015). https://doi.org/10.1016/j.electacta.2015.07.003

[11]   Y. He, W. Kong, W. Wang, T. Liu, Y. Liu, Q. Gong, J. Gao, Modified natural halloysite/potato starch composite films, Carbohydr. Polym. 87 (2012) 2706-2711. https://doi.org/10.1016/j.carbpol.2011.11.057

[12]   Y. Liu, X. Wang, X. Gao, J. Zheng, J. Wang, A. Volodin, Y.F. Xie, X. Huang, B. Van der Bruggen, J. Zhu, High-performance thin film nanocomposite membranes enabled by nanomaterials with different dimensions for nanofiltration, J. Memb. Sci.596 (2020). https://doi.org/10.1016/j.memsci.2019.117717

[13]   M. Rahimnejad, M. Ghasemi, G.D. Najafpour, M. Ismail, A.W. Mohammad, A.A. Ghoreyshi, S.H.A. Hassan, Synthesis, characterization and application studies of self-made Fe 3O4/PES nanocomposite membranes in microbial fuel cell, Electrochim. Acta. 85 (2012) 700-706. https://doi.org/10.1016/j.electacta.2011.08.036

[14]   L. Bellani, L. Giorgetti, S. Riela, G. Lazzara, A. Scialabba, M. Massaro, Ecotoxicityof halloysite nanotube-supported palladium nanoparticles in Raphanus sativus L, Environ. Toxicol. Chem. 35 (2016) 2503-2510. https://doi.org/10.1002/etc.3412

[15]   D. Czarnecka-Komorowska, K. Bryll, E. Kostecka, M. Tomasik, E. Piesowicz, K. Gawdzińska, The composting of PLA/HNT biodegradable composites as an eco-approach to the sustainability, Bull. Polish Acad. Sci. Tech. Sci. 69 (2021).

[16]   L.F. Wang, J.W. Rhim, Functionalization of halloysite nanotubes for the preparationof carboxymethyl cellulose-based nanocomposite films, Appl. Clay Sci. 150 (2017) 138-146. https://doi.org/10.1016/j.clay.2017.09.023

[17]   P. Fathiraja, S. Gopalrajan, M. Karunanithi, M. Nagarajan, M.C. Obaiah, S. Durairaj, N. Neethirajan, Response surface methodology model to optimize concentration of agar, alginate and carrageenan for the improved properties of biopolymer film, Polym.

Bull. (2021). https://doi.org/10.1007/s00289-021-03797-5

[18]   N.A.N. Nik Malek, S.A. Mohd Hanim, Antibacterial Activity of Amine-Functionalized Silver-Loaded Natural Zeolite Clinoptilolite, Sci. Lett. 15 (2021). https://doi.org/10.24191/sl.v15i1.11790

[19]   S. Saedi, M. Shokri, J.W. Rhim, Preparation of carrageenan-based nanocomposite films incorporated with functionalized halloysite using AgNP and sodium dodecyl sulfate, Food Hydrocoll. 106 (2020). https://doi.org/10.1016/j.foodhyd.2020.105934

[20]   M. Akrami-Hasan-Kohal, M. Ghorbani, F. Mahmoodzadeh, B. Nikzad, Development of reinforced aldehyde-modified kappa-carrageenan/gelatin film by incorporation of halloysite nanotubes for biomedical applications, Int. J. Biol. Macromol. 160 (2020) 669-676. https://doi.org/10.1016/j.ijbiomac.2020.05.222

[21]   M. Dash, F. Chiellini, R.M. Ottenbrite, E. Chiellini, Chitosan - A versatile semi-synthetic polymer in biomedical applications, Prog. Polym. Sci. 36 (2011) 981-1014. https://doi.org/10.1016/j.progpolymsci.2011.02.001

[22]   D. Huang, Z. Zhang, Y. Zheng, Q. Quan, W. Wang, A. Wang, Synergistic effect of chitosan and halloysite nanotubes on improving agar film properties, Food Hydrocoll. 101 (2020). https://doi.org/10.1016/j.foodhyd.2019.105471

[23]   S. Roy, J.-W. Rhim, Effect of chitosan modified halloysite on the physical and functional properties of pullulan/chitosan biofilm integrated with rutin, Appl. Clay Sci. 211 (2021) 106205. https://doi.org/10.1016/j.clay.2021.106205

[24]   S. Shankar, S. Kasapis, J.W. Rhim, Alginate-based nanocomposite films reinforcedwith halloysite nanotubes functionalized by alkali treatment and zinc oxide nanoparticles, Int. J. Biol. Macromol. 118 (2018) 1824-1832. https://doi.org/10.1016/j.ijbiomac.2018.07.026

[25]   S. Kouser, S. Sheik, A. Prabhu, G.K. Nagaraja, K. Prashantha, J.N. D'souza, M.K. Navada, D.J. Manasa, Effects of reinforcement of sodium alginate functionalized halloysite clay nanotubes on thermo-mechanical properties and biocompatibility of poly (vinyl alcohol) nanocomposites, J. Mech. Behav. Biomed. Mater. 118 (2021). https://doi.org/10.1016/j.jmbbm.2021.104441

[26]   J.A. Burdick, M.M. Stevens, Biomedical hydrogels, in: Biomater. Artif. Organs Tissue Eng., Elsevier Inc., 2005: pp. 107-115. https://doi.org/10.1533/9781845690861.2.107

[27]   S. Tizchang, M.S. Khiabani, R.R. Mokarram, H. Hamishehkar, N.S. Mohammadi,

Y. Chisti, Immobilization of β-galactosidase by halloysite-adsorption and entrapmentin a cellulose nanocrystals matrix, Biochim. Biophys. Acta - Gen. Subj. 1865 (2021). https://doi.org/10.1016/j.bbagen.2021.129896

[28]   G. Gorrasi, Dispersion of halloysite loaded with natural antimicrobials into pectins: Characterization and controlled release analysis, Carbohydr. Polym. 127 (2015) 47-53.https://doi.org/10.1016/j.carbpol.2015.03.050

[29]   F. Topuz, T. Uyar, Antioxidant, antibacterial and antifungal electrospun nanofibersfor food packaging applications, Food Res. Int. 130 (2020). https://doi.org/10.1016/j.foodres.2019.108927

[30]   J. Joung, A. Boonsiriwit, M. Kim, Y.S. Lee, Application of ethylene scavenging nanocomposite film prepared by loading potassium permanganate-impregnated halloysite nanotubes into low-density polyethylene as active packaging material forfresh produce, LWT. 145 (2021). https://doi.org/10.1016/j.lwt.2021.111309

[31]   S. Md Nor, P. Ding, Trends and advances in edible biopolymer coating for tropicalfruit: A review, Food Res. Int. 134 (2020). https://doi.org/10.1016/j.foodres.2020.109208

[32]   L. Buendía−Moreno, M.J. Sánchez−Martínez, V. Antolinos, M. Ros−Chumillas, L. Navarro−Segura, S. Soto−Jover, G.B. Martínez−Hernández, A. López−Gómez, Active cardboard box with a coating including essential oils entrapped within cyclodextrins and/or hallosyte nanotubes. A case study for fresh tomato storage, Food Control. 107 (2020). https://doi.org/10.1016/j.foodcont.2019.106763

[33]   Q. Li, T. Ren, P. Perkins, X. Hu, X. Wang, Applications of halloysite nanotubes in food packaging for improving film performance and food preservation, Food Control. 124 (2021). https://doi.org/10.1016/j.foodcont.2021.107876

[34]   E. Abdullayev, Medical and health applications of halloysite nanotubes, in: Nat. Miner. Nanotub. Prop. Appl., 2015.

[35]   A. Karewicz, A. Machowska, M. Kasprzyk, G. Ledwójcik, Application of halloysitenanotubes in cancer therapy-A review, Materials (Basel). 14 (2021). https://doi.org/10.3390/ma14112943

[36]   H.Y. Liu, L. Du, Y.T. Zhao, W.Q. Tian, In vitro hemocompatibility and cytotoxicityevaluation of halloysite nanotubes for biomedical application, J. Nanomater. 2015 (2015). https://doi.org/10.1155/2015/685323

[37]   M. Zatorska-Płachta, G. Łazarski, U. Maziarz, A. Foryś, B. Trzebicka, D. Wnuk, K.Chołuj, A. Karewicz, M. Michalik, D. Jamroz, M. Kepczynski, Encapsulation of

Adv. App. of Micro and Nano Clay – Biopolymer-based Composites   Materials Research Forum LLC
Materials Research Foundations **125** (2022) 254-274   https://doi.org/10.21741/9781644901915-11

curcumin in polystyrene-based nanoparticles-drug loading capacity and cytotoxicity,
ACS Omega. 6 (2021). https://doi.org/10.1021/acsomega.1c00867

[38]   P.P. Dandekar, R. Jain, S. Patil, R. Dhumal, D. Tiwari, S. Sharma, G. Vanage, V. Patravale, Curcumin-loaded hydrogel nanoparticles: Application in anti-malarial therapy and toxicological evaluation, J. Pharm. Sci. 99 (2010) 4992-5010. https://doi.org/10.1002/jps.22191

[39]   I. Armentano, M. Dottori, E. Fortunati, S. Mattioli, J.M. Kenny, Biodegradable polymer matrix nanocomposites for tissue engineering: A review, Polym. Degrad. Stab. 95 (2010) 2126-2146. https://doi.org/10.1016/j.polymdegradstab.2010.06.007

[40]   S. Torgbo, P. Sukyai, Bacterial cellulose-based scaffold materials for bone tissue engineering, Appl. Mater. Today. 11 (2018) 34-49. https://doi.org/10.1016/j.apmt.2018.01.004

[41]   S. V. Dorozhkin, Calcium orthophosphate-based biocomposites and hybrid biomaterials, in: J. Mater. Sci., 2009: pp. 2343-2387. https://doi.org/10.1007/s10853-008-3124-x

[42]   B.O. Okesola, S. Ni, B. Derkus, C.C. Galeano, A. Hasan, Y. Wu, J. Ramis, L. Buttery, J.I. Dawson, M. D'Este, R.O.C. Oreffo, D. Eglin, H. Sun, A. Mata, Growth-Factor Free Multicomponent Nanocomposite Hydrogels That Stimulate Bone Formation, Adv. Funct. Mater. 30 (2020) 1-13. https://doi.org/10.1002/adfm.201906205

[43]   F.H. Zulkifli, F.S.J. Hussain, S.S. Zeyohannes, M.S.B.A. Rasad, M.M. Yusuff, A facile synthesis method of hydroxyethyl cellulose-silver nanoparticle scaffolds for skin tissue engineering applications, Mater. Sci. Eng. C. 79 (2017) 151-160. https://doi.org/10.1016/j.msec.2017.05.028

[44]   P. Ezati, J.W. Rhim, M. Moradi, H. Tajik, R. Molaei, CMC and CNF-based alizarinincorporated reversible pH-responsive color indicator films, Carbohydr. Polym. 246 (2020). https://doi.org/10.1016/j.carbpol.2020.116614

[45]   H.J. Kim, S. Roy, J.W. Rhim, Effects of various types of cellulose nanofibers on thephysical properties of the CNF-based films, J. Environ. Chem. Eng. 9 (2021). https://doi.org/10.1016/j.jece.2021.106043

[46]   K. Zhang, T.S. Huang, H. Yan, X. Hu, T. Ren, Novel pH-sensitive films based on starch/polyvinyl alcohol and food anthocyanins as a visual indicator of shrimp deterioration, Int. J. Biol. Macromol. 145 (2020) 768-776. https://doi.org/10.1016/j.ijbiomac.2019.12.159

[47]   H. Kawamoto, Trends in research and development on plastics of plant origin -

fromthe perspective of nanocomposite polylactic acid for automobile use -, Sci. Technol. Trends. 22 (2007) 62-75.

[48]   V. Nimbagal, N.R. Banapurmath, A.M. Sajjan, A.Y. Patil, S. V. Ganachari, Studieson Hybrid Bio-Nanocomposites for Structural Applications, J. Mater. Eng. Perform. (2021). https://doi.org/10.1007/s11665-021-05843-9

[49]   Y. Zhang, A. Tang, H. Yang, J. Ouyang, Applications and interfaces of halloysite nanocomposites, Appl. Clay Sci. 119 (2016) 8-17. https://doi.org/10.1016/j.clay.2015.06.034

[50]   N.P. Risyon, S.H. Othman, R.K. Basha, R.A. Talib, Characterization of polylactic acid/halloysite nanotubes bionanocomposite films for food packaging, Food Packag.

Shelf Life. 23 (2020). https://doi.org/10.1016/j.fpsl.2019.100450

[51]   S.M.M. Meira, G. Zehetmeyer, J.O. Werner, A. Brandelli, A novel active packagingmaterial based on starch-halloysite nanocomposites incorporating antimicrobial peptides, Food Hydrocoll. 63 (2017) 561-570. https://doi.org/10.1016/j.foodhyd.2016.10.013

Adv. App. of Micro and Nano Clay – Biopolymer-based Composites   Materials Research Forum LLC
Materials Research Foundations **125** (2022) 275-301   https://doi.org/10.21741/9781644901915-12

Chapter 12

Kaolinite–Cellulose based Nano–Composites and Applications

Mriganka Sekhar Manna[1]*, Susanta Ghanta[2]

[1]Department of Chemical Engineering, National Institute of Technology Agartala, India-799046

[2]Department of Chemistry, National Institute of Technology Agartala, India-799046

* mrigankamanna602@gmail.com

Abstract

This chapter is intended to detail the properties of kaolinite-cellulose nano-composite, its preparation, characterization, and applications in various fields. The fields of applications cover the conventional areas of application like wastewater treatment, biomedical application, fuel cells as well as in advanced applications e.g. in the fields of packaging, flame retardant, printed electronics, etc. The brisk viewpoints on the challenges to fabricate nanostructured kaolinite, nanostructure cellulose, and their composite as well are discussed. The direction to tackle the said challenges and updates on the recently developed applications of kaolinite-cellulose nano-composites are discussed.

Keywords

Cellulose Nanocrystals, Cellulose Nanofibers, Nano-Cellulose, Nano Fibrillated Cellulose, Nano-Structure Kaolinite, Clay Minerals, Nano-Composite, Adsorption, Wastewater Treatment

Abbreviations:

ATR-FTIR	: Attenuated total reflectance-Fourier transform infrared spectroscopy
BET	: Brunauer–Emmett–Teller
BJH	: Barrett–Joyner–Halenda
CCB	: Clay-cellulose bio-composite
CMF	: Cellulose microfibril
CNC	: Cellulose nanocrystal
CNF	: Cellulose nanofiber
CNTs	: Carbon nanotubes
CTAB	: Cetyltrimethylammonium bromide
DMSO	: Dimethyl sulfoxide

DSC	: Differential scanning calorimetry
EDX	: Electron dispersive X-ray spectrometer
EXK	: Exfoliated kaolinite
KNT	: Kaolinite nanotubes
MFC	: Microfibrillated cellulose
RFID	: Radio frequency identifier
RS	: Reflectance spectroscopy
SEM	: Scanning electron microscope
TEM	: Transmission electron microscope
TGA	: Thermo-gravimetric analyzer
WCA	: Water contact angle
XPS	:X-ray photoelectron spectroscopy

Contents

1.   Introduction

The nano-technology has emerged as a promising area of research and development in the last few decades as it has enormous and wide applications in various fields of applied science and technology. After the successful research and development of nano-particles of a material, a substantial amount of applications of these nanomaterials have been achieved in different needs of mankind such as energy storage, wastewater treatment, drug delivery, etc. Nano-materials are classified as clay-based and polymer-based with specific characteristics and applications as well in the respective fields. The recent trend of the scientific and technological endeavor is the integration of these two kinds of nano-materials to exploit synergistically their specific characteristics for better performance. Polymeric nano-materials can improve the mechanical properties of the nano-composites whereas clay-based nano-materials (kaolinite) improve the surface and rheological properties of the nano-composites. Consequently, the clay-polymer nano-composites show combined and synergistic effects at times. The research and development of clay-polymer-based nanocomposites are increasingly being applied to the various areas of applications. The nanocomposites are used as material to provide a barrier against passage of oxygen, to resist inflammation. It is also used in the biomedical and electrical/electronic applications as well as in the fuel cell [1]. The clay minerals with the specific characteristics are non-toxic, abundant; a low cost that has the capacity of ion exchange. The high surface area per unit volume of the clay minerals makes it the most widely used material for physicochemical interaction such as adsorption, catalytic activity. So, they can be used widely for environmental applications. The majority of the clay minerals are very effective at degrading cationic pollutants, however, they are able to decontaminate the anionic pollutants with various suitable modifications [2]. The compatible combination of different clays with organic material like cellulose may even provide a synergistic effect for decontamination of pollutants like heavy metals and obnoxious compounds like nitrates [3]. Various clay minerals like bentonite, montmorillonite, attapulgite, mica, and kaolinite, etc. are mixed up with the respective biopolymer to enhance various mechanical characteristics of a polymeric material. Various moduli of the clay are improved. Both the strength and stiffness are enhanced. The stability of the three dimensional structure of the

Adv. App. of Micro and Nano Clay – Biopolymer-based Composites   Materials Research Forum LLC
Materials Research Foundations **125** (2022) 275-301   https://doi.org/10.21741/9781644901915-12

clay is increased. The mixing of clay with natural organics also improves the chemical properties of clay composite. The composite becomes more resistant against heat and solvent attack. The wettability and in turn the dyeability of the clay composite are achieved. Other properties like flame retardancy, electrical conductivity of the composite are modified with different types of clay. The ratio of clay to biopolymer plays a major role in the modification of composite [4, 5].

The nano-structure of clay materials (kaolinite) offers a wide range of bio-sorption of inorganic hazardous chemical contaminants, specifically the much challenging heavy metals. Nano-structured materials are with established higher adsorption capacities as compared to the same material of larger particle size. Similarly, nano-cellulose has already been a very promising high-performance bio-adsorbent owing to its interesting and complex structural and functional characteristics towards high adsorption capacity. The nano-cellulose is very strong due to its long fibrous structure. It is hydrophilic in nature due to the presence of hydroxyl groups. It is renewable as it can be naturally produced. The biodegradability of nano-cellulose is one of the advantageous features of the nano-cellulose that permits extensive use of it for structural purposes [6]. The nano-cellulose is considered a sustainable functional material that has drawn increasing interest for the solution of environmental issues for the reason it contains active functional groups leading to specific properties. Nano-cellulose is obtained either by breaking down the naturally originated cellulosic materials mechanically, chemically, and thermally or is manufactured biologically through bacterial action [7]. The alteration of nano-cellulose by various kinds of treatments viz. chemical, thermal treatments towards the up-gradation of efficiency, improvement of desorption capability considering the environmental issues are the recent trends of research and development for the kaolinite-cellulose nano-composites [7].

2.   Inorganic–organic nano–composites

Clay minerals are porous functional inorganics that are surface active due to various functional groups present on the surface. The recent focus is on the kaolinite and its nano-structure for upgradation of its characteristics. On the other hand, organic cellulosic materials that are easily available from natural bacteria and plants are having the similar surface characteristics and are loaded with functional groups to be used in similar type of applications (adsorption, catalysis, etc) as surface active agents. The nano-structure of cellulose evidently shows better performance compared to its micro-structured. Both the kaolinite and cellulose have some drawbacks for their effective applications. Hence, the nano-structures of both the materials are composed to prepare inorganic-organic nano-composite to achieve a synergistic effect for the application as a composite.

2.1　Nano-structured kaolinite

Naturally available and extensively distributed clay minerals, in general, are of layered silicates that have distinguished ability to exchange ions, large surface-active area to exhibit great performance for adsorption [8, 9]. Kaolinite is a very common clay mineral and the major component of the kaolinite is aluminum silicate [10]. The aluminium silicate is bound by hydroxyl groups and forms the crystal structure of kaolinite. to layers of It is one of the most commercially available clay minerals which is widely used as the surface-active reagent. The molecular formula of kaolinite is $Al_2Si_2O_5(OH)_4$ with the following compositions: 39.8 % alumina, 46.3 % silica, and 13.9 % water [11]. The net negative charge is associated with the water molecule and the clay mineral is composed of two sheet layers. One tetrahedral silicon-oxygen $[Si_{2n}O_{5n}]^{2n-}$ lattice is attached to another octahedral alamo-oxygen-hydroxylic $[Al_{2n}(OH)_{4n}]^{2n+}$ lattice [12, 13]. Kaolinite is the principal constituent of kaolin that is formed from the sedimentary rocks by the process known as kaolinization. Kaolinization is the process known as the weathering of rocks i.e. decomposition of feldspars (potassium feldspars), granite, and aluminum silicates [14]. It is found in different colors e.g. white, greyish-white, little colored substance. Kaolinite is chemically a hydrous aluminum silicate. Ceramic can be produced from kaolinite due to its stable chemical structure and good physical properties. The shrinkage is very low during the drying of kaolinite due to its plastic nature. The melting point of kaolinite is as high as 1750°C. Kaolinite is developed by pseudo-hexagonal triclinic crystals with a diameter of 0.2–10 μm and with a thickness of 0.7 nm. The density of the kaolinite is measured to be 2.6 g.cm^{-3}. The ratio of two types of crystals of kaolinite i.e. SiO_4 tetrahedral sheets and $Al(OH)_6$ octahedral sheets are 1: 1 with pseudo-hexagonal symmetry.

Kaolinite has been developed in recent times as surface-active materials for remediation of chemical contaminants owing to its ability to the adsorption of heavy metals from groundwater, wastewater, and soil. Kaolinite is inexpensive and the remediation process by using it is a green technology. At present, kaolinite among other clay minerals such as montmorillonite, sepiolite, attapulgite, is an easily available clay mineral and is studied at length for remediation of environmental contaminants [15]. The surface reactivity of kaolinite is promoted by the different organic and inorganic alternations of the surface structure to enhance its adsorption affinity for a specific ion thereby increasing its selectivity. The same modification often increases its total surface area per unit weight to promote better adsorption capacity [10]. The breaking of the bulk minerals of kaolinite into single sheets (exfoliation) is recently developed as an efficient modification to have better properties of kaolinite [16].

However, the adsorption of only one heavy metal for wastewater treatment has great limitations. The surface properties and the affinity of kaolinite to combine with the other

Adv. App. of Micro and Nano Clay – Biopolymer-based Composites   Materials Research Forum LLC
Materials Research Foundations **125** (2022) 275-301   https://doi.org/10.21741/9781644901915-12

substances are limited by the structure of the crystal. The negative charge as well involves in the clay minerals restricts its applications as a binding agent [17]. Hence, the modification of kaolinite is necessarily important for the introduction of more surface area and a greater number of functional groups so as to have the affinity for binding multiple ions. The modification of kaolinite is required to enhance surface properties. The modified kaolinite has widespread applications based on the functionality introduced [16]. The adsorption affinity specific to an adsorbate is enhanced by the inorganic and organic modification of kaolinite. The surface of the sheets of kaolinite becomes more reactive by the modification [10, 18]. The different layers are separated to individual sheets of kaolinite by the process known as exfoliation. It is very widely used to achieve surface reactivity, more surface area, active functional groups for adsorption of specific adsorbates [19]. Several studies have been recently carried out on the exfoliation of kaolinite for the environmental and technical upgradations of exfoliated kaolinite sheets towards its applications as a surface-active agent specifically as advanced adsorbent [10] . Kaolinite can be of both types i.e., amorphous and crystalline depending on the inner layer structures. The crystalline kaolinite is made of individual active layers of it and the layers are of 1:1 type [20]. The stack arrangement of active layers of kaolinite induces the better capacity for the treatment of heavy metals.

The first step to manufacture nano-structure kaolinite is the separation of different layers. The scrolls of the kaolinite are produced conventionally in the exfoliation. Dimethyl sulphoxide (DMSO) helps in the breaking of the hydrogen bonds connecting two layers of kaolinite. Fifty mL of 80% DMSO is reacted with 15 g solid kaolinite for the purpose of layer separation (exfoliation). The layer separation process is carried for 2 h with the provision of stirring. Methanol is used for the washing of treated layers that transforms the kaolinite to methoxy kaolinite during washing. Now, the cetyltrimethylammonium bromide (CTAB) solution is gradually mixed with these methoxy kaolinite particles for 48 h. The resultant supernatant is separated. Two steps of washing of the supernatant by ethanol and distilled water are performed one by one. The sample of nano-structured kaolinite is dried for 12 h at 60 °C [8].

Han et al. Have reviewed the modified clay-based nanocomposites for heavy metals and organic pollutants [17]. However, the modification of nanostructured kaolinite using acids, calcination, polymers, or surfactants is not enough for the adsorption of more complex recalcitrants like antibiotics, aromatics, and various dyes. The co-presence of multiple contaminants in industrial wastewater prevents the expected performance of adsorbents. Therefore, more rigorous treatment of the kaolinite clay is required to get a satisfactory removal efficiency for separating organic and inorganic contaminants from real wastewater [17]. The active functionalities in kaolinite induce adsorption capacity for the adsorption of various adsorbates, specifically heavy metals. The active functional groups are attached on the internal surface of individual layers of kaolinite clay during the organic modification of kaolinite. The modified kaolinite shows more advanced adsorption capacity with a larger surface area and specific functional groups. Kaolinite shows hydrophilicity due to the hydrated cations present in the space between layers [21]. Hydrophobicity also is achieved by this organic modification of kaolinite. The surface energy of the kaolinite sheets is reduced by this modification resulting in the increase of its affinity towards adsorption of adsorbates e.g., heavy metals. The -SH group attached to the surface of the organically modified kaolinite removes impurities from inside and outside the pores. The attachment of --SH group is advantageous to the stable adsorption of mercury [22].

2.2   Nano-structured cellulose

Cellulose is a structural and functional bio-component of all green plants. It is a carbohydrate in nature and structurally a polymer of various sugar molecules. The chemical structure of cellulose is depicted in Fig. 1. It is naturally produced, hence abundant as raw material. The D-glucose monomers (repeating units) are polymerized by the β-(1-4)-linkage to form a chain of cellulose. Further, different cellulosic chains are linked by different feeble interactions. Different layers of cellulose are linked by van der Waals force and the functional groups (-OH) attached to the different carbon atoms are connected with weak hydrogen bonds to form the crystalline form of cellulose. The assembled repeating units in nano-sized, thread-like agglomerates result as a complex structure of fibers of cellulose [23]. The nano-celluloses have attracted wide-ranging attention as a sustainable advanced biological functional material. Nano-celluloses are available in different structural varieties such as rod-like shaped nano-crystals. Nano-crystals are a highly crystalline form of cellulose. The various properties of nano-cellulose and their contributions to wards the adsorption capacity of it have been elaborated in Table 1. On the other hand, nano-fibres are longer with more entanglement of the cellulose fibres. The other types of cellulose are microfibrillated cellulose and bacterial cellulose. The recently developed advanced applications of nano-cellulose include nano-composites, films, fibres, gels, and aerogels. It can be used for the monitoring of viscosity of a low viscous fluid, barrier material, and fire extinguishing foams. Different forms of cellulose are being used in the manufacturing of filtering membranes as cellulose acetate and energy storage applications. Nevertheless, enough challenges remain for the nano-celluloses to master their interactions and tailor their ability as functional materials.

The conversion of natural lignocellulosic material to nano-cellulose can be accomplished by various processes. The morphological characteristics and all other properties of transformed nano-cellulose greatly depend on the process used. The pre-treatment by

Adv. App. of Micro and Nano Clay – Biopolymer-based Composites   Materials Research Forum LLC
Materials Research Foundations **125** (2022) 275-301   https://doi.org/10.21741/9781644901915-12

chemical reagents such as acid, base and the agent for bleaching purposes are common practices. More efficient pre-treatment of lignocellulose can be achieved by the use of eutectic solvents or ionic liquids. A continuous effort is required for the optimization of a pre-treatment process as it largely depends on the type of raw material (lignocellulose) and other parameters. The environmental issues and the economy of the overall process of nano-cellulose production also decide a pre-treatment process. Nano-fibres are most often separated by mechanical means (fibrillation). On the contrary, hydrolysis of lignocellulose by the aid of acid helps to extract nano-crystals. The yield of nano-cellulose from cellulosic biomass is improved by various treatments following the pre-treatments. The treatments enhance the surface activity of nano-cellulose through the incorporation of functional hydroxyl groups. The introduction of hydroxyl groups imparts a very important balance between hydrophobicity and hydrophilicity of produced nano-cellulose. The demand for nano-cellulosic nanomaterials is continuously increasing towards various applications with the modified properties of it [24].

Various structures of cellulosic material viz. cellulose microfibrils (CMFs) and cellulose nano-fibrils (CNFs) are recently being used in pulp and paper technology. The different shapes and sizes of cellulose with enormous potential have emerged as the widespread use of biopolymer cellulose [25]. Thirty-six cellulose chains arranged in a crystal structure from CNF as an elementary fibril are synthesized from wood and plant materials. The matrices of polysaccharides have distinctive hydrophobic and hydrophilic surfaces to interact and intertwine with themselves to form a strong but elastic property. The bacterial and plant cellulose are initially synthesized and then cellulose microfibrils are produced by the reduction of the number of cellulosic chains from 36 to 18. A recent study on the nanostructure of cellulose and structure-dependent activities reveals that pectin–cellulose interactions are highly prominent as compared to cellulose-xyloglucan interaction. However, computational calculations show that the xyloglucan binds strongly to the hydrophobic surface of cellulose microfibrils. The extensibility of cellulose microfibrils is limited by the rare cellulose-cellulose contacts. The regions of extension are mediated by the presence of a trace amount of xyloglucan [23].

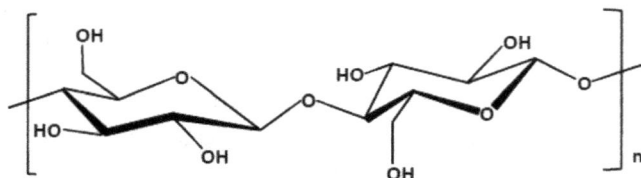

Fig. 1: Chemical structure of the cellulose

Adv. App. of Micro and Nano Clay – Biopolymer-based Composites   Materials Research Forum LLC
Materials Research Foundations **125** (2022) 275-301   https://doi.org/10.21741/9781644901915-12

Table 1: The properties of nano-cellulose towards the adsorption

Property	Advantages	Reference
Surface functionalization	Surface functionalization is executed through oxidation, esterification, etherification, and addition and grafting. It causes an enhancement in adsorption capacity.	[26]
Desorption capacity	High quality of desorption facilitates the reusability of the cellulose as adsorbent. Nano-cellulose needs a process of regeneration without hampering in adsorption capacity.	[27]
Cost effective	It can be produced from several biomass and is cost effective compared to activated carbon.	[28]
Biodegradability	It is not harmful to the environment due to its biodegradability.	[29]
High specific surface area	Many active sites are available for specific functionalization to improve the adsorption capacity.	[30]
Mechanical strength	The stiffness along with cohesion of nano-cellulose offers mechanical strength to the adsorbent offering its trouble free regeneration.	[31]
Surface tension	Low value of surface tension of cellulose increases its water wettability.	[32]
Stability in water	Hydrophilicity of nano-cellulose results in the very minimum chance of fouling by biomaterials and organic materials. The crystalline nano-cellulose reduces the biological and chemical corrosion of cellulose in water.	[33]

2.2.1 The role of cellulose-water interaction for its adsorption properties

The interaction of cellulose with water is believed to be the most prominent reason for the diversified applications of nano-cellulose-based materials. A wood chip soaks a moderate amount of water due to its lignin content that is hydrophobic in nature. On the other hand, a piece of paper soaks water in a higher amount when it comes in contact with water and disintegrates very easily. Apart from the presence of hydrophilic and hydrophobic substances in varying concentrations, crystallinity plays an important role in the soaking of water by cellulose. Single micro-fibril of cellulose does not swell by the interaction with water. But, the network of microfibrils swells as their surfaces are loaded with hygroscopic materials [34]. The crystalline structure of cellulose is mostly impenetrable by water whereas the amorphous structure with a comparatively loose network is susceptible to water. The hemicellulose content of cellulosic material induces the amorphous structure in cellulosic material and uptakes more water compared to crystalline and semi-crystalline microfibrils [35]. The water interaction is revealed by the conventional classical water retention and dynamic vapour sorption techniques. Moreover, sophisticated instrumental analysis, like NMR [36], scattering by neutron [37], thermo-porosimetry, and molecular dynamics simulations [38], have identified the presence of two different forms of H_2O molecules attached to the fibres of cellulose. Both the forms of H_2O are adhered to the nano-pores of microfibrils. Non-freezing water is directly attached to the surface of crystalline structure of cellulose whereas the second species of water is weakly bound in between the micro-fibril network structure of cellulose. Molecular dynamic study reveals the impact of the chirality of microfibrils for water interaction. The difference in the disorder resulting from the twist of chiral microfibrils facilitates water to enter into the compact aggregation of microfibrils [38]. Contrary, the restoration of cellulose swelling does not occur by the removal of water due to irreversible aggregation of microfibrils in the structure of cellulose. Cellulose nanofibers (CNFs) and cellulose nano-crystals are generally manufactured from refined cellulose with compact mass of microfibrils. The disentangled bundles for making individual fibrils and crystallites are made following an isolation technique. The interfaces between the microfibrils generally remain ordered but become disordered upon disintegration of CNFs.

The cellulose–water interaction is considered as an annoyance and is supposed to be a demerit of nano-cellulose for its usefulness. The presence of water hampers the strength of cellulose and is therefore unfavourable for manufacturing cardboard, nano-paper, paper etc by cellulose. Furthermore, many commercial applications of nano-cellulose have also been devalued by the presence of water. However, the same water can actually be beneficial when functional nano-cellulose is prepared. The nano-cellulose hydrogels that are used in wound dressing are manufactured based on the interaction between nano-cellulose and water. Various hydrogels are used in the growth process of biological tissue too. The functional transformation of the hydroxyl groups leads to the variation of the solubility of water in the nano-cellulose [39].

The derivatives of cellulose are not toxic and biocompatible as well. Hence, they are employed in various applications such as emulsifiers, viscosity modifiers, in pharmaceutical and biomedical applications. Dressing of wounds and culture of biological tissues by using cellulose derivatives are very common practices in biomedical applications [29, 40]. The hydrogen bonds along with others interactions among the functional atoms of cellulose endorse noticeable aggregation of the cellulose nano-particle.

Pre-treatment by chemical agents and the functionalization very often promote the introduction of both of the active cations and anions on the surface of nano-cellulose during different modification processes. Various functional groups like carboxylate, sulfate, ester, phosphorylate and quaternary amine present in the nano-celluloses are used for increasing colloidal stability [41]. The tendency of cellulose to interact with water, large amount of active surface per unit volume and the substantial aspect ratio provides interconnected networks for achieving highly viscous suspension of nano-cellulose and hence, the diversified applications [42].

2.2.2 Manufacturing of nano–cellulose

The preparation of nano-cellulose from various cellulosic materials is conducted through appropriate pre-treatment and after treatment for refining the nano-cellulose product. The individual pre-treatment processes are sometimes combined in sequence to get better quality. The pre-treatment processes are elaborated in the following sub-sections:

2.2.2.1 Mechanical treatment

The preparation of nano-structured cellulose (CNF, CNC, etc.) from natural cellulosic material is more often accomplished by mechanical treatment [43]. This treatment is used for the disintegration of cellulose pulp into smaller particles. The separation of prepared CNC is also accomplished by mechanical techniques after cellulose is treated. The cellulosic fiber needs to be delaminated to elementary fibril of dry cellulosic fiber to improve mechanical properties of nano-cellulose after shredding of cellulosic raw material [44]. The mechanical technique for the breaking of cellulose fiber is facilitated with the addition of water to improve delamination by breaking the interfibrillar hydrogen bonding. The aqueous medium also helps prevent reverse coalescence or fibril aggregation. The grinding and homogenization are two extensively used mechanical techniques for the delamination of cellulosic fiber.

2.2.2.2 Chemical treatment

The chemical pretreatment of cellulosic raw material provides fine structure and shorter nano-celluloses as compared to mechanical treatment. The problem of agglomeration is also reduced in chemical treatment. The chemical agents contribute significantly to the defibrillation of nano-cellulose production [45]. The prevalent techniques of chemical treatment involve acid hydrolysis, carboxymethylation, carboxylation, sulphonation, etc. to the modification of surface charge of nano-cellulose. Acid hydrolysis is mostly used to transform cellulose nanocrystal from cellulose. Acid hydrolysis degrades only the amorphous regions in microfibrils and the crystalline regions are unaffected. The CNC obtained by the H_2SO_4 hydrolysis shows better adsorption capacity for Ag(I) over CNC obtained by mechanical treatment [46]. This acid hydrolysis also fabricates bacteria nano-crystals from bacterial cellulose (BC) microfibrils. The chemical pretreatment step is sometimes accomplished to facilitate the delamination of nano-fibrils subsequently to the mechanical disintegration for the production of nano-cellulose. Another method for the hydrolysis of cellulosic material is accomplished by the use of ionic liquid which is considered as the green solvent for green technology. The use of ionic liquid-like 1-butyl-2-methyl imidazolium hydrogen sulfate (BmimHSO$_4$) has become a potential alternative method for the preparation of nano–cellulose. The preparation of nano-cellulose using ionic liquids as both solvent and catalyst is considered as a green technology for very minimum production of hazardous wastes [47].

2.2.2.3 Enzymatic hydrolysis

Mechanical pretreatment is constrained by the high energy cost for the preparation of nano-cellulose whereas the chemical treatment is burdened with environmental concern due to the production of hazardous chemicals drained to the water body and/or requirement of the high cost for waste treatment. Consequently, enzymatic hydrolysis is introduced in addition to either mechanical or chemical treatment or both to reduce the aforementioned problems. Hydrolysis of cellulose with the aid of enzymes is generally conducted prior to the mechanical operation for breaking down the cellulose for the production of nano-cellulose. The endoglucanase enzymes promote delamination of the cell walls in a homogenizer for the disintegration of cellulose [48]. The nano-cellulose (nanofibers) produced by endoglucanase yields better average molar mass and a higher aspect ratio when they are compared with nanofibers produced in acid hydrolysis [49]. Various other enzymes that are used for enzymatic hydrolysis of cellulosic material include cellulases, pectinases, xylanases, and ligninases for the hydrolysis of corresponding substrates.

2.2.3 Challenges to the preparation and application of nano-cellulose

The nano-cellulose is considered as an encouraging adsorption material for the removal of varied contaminants and other novel applications. However, there still remain many challenges for the application of this nanomaterial. The application in large scale demands easy scalability, low production cost and production have to be shifted from laboratory scale to industrial scale. The selection of treatment routes for the commercial production of nano-cellulose from any cellulosic material like wood cellulose has generally detrimental environmental impact. The mechanical disintegration of the cellulose material is found energy-intensive and is not cost-effective for up-scaling. The biological and chemical pre-treatment techniques are under development to tackle this issue. The carboxymethylation process as a chemical pre-treatment has not been proved yet cost-effective as a large quantity of methyl, ethyl and isopropyl alcohols etc are required for the preparation of nano-cellulose such as CNFs [50]. Moreover, the chemical pre-treatment of nano-cellulose results in a substantial problem to the environment because of the mixing of chemicals, especially various acids into the aquatic environment. Continuous research for the development of a better production strategy of nano-cellulose is needed for better economic feasibility with lower environmental impact. The toxicology of various modified nano-cellulose materials is the greatest impediment for its wider applications and marketability. The problems associated with a smaller particle size of nano-cellulose increase with decreasing particle size. The fine nano-sized particle of cellulose is inhaled and is up-taken into cells, blood, and lymph circulation systems and promotes health hazards. Most nano-cellulose-based materials are comparatively non-toxic in general. But, exposure to high concentration nano-cellulose has a negative effect on the viability and proliferation of living cells.

The preparation and application of nano-cellulose are considered as the landmark in the domain of material science and engineering. The nano-cellulose has achieved global attention with potential applications in packaging materials for food and beverages, electronic and optical appliances, and regeneration of biological tissue in advanced biotechnology, etc. The inbuilt large surface area, scope to modify the surface area with functionalization, environmental benignity with low toxicity, and renewability of nano-cellulose offer a promising utilization of it. Remediation of contaminants from wastewater and removal of carbon dioxide from the air are accomplished by the use of nano-cellulose. The specific applications of nano-cellulose for catalysis, membrane based separation and as an adsorbent are burdened with various challenges to be addressed. The contaminants that are separated from the environment by the application of cellulose based nano-composite include heavy metals, cationic and anionic dyes, carbon dioxide etc. The nano-material shows its activity against various malicious microbes too. An additional effort for the commercialization of nano-cellulose based nano-composite for the application in various fields is elaborated in the following sections.

2.3   Kaolinite–cellulose nano-composite

The combination of the single sheet of kaolinite with cellulose fibers makes excellent biopolymers with exquisite surface characteristics. The novel composite exhibits the synergistic effect for encouraging adsorption capacity [8, 19]. Cellulose, polysaccharide polymer is a natural biopolymer that plays an important role in numerous industrial, medical, and environmental applications [40]. The preparation of composite between cellulose and kaolinite and the characterization of produced composites are elaborated in the following sections.

2.3.1 Preparation

The preparation of nano-structured material on a large scale is difficult. The small scale production of exfoliated kaolinite and cellulose composite is prepared in the following method. The ingredients of the composite preparation include raw kaolinite, crystalline cellulose fibers, and the reagents like methanol, CTAB, and DMSO. CTAB, DMSO, and methanol are used for the disintegration of raw kaolinite. The nanocomposite of the exfoliated kaolinite–cellulose fiber (EXK/CF) is produced on a small scale by taking only 6 g of single-layered sheets of kaolinite. The sheets are initially dispersed into 30 mL of distilled water. The dispersion is facilitated with stirring of the mixture at 500 rpm for 2 h. The vigorous mixing is provided by sonication at a power of 240 W for 2 h. The solution of cellulose fiber (3 g cellulose fiber in 30 mL water) is mixed with a supernatant of previously treated kaolinite. The mixture is now homogenized effectively for a day. The homogenized mixture is finally undergone by sonication waves at 240 W once again to be used as the kaolinite-cellulose nano-composite for application as novel adsorbent. The kaolinite-cellulose nanocomposite made by the dispersion of exfoliated nano-structured kaolinite in nano-cellulose is depicted in Fig. 2.

The melt-processed polymer-clay nano-composites are easier to prepare when the clay content is kept below 5 w%. The nano-composite between microfibrillated cellulose (MFC) and kaolinite is prepared on a pilot scale. The nano paper structure of thickness of around 100 µm is continuously produced [51]. The MFC suspension is of colloidal nature and can easily be disintegrated from chemical wood fibre pulp. Such disintegrated MFC can be mixed with nanostructure particles of kaolinite. The processing of nano-composite such as dewatering, retention of small particles is optimized for laboratory scale preparation of kaolinite-cellulose nano-composite. The monitoring of density and certain mechanical properties of fabricated composite is also important and is accomplished by their measurements analytically. The sheet structure of the composite for manufacturing of different quality papers is obtained in web form. Clay-cellulose bio-composite (CCB) is prepared by applying rotation and pressure during the heating to the mixture of single layered clay tubules and shredded cellulose. The negative charge is introduced onto the surface of the prepared CCB at high temperature in the presence of sodium hydroxide. This modification of CCB enables it for the adsorption of cationic adsorbates. [52]. Composites with varying compositions of in-plane oriented nano-cellulose, kaolinite, and wood fibres are successfully produced at a pilot scale based on the targeted application of the prepared composites.

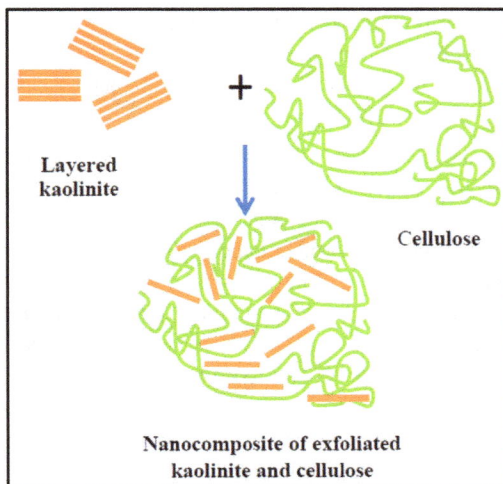

Fig. 2: Exfoliated nano-structured kaolinite dispersed in nano-cellulose to form nanocomposite of cellulose and kaolinite

2.3.2 Characterization

The efficiency of the preparation of composite is investigated in several analytical instruments to meet various properties of the composite. The process of preparation is modified accordingly in an iterative process to finalize the composite prior to its applications. All the possible analytical measurements collectively provide the required quality of the composite. The PAN analytical x-ray diffractometer is used for investigating the structural properties of prepared nano-composite (EXK/CF). The crystallinity of the composite is also measured by the same instrument. On the other hand, a Bruker spectrometer is used to find the prevalent functional groups in the cellulose and kaolinite and their alternations in the process of integration of cellulose and kaolinite. A scanning electron microscope (SEM) and a transmission electron microscope (TEM) are used to measure and study the external surface morphologies and internal features of the prepared composite, respectively. The Barrett–Joyner–Halenda (BJH) is used to find the distribution of pore size of nano-composite whereas Brunauer–Emmett–Teller (BET) methods is used to determine the total surface area of all the pores [16].

Apart from the very general characterizations of nano-composites, the specific characterization of nano cellulose is necessary for several specific applications. Mechanical properties like various stress coefficients are found out in uniaxial tension. The cone

Adv. App. of Micro and Nano Clay – Biopolymer-based Composites   Materials Research Forum LLC
Materials Research Foundations **125** (2022) 275-301   https://doi.org/10.21741/9781644901915-12

calorimetric measurement provides the fire retardancy of cellulosic composite material. The fire retardant paper board is an example of bio-composite thermoset polymer. This thermoset has a potential application as moderately strong mechanical structures and stronger mats [51].

The functionality of the composite of silica-kaolinite network on cellulose fibres is improved by the induction of UV radiation. The silica–kaolinite is put on the surface of cotton by using succinic acid (SA). The SA acts as an agent for cross-linking between the cotton and the silica-kaolinite. On the other hand, sodium hypophosphite (SHP) is used as a catalyst. The thermal stability of cellulose fibre is investigated using different analytical instruments to confirm the role of silica-kaolinite to enhance the thermal resistance of cellulose. The interactions between the ingredients of composites and the cellulose functional groups on the surface of composite are investigated in attenuated total reflectance-Fourier transform infrared spectroscopy (ATR-FTIR). The increase of thermal stability by providing a coating of silica-kaolinite increases the utility of cotton garments with better safety against the fire accident [15].

The characterization of the morphology of the cellulose-kaolinite nano-composite is best accomplished in scanning electron microscopy (SEM) coupled with energy-dispersive x-ray spectroscopy (EDS). On the other hand, the chemical composition and characteristics are investigated in X-ray photoelectron spectroscopy (XPS). The crystal structure of composite is measured by X-ray diffraction (XRD) method. The pore space characteristics along with the density composite are determined in mercury porosimetry [53]. The efficiency of the synthesized nano-composite for the adsorption capacity to hazardous heavy metals such as Pb(II) and Cd(II) can be tested by different characterization methods. The adsorption kinetics of cadmium (II) and lead onto synthesized nano-composite have been investigated. The kinetics for both the metals are observed to be pseudo-second-order. The adsorption capacity (Langmuir) of composite has been determined as 115.96 mg g^{-1} for cadmium whereas that for lead is found as 389.78 mg g^{-1}. The adsorption study in continuous mode can also be performed to investigate the efficiency of synthesized nano-composite [52].

2.4   Applications of the kaolinite-cellulose nano composite

The kaolinite-cellulose nano-composite is being used in various fields of science and technology. The conventional uses are in the treatment of ground water as well as wastewater, and in the drug delivery by the process of adsorption. The advanced applications of this composite in the diversified areas of science and technology along with the conventional applications are elaborated in the following subsections.

Adv. App. of Micro and Nano Clay – Biopolymer-based Composites   Materials Research Forum LLC
Materials Research Foundations **125** (2022) 275-301   https://doi.org/10.21741/9781644901915-12

2.4.1 Wastewater treatment

Rapid urbanization and industrialization have prompted water contamination by the trace quantity of various hazardous heavy metals such as Hg, Cu, Pb, Cd, As, and Cr ions. Heavy metal contamination has a very detrimental effect on the general health of humans. The adsorption technique has arisen as one of the efficient and cost-effective separation processes for the trace quantity of heavy metals. The crude clays and the improved clays after modification are used to remove the heavy metals from wastewater [20]. At the same time, cellulose-based adsorbents materials can replace conventional activated carbon for purification of wastewater in regard to the cost of the purification process.

Unmodified cellulose is a very poor adsorbent. With the modification of the cellulose, the adsorption properties such as functionality, hydrophilicity, surface area, aspect ratio, particle size, and chemical accessibility are enhanced. Nano-structured cellulose has a high amount of affinity towards heavy metals and organic pollutants. Hence, nano cellulose is considered a good natural biomaterial adsorbent for the treatment of wastewater. The surface area, crystallinity, and a number of functional moieties of unmodified cellulose are enhanced by the transformation of the cellulose particle size to the nano level. The amine group ($-NH_2$) is introduced to the surface of nano-cellulose to obtain magnetite properties in the cellulose. The transformed nano-cellulose after functionalization is very efficient adsorbent for the separation of carcinogenic arsenic from wastewater [54]. The iron coordination of the amino group generates active sites on the nano-cellulose surface for the adsorption. The number of active sites largely depends on the particle size of adsorbent. The smaller the size of nano-cellulose particles, the higher is the number of active sites. The nano-composite of nanostructured kaolinite and cellulose has synergistic effects on its adsorption capacity for the treatment of wastewater.

2.4.2 Drug delivery

The nano-composites are used as the carrier of certain drugs because of their surface and rheological properties. The multi-particulate system provides a slow and steady drug release with improved mechanical properties. The kaolinite-cellulose nano-composite in drug delivery systems has been a potential application of nanotechnology. Another advanced application of this nano-composite is in tissue engineering. The regeneration of nerve growth is promoted by this composite in a biological living system. The nano-particle drug delivery provides an improved half-life, better biocompatibility, and the lowest immunogenicity. The membrane barrier for the inclusion of drugs is reduced by the nanostructure of the composite. The recently developed smart material that alters its own structure and function in response to the environment can be used for better efficiency when combined with this nano-composite of kaolinite and cellulose nano-composite.

The 5-fluorouracil ($C_4H_3FN_2O_2$) abbreviated as 5FU hereafter is a widely useful drug for chemotherapies against various cancer types. The 5FU drug is not selective for cancer cells. Hence, the same drug is effective for cancer in breast, rectum, pancreas etc.The dosage of this drug is very critical otherwise it negatively affects the non-cancerous cells and tissues [55, 56]. Very critical impact to the health arises from the overdose of the drug. The damage of the kidney, heart, and intestinal tract may happen. The hematological as well as dermatological reactions may occur. Consequently, a proper carrier is necessary for controlling the dose of 5FU to achieve higher therapeutic efficiency and lower side effects [57]. The synthesized nano-level kaolinite nanotubes (KNT) acts as the carrier of 5FU. The encapsulation capacity reaches approximately 103 mg 5FU per gm of KNT. The loading capacity is determined through the characterization of samples by FT-IR and XRD. The synthetic KNTs loaded by 5FU (KNTs/5FU) facilitate the controlled release of the drug for 60 h. The colonic and intestinal fluid can release the drug at the releasing percentages of 91% and 74.3%, respectively. The L929 cell lines are utilized for the studies to check the cell viability. Substantial cell viability and non-toxicity are found. Hence, the KNTs may be applicable as an efficient drug carrier for 5FU that is used for the treatment of cancer [8].

2.4.3 Packaging

The nano-composite of cellulose nanofiber (CNF) and kaolinite are mixed with the polypropylene (PP) by a method known as melt-mixing. The constant composition of CNF (1 wt. %) is taken and it is mixed with a varying percentage of kaolinite in the range 1-5wt %. This polypropylene-based composite shows the best synergistic effect of kaolinite and cellulose for the 3wt% of kaolinite in the composite. The mechanical properties of the modified polypropylene such as tensile and flexural moduli are improved by incorporating the nano range of cellulose-kaolinite composite. The tensile modulus of PP is enhanced by 51% whereas flexural modulus is increased by 26% with the inclusion of kaolinite-cellulose nano-composite in comparison with unmodified PP [58]. Moreover, the oxygen permeability of this blended PP is reduced by 28% without compromising with the water vapour permeability that improves the barrier property of a packaging material (PP). The fabricated PP-kaolinite-CNF nano-composite is better packaging material, especially for food materials. The PP is converted to a better packaging material with improved mechanical strength and more resistance against the passage of O_2 through it.

2.4.4 Flame retardant

The thermal stability of cellulose is studied by many investigators and it is found that the cellulose fibre is highly combustible due to its very low thermal resistance. Hence, the dresses made of cellulose are unsafe against the flame and fire. So is the unsafe the congested public places like school, theatre etc. [59, 60]. Consequently, nano-particle clay such as kaolinite is extensively used to improve thermal resistance of cellulose. The manufacturing of thermally stable kaolinite-cellulosic nano-composite is possible owing to the multiple effects of nano-composite such as high thermal stability and insulation properties [60, 61]. The embedding of kaolinite on cellulose fibre is evidently possible for coating of nano-composite on cellulosic textile materials to be used as the flame retardant.

The investigation of the retardant property of coated cellulose shows that the principle of action lies in the physical effects such as the chemical nature of the kaolinite, its particle size, porosity, orientation, density of crystal, and the presence of -OH groups on the surface of cellulose. Some additives are used to promote the formation of a shallow carbonaceous coating on the surface of cellulose. The release of free radicals by the pyrolysis of cellulose fibres with application of heat also facilitates the carbonaceous coating. The coating provides extra resistance to the transfer of mass and heat between the substrate and the gas phase [62]. Kaolinite exhibits intumescent action (swelling up to protect against heat) as it forms an expanded layer of carbon-mineral that is capable of shielding the cellulosic material for flame extinguishing [63]. The functionalization of cellulosic fibres with a kaolinite-TiO_2 nano-hybrid composite is accomplished through a solvo-thermal process to improve the flame retardant properties of cellulose. The flame retardant property without compromising resistance to washing the cellulosic fabric is one of the desired qualities of garments and textiles. A covering by the kaolinite nanostructure on the treated cotton fibres is confirmed by the SEM analysis. This coated layer does not propagate flame when it comes to the effect of fire and provides self-extinguishing characteristics immediately after the heat source is removed. The functionalization of the fabric reduces the thermal degradation of the cellulosic fibre by 41% [64]. Insignificant resistance against the washing of cellulosic fabric is observed whereas mechanical strength of the fabric is almost retained.

2.4.5 Printed electronics

The nano-composite of cellulose and kaolinite has a very good potential to be used as electronic material in printing electronics. Printing electronics, a specific and recently developed branch of electronics that fabricates and applies various advanced organic and inorganic materials. The materials are used to create a discrete component of a device or the entire device by sequential and/or parallel combination of material to be used in the technologies like screen printing or direct writing. The conductive traces, passive and active electronic components are directly fabricated on the kaolinite-cellulose nano-composite to manufacture electronic devices like electronic circuits, displays, sensors, radio frequency identifiers (RFID), etc. Printing electronics may be considered to be an emerging technology for food packaging in the time to come as it offers new functionality to the packaging.

The use of the kaolinite-cellulose composite is advanced by its application in printed electronics. The micro and nanoscale cellulosic fibrils are mixed with kaolinite in various proportions to regulate the properties of the composite to be used in various applications as printing materials. The quality of nano-cellulose and that of kaolinite; and the treatment processes in the making of composite decide the quality of printing material. The properties of the composite for use as printed electronics include porosity, surface smoothness, mechanical properties, and density. The high proportion of kaolinite (nearly 80 wt%) provides better control on shrinkage during drying of the composite. Also, the dimensional stability of the composite is facilitated by a high proportion of kaolinite in the composite. In contrast, the higher amount of nano-cellulose along with plasticizers provides a better quality composite in terms of elasticity [65].

The performance of devices like transistors made from kaolinite-cellulose nano-composite strongly depends on the raw materials used. The quality of the semiconductor layer depends upon the porosity, polarity, and absorption properties of the upper surface of the composite. The fabricated kaolinite-cellulose nano-composite can also be utilized in novel applications such as in sensors and simple displays [65].

Conclusions

The chapter demonstrates the synergistic effect of the nano-composite made from nanostructured inorganic clay material i.e. kaolinite and nanostructured organic cellulose. The exfoliated sheets of kaolinite and the cellulose fibers are mixed with various pretreatments and post-treatments to obtain the quality composite. The various properties of the composite, specific to a particular application are measured by the analytical instruments. The presence of functional groups are incorporated by the chemical treatment to increase the surface reactivity of the composite. On the other hand, the materials (both raw materials and reagents) and specific methods contribute to the available surface of composite for adsorption. However, further research is needed to optimize the mechanical properties with respect to long time durability. The nano-composite has the potential to be a better green nanomaterial with several outstanding useful characteristics for widespread applications as a material of surface-active agent. The diversified applications of the composite in the field of engineering, general science, material science, and biomedical engineering exhibit a high potential for evolving industries. The further development in the process optimization for fabrication, better environmental aspects in both fabrication and application, and minimization of cost and energy consumption towards its manufacturing will certainly help in the living standard of humans.

References

[1] P.V.K. Kumari, Y.S. Rao, S. Akhila, Role of nanocomposites in drug delivery, GSC Biol. Pharm.Sci. 8 (2019) 94-103.https://doi.org/10.30574/gscbps.2019.8.3.0150

[2] C. Lazaratou, D. Vayenas, D. Papoulis, The role of clays, clay minerals and clay-based materials for nitrate removal from water systems: A review, Appl. Clay Sci. 185 (2020) 105377.https://doi.org/10.1016/j.clay.2019.105377

[3] M. Borjesson, G. Westman, Crystalline nanocellulose-preparation, modification, and properties, in: M. Poletto, H.L.O. Junior (Eds.), Cellulose-Fundamental Aspects and Current Trends, Rijeka, Croatia, 2015, pp.159-191.https://doi.org/10.5772/61899

[4] M. P. Gashti, S. Moradian, Effect of nanoclay type on dyeability of polyethylene terephthalate/clay nanocomposites, J. Appl. Polym. Sci.125 (2012) 4109-4120.https://doi.org/10.1002/app.35493

[5] M. Parvinzadeh, S. Moradian, A. Rashidi, M.-E. Yazdanshenas, Effect of the addition of modified nanoclays on the surface properties of the resultant polyethylene terephthalate/clay nanocomposites, Polym. Plast. Technol. Eng. 49 (2010) 874-884.https://doi.org/10.1080/03602551003664628

[6] M.N.F. Norrrahim, N.A.M. Kasim, V.F. Knight, M.S.M. Misenan, N. Janudin, N.A.A. Shah, N. Kasim, W.Y.W. Yusoff, S.A.M. Noor, S.H. Jamal, K.K. Ong, W.M.Z.W. Yunus, Nanocellulose: a bioadsorbent for chemical contaminant remediation, RSC Adv.11 (2021) 7347-7368.https://doi.org/10.1039/D0RA08005E

[7] K.P.Y. Shak, Y.L. Pang, S.K. Mah, Nanocellulose: Recent advances and its prospects in environmental remediation, Beilstein J. Nanotechnol. 9 (2018) 2479-2498.https://doi.org/10.3762/bjnano.9.232

[8] M.R. Abukhadra, A.F. Allah, Synthesis and characterization of kaolinite nanotubes (KNTs) as a novel carrier for 5-fluorouracil of high encapsulation properties and controlled release, Inorg. Chem. Commun. 103 (2019) 30-36.https://doi.org/10.1016/j.inoche.2019.03.005

[9] M.R. Abukhadra, M. Mostafa, Effective decontamination of phosphate and ammonium utilizing novel muscovite/phillipsite composite; equilibrium investigation and realistic application, Sci. Total Environ. 667 (2019) 101-111.https://doi.org/10.1016/j.scitotenv.2019.02.362

[10] M. Shaban, M.I. Sayed, M.G. Shahien, M.R. Abukhadra, Z.M. Ahmed, Adsorption behavior of inorganic-and organic-modified kaolinite for congo red dye from water,

kinetic modeling, and equilibrium studies, J. Sol-Gel Sci. Technol. 87 (2018) 427-441.https://doi.org/10.1007/s10971-018-4719-6

[11] G.Varga, The structure of kaolinite and metakaolinite, Epitoanyag, 59 (2007) 6-9.https://doi.org/10.14382/epitoanyag-jsbcm.2007.2

[12] S.S. Ray, M. Okamoto, Polymer/layered silicate nanocomposites: a review from preparation to processing, Prog. Polym. Sci. 28 (2003) 1539-1641.https://doi.org/10.1016/j.progpolymsci.2003.08.002

[13] J. Herney-Ramirez, M.A. Vicente, L.M. Madeira, Heterogeneous photo-fenton oxidation with pillared clay-based catalysts for wastewater treatment: a review, Appl. Catal. B: Environ. 98 (2010) 10-26.https://doi.org/10.1016/j.apcatb.2010.05.004

[14] R. Ismail, W. Almaqtri, M. Hassan, Kaolin and bentonite catalysts efficiencies for the debutylation of 2-tert-butylphenol, Chem. Int. 7 (2021) 21-29.

[15] M.P. Gashti, A. Elahi, M.P. Gashti, UV radiation inducing succinic acid/silica-kaolinite network on cellulose fiber to improve the functionality, Compos. B. Eng. 48 (2013) 158-166.https://doi.org/10.1016/j.compositesb.2012.12.002

[16] M.R. Abukhadra, A. AlHammadi, A.M. El-Sherbeeny, M.A. Salam, M.A. El-Meligy, E.M. Awwad, M. Luqman, Enhancing the removal of organic and inorganic selenium ions using an exfoliated kaolinite/cellulose fibres nanocomposite, Carbohydr. Polym. 252 (2021) 117163.https://doi.org/10.1016/j.carbpol.2020.117163

[17] H. Han, M.K. Rafiq, T. Zhou, R. Xu, O. Mašek, X. Li, A critical review of clay-based composites with enhanced adsorption performance for metal and organic pollutants, J. Hazard. Mater. 369 (2019) 780-796.https://doi.org/10.1016/j.jhazmat.2019.02.003

[18] D. Tan, P. Yuan, F. Annabi-Bergaya, D. Liu, H. He, High-capacity loading of 5-fluorouracil on the methoxy-modified kaolinite, Appl. Clay Sci. 100 (2014) 60-65.https://doi.org/10.1016/j.clay.2014.02.022

[19] M.R. Abukhadra, B.M. Bakry, A. Adlii, S.M. Yakout, M.E. El-Zaidy, Facile conversion of kaolinite into clay nanotubes (KNTs) of enhanced adsorption properties for toxic heavy metals (Zn^{2+}, Cd^{2+}, Pb^{2+}, and Cr^{6+}) from water, J. Hazard. Mater. 374 (2019) 296-308.https://doi.org/10.1016/j.jhazmat.2019.04.047

[20] S. Gu, X. Kang, L. Wang, E. Lichtfouse, C. Wang, Clay mineral adsorbents for heavy metal removal from wastewater: a review, Environ. Chem. Lett. 17 (2019) 629-654.https://doi.org/10.1007/s10311-018-0813-9

[21] B. Sarkar, R. Rusmin, U.C. Ugochukwu, R. Mukhopadhyay, K.M. Manjaiah, Modified clay minerals for environmental applications, in: M. Mercurio, B. Sarkar, A. Langella (Eds.) Modified Clay and Zeolite Nanocomposite Materials, Elsevier, 2019, pp. 113-127.https://doi.org/10.1016/B978-0-12-814617-0.00003-7

[22] Y. Wang, S. Li, H. Yang, In situ stabilization of some mercury-containing soils using organically modified montmorillonite loading by thiol-based material, J. Soils Sediments, 19 (2019) 1767-1774.https://doi.org/10.1007/s11368-018-2150-9

[23] D.J. Cosgrove, Re-constructing our models of cellulose and primary cell wall assembly, Curr. Opin. Plant Biol. 22 (2014) 122-131.https://doi.org/10.1016/j.pbi.2014.11.001

[24] K. Dhali, M. Ghasemlou, F. Daver, P. Cass, B. Adhikari, A review of nanocellulose as a new material towards environmental sustainability, Sci. Total Environ. (2021) 145871.https://doi.org/10.1016/j.scitotenv.2021.145871

[25] M. He, G. Yang, B.-U. Cho, Y.K. Lee, J.M. Won, Effects of addition method and fibrillation degree of cellulose nanofibrils on furnish drainability and paper properties, Cellulose, 24 (2017) 5657-5669.https://doi.org/10.1007/s10570-017-1495-3

[26] R. Ilyas, S. Sapuan, M. Sanyang, M. Ishak, Nanocrystalline cellulose reinforced starch-based nanocomposite: A review, 5th Postgraduate seminar on natural fiber composites, Universiti Putra Malaysia Serdang, Selangor, 2016, pp. 82-87.

[27] L. Mo, H. Pang, Y. Tan, S. Zhang, J. Li, 3D multi-wall perforated nanocellulose-based polyethylenimine aerogels for ultrahigh efficient and reversible removal of Cu (II) ions from water, Chem. Eng. J. 378 (2019) 122157.https://doi.org/10.1016/j.cej.2019.122157

[28] H. Voisin, L. Bergström, P. Liu, A.P. Mathew, Nanocellulose-based materials for water purification, Nanomaterials, 7 (2017) 57.https://doi.org/10.3390/nano7030057

[29] N. Mahfoudhi, S. Boufi, Nanocellulose as a novel nanostructured adsorbent for environmental remediation: a review, Cellulose, 24 (2017) 1171-1197.https://doi.org/10.1007/s10570-017-1194-0

[30] N. Lin, J. Huang, A. Dufresne, Preparation, properties and applications of polysaccharide nanocrystals in advanced functional nanomaterials: a review, Nanoscale, 4 (2012) 3274-3294.https://doi.org/10.1039/c2nr30260h

[31] R. Rusli, S.J. Eichhorn, Determination of the stiffness of cellulose nanowhiskers and the fiber-matrix interface in a nanocomposite using Raman spectroscopy, Appl. Phys. Lett. 93 (2008) 033111.https://doi.org/10.1063/1.2963491

Adv. App. of Micro and Nano Clay – Biopolymer-based Composites   Materials Research Forum LLC
Materials Research Foundations **125** (2022) 275-301   https://doi.org/10.21741/9781644901915-12

[32] H. Ibrahim, N. Sazali, I. Ibrahim, M.S. Sharip, Nano-structured cellulose as green adsorbents for water purification: A mini review, J. Appl. Membr. Sci. Technol. 23 (2019).https://doi.org/10.11113/amst.v23n2.154

[33] T. Nguyen, F.A. Roddick, L. Fan, Biofouling of water treatment membranes: a review of the underlying causes, monitoring techniques and control measures, Membranes, 2 (2012) 804-840.https://doi.org/10.3390/membranes2040804

[34] C. Driemeier, J. Bragatto, Crystallite width determines monolayer hydration across a wide spectrum of celluloses isolated from plants, J. Phys. Chem. B, 117(1) (2013) 415-421.https://doi.org/10.1021/jp309948h

[35] T.C. Maloney, H. Paulapuro, The formation of pores in the cell wall, J. Pulp Pap. Sci. 25 (1999) 430-436.

[36] E.L. Lindh, C. Terenzi, L. Salmén, I. Furo, Water in cellulose: evidence and identification of immobile and mobile adsorbed phases by 2 H MAS NMR, Phys. Chem. Chem. Phys. 19 (2017) 4360-4369.https://doi.org/10.1039/C6CP08219J

[37] D. Vural, C. Gainaru, H. O'Neill, Y. Pu, M.D. Smith, J.M. Parks, S.V. Pingali, E. Mamontov, B.H. Davison, A.P. Sokolov, A.J. Ragauskas, J.C. Smith, L. Petridis, Impact of hydration and temperature history on the structure and dynamics of lignin, Green Chem. 20 (2018) 1602-1611.https://doi.org/10.1039/C7GC03796A

[38] A. Paajanen, S. Ceccherini, T. Maloney, J.A. Ketoja, Chirality and bound water in the hierarchical cellulose structure, Cellulose, 26 (2019) 5877-5892.https://doi.org/10.1007/s10570-019-02525-7

[39] N. Grishkewich, N. Mohammed, J. Tang, K.C. Tam, Recent advances in the application of cellulose nanocrystals, Curr. Opin. Colloid Interface Sci. 29 (2017) 32-45.https://doi.org/10.1016/j.cocis.2017.01.005

[40] J.N. Putro, S.P. Santoso, F.E. Soetaredjo, S. Ismadji, Y.-H. Ju, Monitoring, management, nanocrystalline cellulose from waste paper: adsorbent for azo dyes removal, Environ. Nanotechnol. Monit. Manag. 12 (2019) 100260.https://doi.org/10.1016/j.enmm.2019.100260

[41] J. Luo, K. Huang, X. Zhou, Y. Xu, Elucidation of oil-in-water emulsions stabilized with celery cellulose, Fuel, 291 (2021) 120210.https://doi.org/10.1016/j.fuel.2021.120210

[42] S. Jodeh, O. Hamed, A. Melhem, R. Salghi, D. Jodeh, K. Azzaoui, Y. Benmassaoud, K. Murtada, Magnetic nanocellulose from olive industry solid waste for the effective

removal of methylene blue from wastewater, Environ. Sci. & Pollut. Res. 25 (2018) 22060-22074.https://doi.org/10.1007/s11356-018-2107-y

[43] A.W. Carpenter, C.-F. de Lannoy, M.R. Wiesner, Cellulose nanomaterials in water treatment technologies, Environ. Sci. Technol. 49 (2015) 5277-5287.https://doi.org/10.1021/es506351r

[44] O. Nechyporchuk, M.N. Belgacem, J. Bras, Production of cellulose nanofibrils: A review of recent advances, Ind. Crops Prod. 93 (2016) 2-25.https://doi.org/10.1016/j.indcrop.2016.02.016

[45] M. Fan, D. Dai, A.Yang, High strength natural fiber composite: defibrillation and its mechanisms of nano cellulose hemp fibers, Int. J. Polym. Mater. 60 (2011) 1026-1040.https://doi.org/10.1080/00914037.2010.551347

[46] P. Liu, H. Sehaqui, P. Tingaut, A. Wichser, K. Oksman, A.P. Mathew, Cellulose and chitin nanomaterials for capturing silver ions (Ag+) from water via surface adsorption, Cellulose, 21 (2014) 449-461.https://doi.org/10.1007/s10570-013-0139-5

[47] X.Y. Tan, S.B. Abd Hamid, C.W. Lai, Preparation of high crystallinity cellulose nanocrystals (CNCs) by ionic liquid solvolysis, Biomass Bioenergy, 81 (2015) 584-591.https://doi.org/10.1016/j.biombioe.2015.08.016

[48] M. Pääkkö, M. Ankerfors, H. Kosonen, A. Nykänen, S. Ahola, M. Österberg, J. Ruokolainen, J. Laine, P.T. Larsson, O. Ikkala, T. Lindstrom, Enzymatic hydrolysis combined with mechanical shearing and high-pressure homogenization for nanoscale cellulose fibrils and strong gels, Biomacromolecules, 8 (2007) 1934-1941.https://doi.org/10.1021/bm061215p

[49] M. Henriksson, G. Henriksson, L. Berglund, T. Lindström, An environmentally friendly method for enzyme-assisted preparation of microfibrillated cellulose (MFC) nanofibers, Eur. Polym. J. 43 (2007) 3434-3441.https://doi.org/10.1016/j.eurpolymj.2007.05.038

[50] J. Feng, S.T. Nguyen, Z. Fan, H.M. Duong, Advanced fabrication and oil absorption properties of super-hydrophobic recycled cellulose aerogels, Chem. Eng. J. 270 (2015) 168-175.https://doi.org/10.1016/j.cej.2015.02.034

[51] D.O. Castro, Z. Karim, L. Medina, J.-O. Häggström, F. Carosio, A. Svedberg, L. Wågberg, D. Söderberg, L.A. Berglund, The use of a pilot-scale continuous paper process for fire retardant cellulose-kaolinite nanocomposites, Compos. Sci. Technol. 162 (2018) 215-224.https://doi.org/10.1016/j.compscitech.2018.04.032

[52] E. Abu-Danso, S. Peräniemi, T. Leiviskä, T. Kim, K.M. Tripathi, A. Bhatnagar, Synthesis of clay-cellulose biocomposite for the removal of toxic metal ions from aqueous medium, J. Hazard. Mater. 381 (2020) 120871.https://doi.org/10.1016/j.jhazmat.2019.120871

[53] D. Wanna, C. Alam, D.M. Toivola, P. Alam, Bacterial cellulose-kaolin nanocomposites for application as biomedical wound healing materials, Adv. Nat. Sci: Nanosci. Nanotechnol. 4 (2013) 045002.https://doi.org/10.1088/2043-6262/4/4/045002

[54] K. Taleb, J. Markovski, Z. Veličković, J. Rusmirović, M. Rančić, V. Pavlović, A. Marinković, Arsenic removal by magnetite-loaded amino modified nano/microcellulose adsorbents: Effect of functionalization and media size, Arab. J. Chem. 12 (2019) 4675-4693.https://doi.org/10.1016/j.arabjc.2016.08.006

[55] F. Tanaka, T. Fukuse, H. Wada, M. Fukushima, The history, mechanism and clinical use of oral 5-fluorouracil derivative chemotherapeutic agents, Curr. Pharm. Biotechnol. 1 (2000) 137-164.https://doi.org/10.2174/1389201003378979

[56] D. Tan, P. Yuan, D. Liu, P. Du, Surface modifications of halloysite, in: P. Yuan, A. Thill, F. Bergaya (Eds.), Developments in clay science, Elsevier 2016, pp. 167-201.https://doi.org/10.1016/B978-0-08-100293-3.00008-X

[57] S.P. Chandran, S.B. Natarajan, S. Chandraseharan, M.S.B.M. Shahimi, Nano drug delivery strategy of 5-fluorouracil for the treatment of colorectal cancer, J. Cancer Res. Pract. 4 (2017) 45-48.https://doi.org/10.1016/j.jcrpr.2017.02.002

[58] B.N. Jung, H.W. Jung, D. Kang, G.H. Kim, J.K. Shim, Synergistic effect of cellulose nanofiber and nanoclay as distributed phase in a polypropylene based nanocomposite system, Polymers, 12 (2020) 2399.https://doi.org/10.3390/polym12102399

[59] C.Q. Yang, Q. He, R.E. Lyon, Y. Hu, Investigation of the flammability of different textile fabrics using micro-scale combustion calorimetry, Polym. Degrad. Stab. 95 (2010) 108-115.https://doi.org/10.1016/j.polymdegradstab.2009.11.047

[60] R. Kozłowski, M. Władyka-Przybylak, Flammability and fire resistance of composites reinforced by natural fibers, Polym. Adv. Technol. 19 (2008) 446-453.https://doi.org/10.1002/pat.1135

[61] M. Moniruzzaman, K.I. Winey, Polymer nanocomposites containing carbon nanotubes, Macromolecules, 39 (2006) 5194-5205.https://doi.org/10.1021/ma060733p

[62] S.P. da Silva Ribeiro, L. dos Santos Cescon, R.Q.C.R. Ribeiro, A. Landesmann, L.R. de Moura Estevão, R.S.V. Nascimento, Effect of clay minerals structure on the

polymer flame retardancy intumescent process, Appl. Clay Sci. 161 (2018) 301-309.https://doi.org/10.1016/j.clay.2018.04.037

[63] C.R.S. de Oliveira, M.A. Batistella, L.A. Lourenço, S.M.d.A.G. Ulson, A.A.U. de Souza, Cotton fabric finishing based on phosphate/clay mineral by direct-coating technique and its influence on the thermal stability of the fibers, Prog. Org. Coat. 150 (2021) 105949.https://doi.org/10.1016/j.porgcoat.2020.105949

[64] C.R.S. de Oliveira, M.A. Batistella, S.M.d.A.G. Ulson, A.A.U. de Souza, Functionalization of cellulosic fibers with a kaolinite-TiO2 nano-hybrid composite via a solvothermal process for flame retardant applications, Carbohydr. Polym. 266 (2021) 118108.https://doi.org/10.1016/j.carbpol.2021.118108

[65] K. Torvinen, F. Pettersson, P. Lahtinen, K. Arstila, V. Kumar, R. Österbacka, M. Toivakka, J.J. Saarinen, Nanoporous kaolin-cellulose nanofibril composites for printed electronics, Flex. Print. Electron. 2 (2017) 024004.https://doi.org/10.1088/2058-8585/aa6d97

Adv. App. of Micro and Nano Clay – Biopolymer-based Composites   Materials Research Forum LLC
Materials Research Foundations **125** (2022) 302-322   https://doi.org/10.21741/9781644901915-13

Chapter 13

Montmorillonite-Cellulose based Nano-Composites and Applications

Purnima Baruah, Debajyoti Mahanta*

Department of Chemistry, Gauhati University, Guwahati, Assam, India

debam@gauhati.ac.in

Abstract

Exploration of different types of polymer-clay nanocomposites have already crossed several decades. In recent times, more emphasis is given in the research of different bio-based polymer nanocomposites as they exhibit eco-friendly biodegradable behavior and biocompatible features. Cellulose is the most copiously available bio-macromolecule with interesting functional, chemical, mechanical and biological properties. Incorporation of montmorillonite clay as nanofillers into cellulose matrix results in significant modification and reinforcement in various properties of the polymer enlarging its applicability. This chapter brings forth a concise account on the development of different montmorillonite-cellulose based nano-composites as prospective materials for multiple biomedical and engineering applications.

Keywords

Cellulose, Montmorillonite, Nanocomposites, Biodegradable, Wound-Healing, Bioplastic, Flame-Retardant

Contents

1.   Introduction

In order to fulfill the huge demand of newer and smarter materials, composite materials have attracted the researchers for varities of applications. Composite materials not only have the properties of the constituent parent materials, but in many cases, enhancement of certain properties is observed due to synergistic effect. Nano-composites contain fillers or additives in nanometer range well-dispersed in the matrix. Nowadays, research on different nano-composites has gained tremendous attention because nano-fillers do not impart disadvantages like bulkiness, opaqueness, weight gain and brittleness which are generally imparted by micro-fillers or other bigger fillers [1, 2]. Polymeric materials have been extensively used for designing different nano-composite materials due to their flexibility, light weight, easy processability and high mechanical strength. Synthetic polymers derived from fossil fuel based resources are mostly non-biodegradable and their utilization for years has been responsible for alarming condition of environmental pollution and global warming. In order to control this frightening situation, utilization of environmentally benign renewable biomass is more preferred for various applications nowadays.

Cellulose is the most abundantly available natural organic compound as well as bio-macromolecule having the general formula $(C_6H_{10}O_5)_n$ [3]. It is a polysaccharide consisting of few hundreds to thousands of linearly connected glucose molecules. It is the principal building material of primary cell wall in green plants and encompasses about 33% of all plant matters. As per estimate, the natural annual production of cellulose is about 1.5×10^{12} t throughout the globe [4]. In 1838, French chemist Anselme Payen discovered cellulose and determined its chemical formula isolating it from plant matter. Like cellulose, clays too are naturally occurring materials formed due to disintrigation and chemical weathering of rock materials on the surface of earth. Significant number of research works are available which reported advantageous impact of nanoclay- dispersion in a polymer matrix not only on its mechanical performance and shielding behavior but also on its thermal stability and biodegradable nature [5-8]. They exhibit layered structures and based on the chemical composition of the layers clays are broadly categorized into four major groups – kaolinite, smectite, illite and chlorite. Montmorillonite (MMT) is a naturally available clay material of phyllosilicate type and an important member of the major clay group- 'smectite' [9]. Clay minerals are comparatively low-cost materials. Owing to a number of advantageous characteristics like high surface area, mechanical as well as chemical stability, cationic exchange capacity, biocompatibility etc clays are regarded as preferential alternatives for various applications [10]. If clays are classified on the basis of arrangement of tetrahedral and octahedral sheets in their layered composition, then MMT comes under 2:1 group indicating that each layer in the MMT structure consists of two tetrahedral sheets and one octahedral sheet. The octahedral sheet containing O-Al(Mg)-O unit is sandwiched between two tetrahedral sheets containing O-Si-O unit [11]. This particular clay is highly abundant in nature which has been broadly used for removing different contaminants from aqueous media for years [12,13]. MMT can be modified through ionic substitution to get desired structure and properties- small cations like Fe^{3+} and Al^{3+} can substitute Si^{4+} and can coordinate with oxygen in tetrahedral fashion. Bigger cations like K^+, Na^+, Cs^+ etc can be placed in between the layers. Al^{3+} ion can be replaced by Mg^{2+}, Fe^{2+}, Li^{1+}, Cu^{2+} etc. ions in the octahedral sheet [14].

From recent advancement in scientific research, it is evident that the use of MMT as nanofiller to get cellulose based nano-composites has a sparkling effect on effectiveness of the material. Research works are going on to investigate applicability of MMT-cellulose based nano-composites in environmental, biomedical and engineering fields. In this chapter we will discuss MMT-cellulose based nano-composites prepared by different research groups using different sources of cellulose, their properties, methods of preparation and their applications in diverse fields.

2.   Characterization

Scanning electron microscopy (SEM) is used to scrutinize the morphological features of MMT-cellulose based nano-composites whereas Fourier transform infrared spectroscopic technique (FTIR) is used for determining the functional groups and chemical compositions. Their thermostability is investigated by thermogravimetric analysis at different heating rates both in nitrogen and in air atmosphere. Mechanical properties are examined using universal testing machine (UTM). X-ray Diffraction Analysis (XRD) is used for

Adv. App. of Micro and Nano Clay – Biopolymer-based Composites   Materials Research Forum LLC
Materials Research Foundations **125** (2022) 302-322   https://doi.org/10.21741/9781644901915-13

investigating the crystallinity, inter-layer spacing etc. Rheometer is used for viscosity measurement of these polymeric nano-composites.

3.   Factors affecting properties of MMT-cellulose based nano-composites

Nanocomposites mostly exhibit superior structural, mechanical, thermal, biological performances. The properties of any kind of polymer-clay nanocomposite are influenced by the arrangement of the clay particles in the polymeric matrix [15,16]. The aspect ratio of the clay particles, their dispersion, intercalation and exfoliation within the polymer matrix etc. are the most imperative factors which decide the property of the composite materials [17].

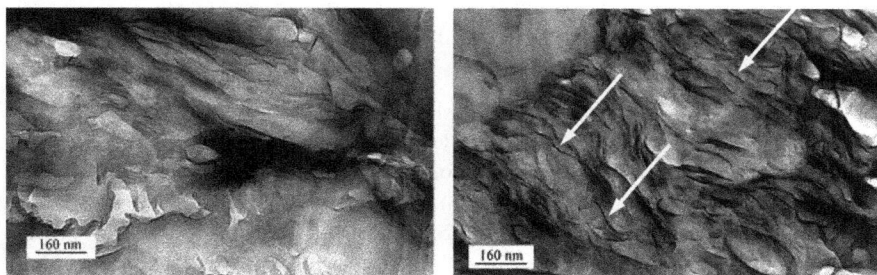

Fig.1: TEM images of MMT-cellulose nanocomposites: (a) 3%MMT-A and (b) 3%MMT-B (extracted from [18]).

Xu et al. synthesized cellulose nanofibrils (CNF)/MMT nanocomposite with CNFs having quaternary ammonium cations (Q-CNF) and carboxylate groups (TO-CNF) and MMT nanoplatelets with negative charge (overall) at the surface and examined the consequence of correlation between oppositely charged nanofibrils and MMT nanoplatelets on mechanical performance and flame-resistance behavior of the nanocomposites. Their study showed that volume fraction of MMT and its interfacial correlation with nanofibrils are important factors for getting desirable properties of nanocomposites. Strong interaction of cationically charged Q-CNF with anionically charged MMT platelets offered synergic improvement in the mechanical properties in case of Q-CNF/MMT nanocomposite. Such synergistic effect was not observed in case of TO-CNF/MMT nanocomposites. On the other hand, TO-CNF/MMT exhibited higher flame retardancy than Q-CNF/MMT as the former can hinder diffusion and transportation of oxygen more through its intercalated structure [19].

The effects of solvent on the properties of the cellulose acetate/ sodium montmorillonite nanocomposites were investigated by *Romero et. al.* in 2009 [20]. They had prepared nanocomposites employing solution intercalation method with both pure and mixed solvents. Two pure solvents- acetone and acetic acid and two mixed solvents- acetone/water and acetic acid/water were used during their investigation. Different values were obtained for the solubility parameters of the same nanocomposite depending on the solvent type and it affected the morphology as well as thermal and mechanical behaviour of the nanocomposite. Thus, solvent plays prime role in regulating the morphology and it can be used as an important processing parameter to get the nanocomposite with desirable range of properties [21].

4.   Methods of preparation of montmorillonite-cellulose based nano-composites
4.1   In-situ method

The in-situ method is used for preparing MMT- cellulose nanocomposites using bacterial cellulose. Here, MMT suspension is added to the bacterial culture medium where MMT nanoparticles get infused into the structure of bacterial cellulose during the formation of cellulose sheet. For this purpose, a homogeneous dispersion of MMT is prepared first in distilled water using magnetic stirring for prolonged period. In order to prevent agglomeration of MMT particles, the dispersion is sonicated for few minutes. These MMT nanoparticles are then added into the culture medium either simultaneously with inoculation or before inoculation.

4.2   Ex-situ method or immersion method

In this method, cellulose sheets are prepared separately and then MMT nanoparticles are incorporated into the already prepared sheets. For that MMT particles are homogeneously dispersed in distilled water using magnetic stirring. Cellulose sheets are then immersed into that suspension of MMT clay particles which results in the formation of nanocomposite. The significant characteristic of immersion method is that no overall structural change is observed in the cellulose sheets after formation of nanocomposite [22].

5.   Application

MMT-cellulose based nano-composite materials have been utilized in diverse kinds of applications.

5.1   As adsorbents of pollutants

Over the years, MMT-cellulose based nano-composite materials established themselves as effective adsorbents for a variety of environmental pollutants. Cr (VI) is a major water

Adv. App. of Micro and Nano Clay – Biopolymer-based Composites   Materials Research Forum LLC
Materials Research Foundations **125** (2022) 302-322   https://doi.org/10.21741/9781644901915-13

pollutant across the world which can be successfully removed by using such nano-composite adsorbents. Composite of sodium montmorillonite (NaMMT) with cellulose was found to be highly effective in removal of Cr(VI) from industrial waste water. The removed Cr(VI) ions from polluted water gets adsorbed as bichromate anion on the surface of the NaMMT/cellulose biosorbent. Sodium hydroxide (NaOH) eluent can be used to regenerate this biosorbent for reusing in quantitative elimination of Cr (VI) content from water medium [23].

MMT-cellulose acetate composite is another appealing material for treating waste water effluents. Water pollution caused by various toxic dyes is a major environmental problem in today's world. Eosin Yellow is one such toxic dye causing severe water pollution with potential threat to aquatic living beings as well as humans. This dye can be effectively removed from waste water using cellulose acetate-MMT composite adsorbent. The biodegradability of this composite further enhanced its importance minimizing the problem of waste management [24]. In 2019, Wang et. al reported MMT-cellulose hydrogels as excellent adsorbent of methylene blue (MB). They had prepared three cellulose/montmorillonite hydrogels CCH-5, CCH-10 and CCH-20 having MMT concentration of 5 wt%, 10 wt%, 20 wt% respectively. The scanning electron micrographs shown in Fig. 2 reveals that even at higher concentrations, the MMT nanoparticles are fully dispersed without agglomeration and well-intercalated in the hydrogel network [25].

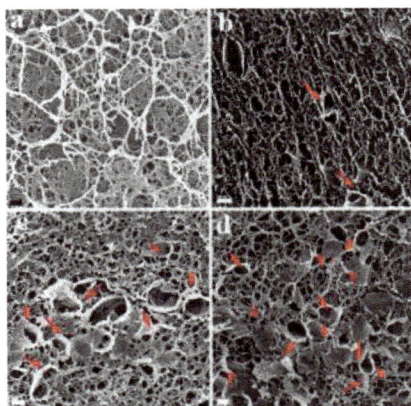

Fig. 2: SEM micrographs of a) regenerated cellulose hydrogel; MMT-cellulose hydrogels with different wt% of MMT: b) CCH-5; c) CCH-10; d) CCH-20 (Extracted from [25])

Moreover, it was detected that the dye adsorption ability of MMT-cellulose nano-composite hydrogels gradually elevated with the rising pH values reaching highest at neutral point.

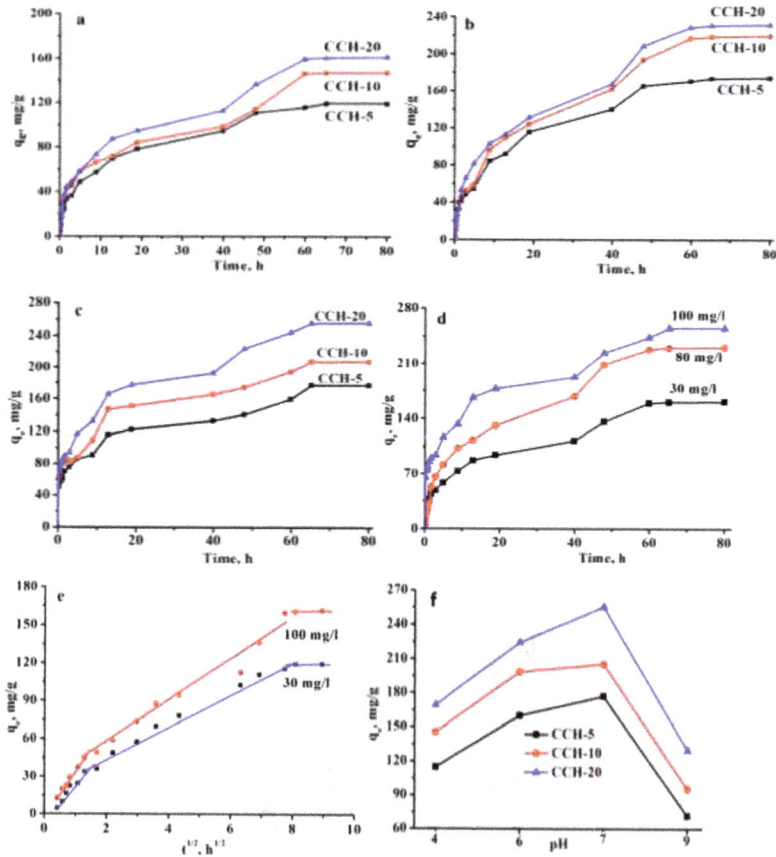

Fig. 3: Performance of MMT-cellulose hydrogels in adsorption of a) 30 mg/L, b) 80 mg/L, c) 100 mg/L MB ; d) adsorption capacity of CCH-20 at different concentration of MB; e) intra-particle diffusion model of CCH-5 at various dye concentration and f) variation of the adsorption capacity of MMT-cellulose hydrogel with respect to pH change (Extracted from [25]).

Adsorbents having bead like structures are preferred to powder adsorbents in the practical field owing to their ease of separation from the aqueous medium. Cellulose derived from sugarcane bagasse was used by Pan et. al. to prepare beads of MMT-cellulose nano-composite for removal of Auramine O (cationic) as well as Amido black (anionic). In order to increase the extent of adsorption by increasing specific surface area calcium carbonate powder was applied on the beads as pore-forming agent. The resultant mesoporous beads were then grafted with tetraethylenepentamine so that they can capture both anionic and cationic dyes [26].

Incorporation of metal into polymer/clay composite materials has significant positive effect on their adsorption efficiency. Biodegradable polymer carboxymethyl cellulose or cellulose gum is well-known for its outstanding emulsifying properties which interact with MMT improving its functionality. Carboxymethylcellulose-MMT composite (Zr-CMC-MMT) encapsulated with Zr(IV) was successfully utilized by P. Sirajudheen et. al. for the elimination of Reactive Red 2 (RR) and Acid Orange 7 (AO) dyes from aqueous solution. Both of these dyes belong to azo dye- a very harmful class of dye responsible for skin irritation, dermatitis and even cancer [27].

5.2   Biomedical application

For biomedical applications we need ultra-pure materials. All kinds of naturally available cellulose are not pure in nature. In case of plant fiber, cellulose remains in mixed state with lignin and hemicelluloses. Cellulose must be isolated first from this mixture using various chemical reagents like sodium chlorite and potassium hydroxide etc. for application in different fields. Bacterial cellulose (BC) is an unmixed form of cellulose formed by specific bacterial strains like Rhizobium, Agrobacterium, Acetobacter xylinum, Gluconoacetobater xylinus, Acetobacter etc. They possess outstanding inherent characteristics like excellent water bearing capacity, higher tensile strength and crystallinity along with fine fibrous network. Moreover, BC is highly flexible which can be casted into a variety of shapes as per requirement. All of these factors make BC superior to plant-based cellulose polymers. *M. Ul-Islam et. al.* prepared MMT-BC nano–composites for the first time via ex-situ impregnation method.

Like BC, MMT also exhibits wound healing ability when applied on injured area as wet paste. But, when both of these two components are combined as composite, it performs better than individual components. The absorbed water molecules are held tightly on the MMT molecules present on the surface of the composite lowering their rate of evaporation. Due to slow water release rate (WRR), the composite material is capable of staying wet for prolonged period making it more efficient for biomedical application as wound healing or dressing material [28]. The weight profile shown in Fig. 5 represents WRR of of pristine BC and BC–MMT nano-composites.

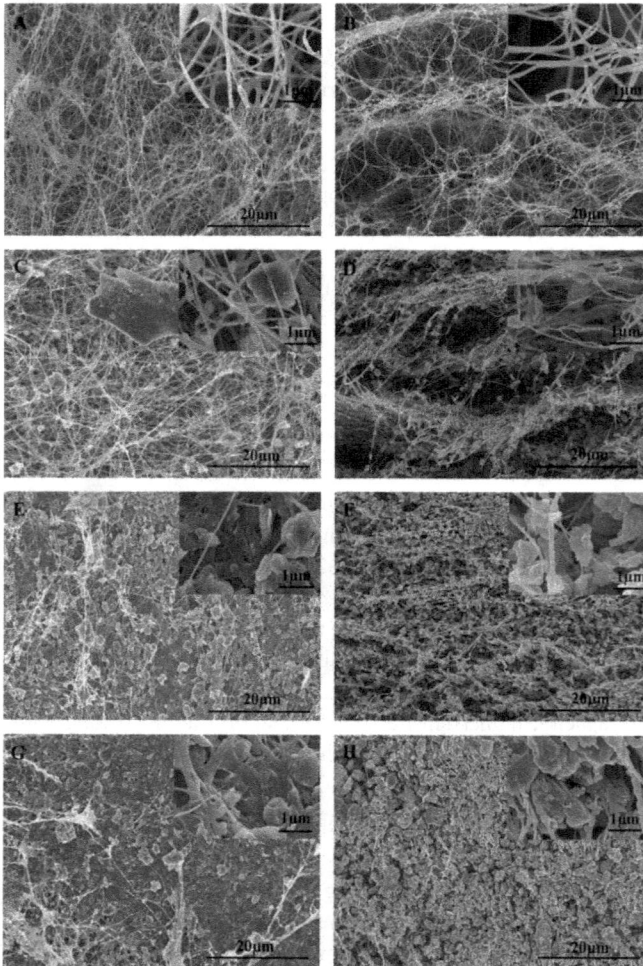

Fig. 4: SEM images of A) surface of pure BC; B) cross section of pure BC; C) surface of BC–MMT1; D) cross section of BC–MMT1; E) surface of BC–MMT2; F) cross section of BC–MMT2 and G) surface of BC–MMT3; H) cross section of BC–MMT3. (Extracted from [28]).

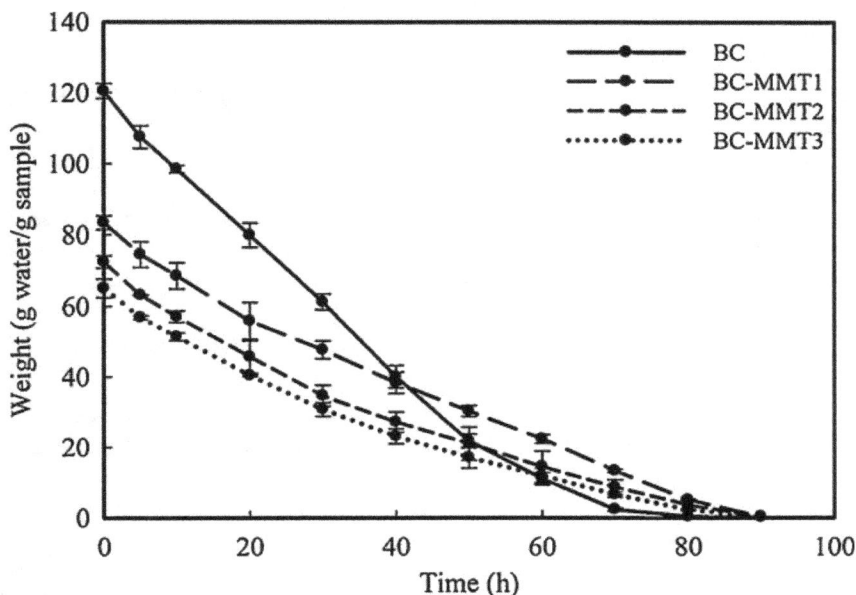

*Fig. 5. Rate of releasing water by pristine BC and three BC–MMT nano-composites.
(Extracted from [28]).*

Though BC is non-toxic and biocompatible, it cannot resist bacterial infection for unavailability of antimicrobial property. In order to exploit tissue regeneration ability of BC and wound healing capacity of MMT, *W. Sajjad et. al.* prepared different modified MMT-BC nanocomposites in 2018. They synthesized BC using bacterial strain *Gluconacetobacter xylinus* cultured in aqueous solution having 0.5% yeast, 0.5% peptone, 2% glucose, 0.115% citric acid and 0.27% Na_2HPO_4. Ca, Cu and Na intercalated modified MMTs were synthesized using ion exchange methods to incorporate antimicrobial activity into MMT. For that, they had used three seperate 100 ml aqueous solutions of 0.2 $molL^{-1}$ $CaCl_2$, $CuSO_4$ and NaCl in each of which 10 g MMT was stirred vigorously at 60^0C for 6 hours [29]. Calcium (Ca) salts increase pH which results in denaturation of cell proteins facilitating bacterial cell death [30]. Denaturation is nothing but alteration of the shape of protein by externally applied agents like heat, acid, base etc. Copper (Cu) causes peroxidation of lipids present in bacterial cell-membranes along with denaturation [31].

Sodium (Na) ions inhibit secretion of certain enzymes within microbial cell and increases cell permeability which ultimately leads to death of the pathogen [32, 33]. All of the MMT-BC composites were exposed to gram negative *C. fruendii, P. aeruginosa, E. coli, S. typhimurium* and gram-positive *S. aureus,* Methicillin-resistant *S. aureus* for examining their antimicrobial activities. From the investigation, it was found that Ca, Cu and Na-modified MMT-BC nanocomposites exhibit significantly enhanced antimicrobial, wound healing and tissue regeneration capability with respect to pristine BC and non-modified MMT-BC composites. All of these were found to be potent material for artificial tissue generation and burn skin substitution. Analysis of digital photographs of the wound area as well as area-measurement of the wound portion at regular intervals demonstrated outstanding wound healing capacity of modified MMT-BC nanocomposites. Wound area reduction was found to be maximum for Cu-MMT-BC treated groups and minimum for Ca-MMT-BC treated groups [29].

In 2017, *Demircan et. al.* reported use of organically modified MMT-cellulose nanocomposite for quorum-sensing inhibition [34]. Quorum-sensing is the unique ability of different bacteria to control the appearance of specific behavioral genes regarding the population density of bacterial cell. They secrete some small signaling molecules called 'auto-inducers' which undergo diffusion into the surrounding environment and these molecules are responsible for coordinating the process of quorum-sensing. Thus, quorum-sensing can be affected by detecting the auto-inducers and by interfering in their production. Any external modulation that can inhibit the procedure of quorum-sensing is termed as anti-quorum-sensing agent [35]. Generally small molecules like farnesol, essential oils, furanones etc. exhibit anti-quorum-sensing effect [36,37]. But high leaching potential of these tiny molecules impart negative effect not only on the anti-quorum-sensing process but also on health and environment. Therefore, polymeric substitutes are more preferred now a days for such biomedical applications [38,39]. Demircan et. al used octadecylamine (ODA)-modified MMT as nanofillers to prepare nanocomposite with RC. The scheme representing the preparative method of RC and nanocomposites of RC with modified MMT is shown in Fig. 6 [34].

They used 4.3 w% cellulose solution in LiCl/DMA coagulated in ethanol bath for preparation of RC. DMA is used to disperse ODA-MMT to prepare nanocomposites. were made in. The dispersions were then mixed with 4.3 w% cellulose solution to prepare three nanocomposites having ODA-MMT content 1, 3, and 5 wt %. Final products (CN_1, CN_3 and CN_5) were gained after coagulation in ethanol bath, washing and drying. Purple pigment producing *C. violaceum* (Gram-negative) was selected for anti-quorum-sensing activity tests of pristine cellulose, ODA-MMT, CN_1, CN_3 and CN_5 using two different methods- Disc Diffusion Test and UV–vis Quantification of Violacein Inhibition [34,40].

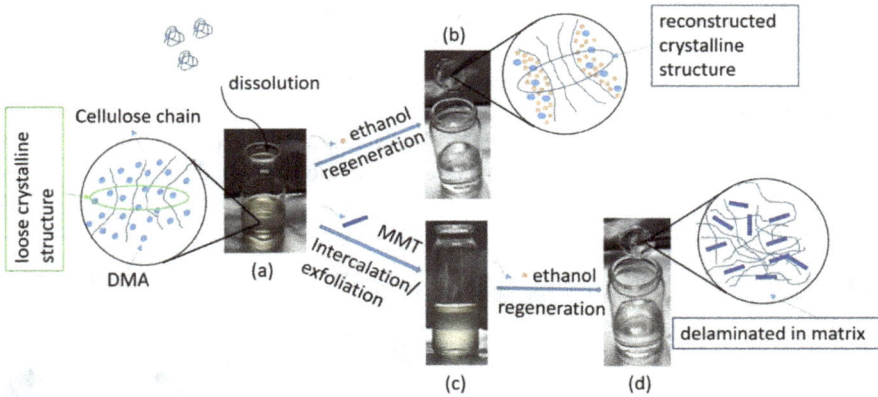

Fig. 6: Scheme for preparing RC and nanocomposites of RC with modified MMT. (Extracted from [34]).

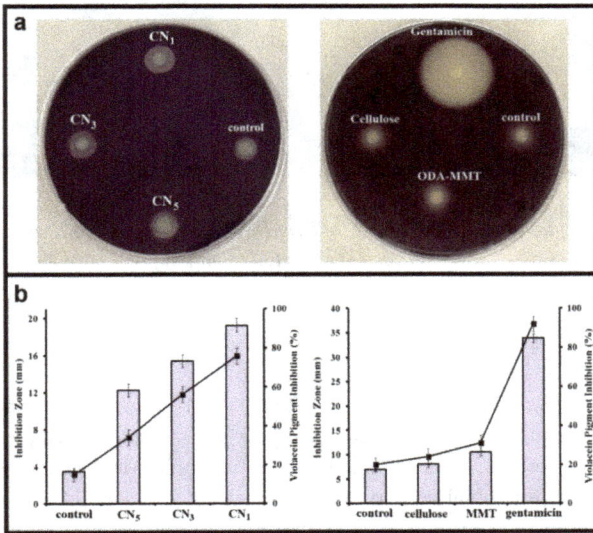

Fig. 7: Study of anti-QS activity: (a) Image of samples in plate diffusion test. (b) Gray bars represent the zone of inhibition in disc diffusion test and black squares are quantitative representation of violacein inhibition by UV−vis technique. (Extracted from [34]).

All of the three nanocomposites (CN_1, CN_3 and CN_5) exhibited notably larger zones of pigment inhibition than pure cellulose and ODA-MMT as displayed in Fig. 7a). Moreover, the zone of pigment inhibition decreased with increasing amount of ODA-MMT in the nanocomposites. Largest zone of inhibition observed in case of CN_1 validates the fact that anti-quorum-sensing activity does not depend on concentration, but depends on exfoliation of ODA-MMT nanoparticles in the polymeric matrix. Violacein inhibition using UV−vis technique revealed that CN_1 possesses anti-quorum-sensing activity comparable to that of gentamicin quantitatively (Fig. 7b).

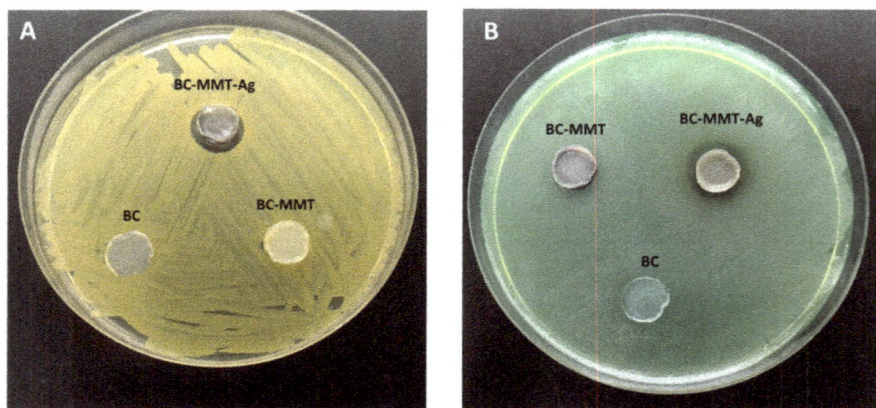

Fig. 8: Antibacterial activity BC, BC-MMT and BC-MMT-Ag on A) S. aureus and B) P. aeruginosa (Extracted from [41]).

In 2020, *M. Horue et. al.* reported bacterial cellulose–silver montmorillonite nanocomposites (BC-MMT-Ag) as potential scaffold for wound healing. They had used cellulose derived from *Komagataeibacter xylinus*. MMT-Ag was prepared by dispersing MMT in 100 mM aqueous $AgNO_3$ solution [41]. Ex situ method was employed to prepare nanocomposites where BC plates were dispersed in stable MMT-Ag suspension [42]. Investigation of antimicrobial activity revealed that not only pristine BC, but also BC-MMT didn't exhibit such property against *S. aureus* and *P. aeruginosa* bacteria whereas BC-MMT-Ag exhibited potential antimicrobial property owing to the presence of silver (Fig. 8) [41]. MMT was incorporated in the composite to administer controlled release of silver. Their study revealed that the release rate of silver is very high in case of BC-Ag which leads to certain serious side effects. Controlled release of silver from BC-MMT-Ag

Adv. App. of Micro and Nano Clay – Biopolymer-based Composites   Materials Research Forum LLC
Materials Research Foundations **125** (2022) 302-322   https://doi.org/10.21741/9781644901915-13

signifies the importance of MMT in the composite for potent antimicrobial use in case of wound healing. Live and dead assays were carried out using two interacting dyes -Syto®9 and propidium iodide. First one binds to living cells imparting fluorescent green color whereas second one combines with dead cells producing red colored spots. When pristine BC matrix was exposed to bacterial cultures only green fluorescent color was observed. This indicates the viability of microbial cells and confirms that we cannot access antimicrobial performance from pristine BC. Conversely, when BC-MMT-Ag nano-composite was exposed to these bacterial cell cultures, red spots were observed confirming the antimicrobial efficiency of the nano-composites (Fig. 9) [41].

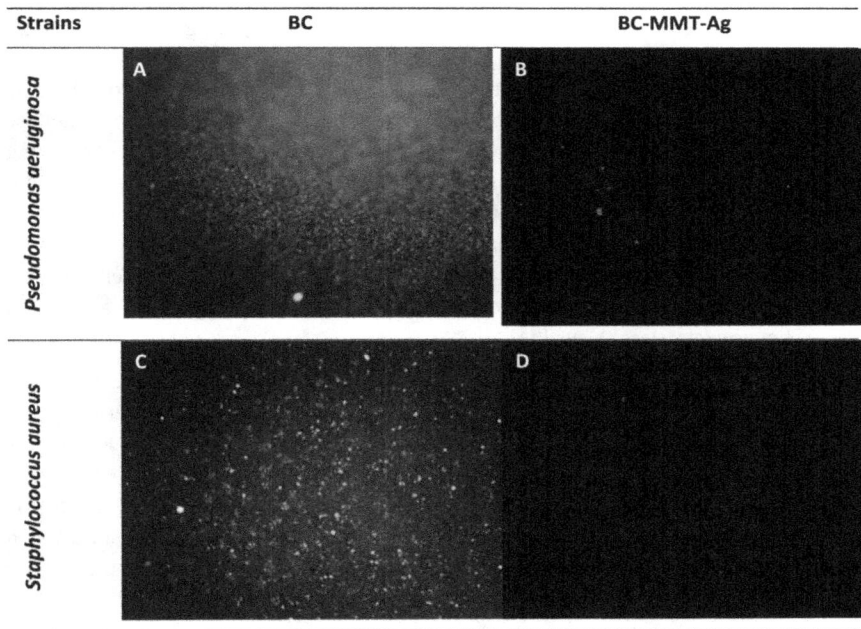

Fig. 9: Performance of BC and BC-MMT-Ag biofilms in live and dead assay after one day bacterial culture. (Extracted from [41]).

5.3   As superior bio-based plastic material with gas permeability and flame retardant behavior:

Extensive use of petroleum based synthetic polymers for packaging essential items like food, beverage, pharmaceuticals etc. have created serious threat to the planet. A major portion of household waste is plastic which usually bring to an end in landfills. Most of the frequently used plastics like polyethylene (PE), poly(ethylene terephthalate) (PET), polypropylene (PP), polystyrene (PS) and poly(vinyl chloride) (PVC) are non-biodegradable in nature and their mounting accumulation in the environment has been a principal reason of soil, air and water pollution [43]. In order to handle this situation, research community has been in search of biodegradable substitute of conventional plastics. Use of cellulose for synthesis of thermoplastic or thermosetting plastic was hardly known till the first decade of 21st century. In 2013, *Wang et. al.* successfully synthesized a new class of bioplastics using regenerated cellulose. Regenerated cellulose is obtained by modification of natural cellulose into more soluble form. In this work, natural cellulose obtained from cotton pulp was first dissolved in a pre-cooled aqueous solution comprising of 7 wt% NaOH and 12 wt% urea. Similarly, transparent solution of cellulose gained from bamboo and wood pulp was made by dissolving it in a aqueous solution comprising from 4.6 wt% LiOH and 15 wt% urea. Each of these solutions was then spread on glass mold and regenerated in the form of cellulose hydrogel with non-solvent and subsequent washing with water. Finally bioplastic of cellulose was prepared by hot-pressing the sandwiched hydrogel between stainless sheets [44].

In order to get superior plastic, the same research group mixed MMT with cellulose and reported a cellulose/MMT bio-based plastic in the year 2015 having flame retardancy and gas barrier properties [45]. Flame retardancy of the cellulose/MMT nanocomposites was examined by performing flammability test. The respective parameter is called limiting oxygen index(LOI) which can be defined as the minimum concentration of oxygen that is required for combustion of a polymer. The target sample is allowed to combust passing a gaseous mixture of oxygen and nitrogen and LOI is calculated by decreasing the oxygen level gradually until a critical level is reached. The higher the value of LOI, the lower is the flammability of the material. From the investigation, Wang et. al reported that LOI values of cellulose/MMT nano-composite plastics were increased with increasing amount of incorporated MMT content (0, 5, 10, 15, 20 wt%). Thus, flame retardacny of the composite plastic became stronger with increase of MMT percentage. Furthermore, LOI values of the cellulose/MMT nanocomposite plastics are higher than that of synthetic plastics- polypropylene (19 %) and polyethylene (17.8%) reported earlier [46]. Gas permeability of prepared plastics were investigated against four different gases- N_2, H_2,

CO_2 and CH_4. The permeability values were found to be decreased with increasing amount of incorporated MMT in case of each of the applied gas.

In 2019, *L. Wang et. al.* also reported a flame retardant cellulose nanofibril/sodium montmorillonite aerogel having enhanced mechanical properties using boric acid and melamine-formaldehyde resins as cross-linking agents [47]. Introduction of boric acid into the porous lamellar structure of the aerogel narrowed down the interspacing of layers whereas introduction of melamine-formaldehyde resulted in formation of polymeric fibrils connecting these layers. Significant increase in LOI values was observed indicating poor flammability of the aerogels owing to the synergistic effect of boron and nitrogen containing cross-linkers. Besides, these two crosslinkers had enhanced the mechanical properties of the cellulose nanofibril/sodium montmorillonite aerogels paving the way for their application in different engineering fields.

Conclusion

In this chapter we have discussed the value of cellulose-MMT nanocomposites and their properties, preparative methods and applications in different fields. These nanocomposites can be fabricated in different ways to get different desirable properties like superior mechanical strength, high flame registance, flexibility, gas permeability, wound healing properties etc. and can be used in diverse fields. We wish to conclude this chapter with the hope that research community will be engaged more in expanding the domain of this particular category of green materials to some new directions so that sustainable development can be achieved further.

References

[1] V. Mittal, Polymer layered silicate nanocomposites: a review, Materials 2 (2009) 992-1057. https://doi.org/10.3390/ma2030992

[2] S. Pavlidou, C. D. Papaspyrides, A review on polymer-layered silicate nanocomposites, Prog. Polym. Sci. 33 (2008) 1119-1198. https://doi.org/10.1016/j.progpolymsci.2008.07.008

[3] H. Kargarzadeh, M. Mariano, J. Huang, N. Lin, I. Ahmad, A. Dufresne, Recent developments on nanocellulose reinforced polymer nanocomposites: A review, Polymer. 132 (2017) 368-93. https://doi.org/10.1016/j.polymer.2017.09.043

[4] D. Klemm, B. Heublein, H. P. Fink, A. Bohn, Cellulose: fascinating biopolymer and sustainable raw material, Angew. Chem. Int. Ed. 36 (2005) 3358-3393. https://doi.org/10.1002/anie.200460587

[5] S. S. Ray, M. Okamoto, Polymer/Layered Silicate Nanocomposites: A Review from Preparation to Processing, Prog. Polym. Sci. 28 (2003) 1539-1641. https://doi.org/10.1016/j.progpolymsci.2003.08.002

[6] D. Paul, L. M. Robeson, Polymer nanotechnology: Nanocomposites, POLYMER 49 (2008) 3187-3204. https://doi.org/10.1016/j.polymer.2008.04.017

[7] S. S. Ray, M. Bousmina, K. Okamoto, Structure and Properties of Nanocomposites Based on Poly(butylene succinate-co-adipate) and Organically Modified Montmorillonite, Macromol. Mater. Eng. 290 (2005) 759-768. https://doi.org/10.1002/mame.200500203

[8] I. Algar, C. Garcia-Astrain, A. Gonzalez, L. Martin, N. Gabilondo, A. Retegi, A. Eceiza, Improved Permeability Properties for Bacterial Cellulose/ Montmorillonite Hybrid Bionanocomposite Membranes by In-Situ Assembling, J. Renew. Mater. 4 (2016) 57-65. https://doi.org/10.7569/JRM.2015.634124

[9] F. UDDIN, Clays, nanoclays, and montmorillonite minerals, Metall. Mater. Trans. A 39A (2008) 2804-2814. https://doi.org/10.1007/s11661-008-9603-5

[10] K. Strawhecker, E. Manias, Structure and properties of poly (vinyl alcohol)/Na-montmorillonite nanocomposites, Chem. Mater. 12 (2000) 2943-2949. https://doi.org/10.1021/cm000506g

[11] M. A. Siddiqui, Z. Ahmed, Mineralogy of the Swat kaolin deposits, Pakistan, Arab. J. Sci. Eng. 30 (2005) 195-218.

[12] H. S. Qian, S. H. Yu, L. B. Luo, J. Y. Gong, L. F. Fei, Synthesis of uniform Te-Carbon-rich composite nanocables with photoluminescence properties and carbonaceous nanofibers by the hydrothermal carbonization of glucose, Chem. Mater. 18 (2006) 2102-2108. https://doi.org/10.1021/cm052848y

[13] X. Q. Qian, F. C. Hu, F. Y. Tian, D. F. Hou, D. S. Li, Equilibrium and kinetic studies on MB adsorption by ultrathin 2D MoS2 nanosheets, Rsc Adv. 6 (2016) 11631-11636. https://doi.org/10.1039/C5RA24328A

[14] Z. Yang, W. Wang, X. Tai, G. Wang, Preparation of modified montmorillonite with different quaternary ammonium salts and application in Pickering emulsion, New J. Chem. 43 (2019) 11543-11548. https://doi.org/10.1039/C9NJ01606F

[15] A. Okada, A. Usuki, Twenty years of polymer-clay nanocomposites, Macromol. Mater. Eng. 291 (2006) 1449–1476. https://doi.org/10.1002/mame.200600260

[16] J. L. Suter, D. Groen, P. V. Coveney, Chemically specific multiscale modeling of clay-polymer nanocomposites reveals intercalation dynamics, tactoid self-assembly

and emergent materials properties, Adv. Mater. 27 (2015) 966–984.
https://doi.org/10.1002/adma.201403361

[17] A. J. Benítez, A. Walther, Cellulose nanofibril nanopapers and bioinspired
nanocomposites: a review to understand the mechanical property space, J. Mater.
Chem. A 5 (2017) 16003–16024. https://doi.org/10.1039/C7TA02006F

[18] P. Cerruti, V. Ambrogi, A. Postiglione J. Rychly, L. M. Rychla, C. Carfagna,
Morphological and thermal properties of cellulose–montmorillonite nanocomposites,
Biomacromolecules 9 (2008) 3004-3013. https://doi.org/10.1021/bm8002946

[19] D. Xu, S. Wang, L. A. Berglund, Q. Zhou, Surface Charges Control the Structure
and Properties of Layered Nanocomposite of Cellulose Nanofibrils and Clay Platelets,
ACS Appl. Mater. Interfaces 13 (2021) 4463–4472.
https://doi.org/10.1021/acsami.0c18594

[20] R. B. Romero, C. A. Paula Leite, M. C. Gonçalves, The effect of the solvent on the
morphology of cellulose acetate/montmorillonite nanocomposites, Polymer 50 (2009)
161-170. https://doi.org/10.1016/j.polymer.2008.10.059

[21] R. Krishnamoorti, J. Ren, A. S. Silva, Linear viscoelasticity of disordered
polystyrene–polyisoprene block copolymer based layered-silicate nanocomposites,
Macromolecules 33 (2000) 3739-3946. https://doi.org/10.1021/ma992091u

[22] N. Khodamoradi, V. Babaeipour, M. Sirousazar, Bacterial cellulose/montmorillonite
bionanocomposites prepared by immersion and in-situ methods: structural,
mechanical, thermal, swelling and dehydration properties, Cellulose 13 (2019) 7847 -
7861. https://doi.org/10.1007/s10570-019-02666-9

[23] S. K. Kumar, S. Kalidhasan, V. Rajesh, N. Rajesh, Application of Cellulose-Clay
composite biosorbent toward the effective adsorption and removal of chromium from
industrial wastewater, Ind. Eng. Chem. Res. 51 (2012), 58-69.
https://doi.org/10.1021/ie201349h

[24] M. Goswami, A. Das, Chemistry, Medicine Carbohydrate polymersSynthesis and
characterization of a biodegradable Cellulose acetate-montmorillonite composite for
effective adsorption of Eosin Y., 206 (2019)
https://doi.org/10.1016/j.carbpol.2018.11.040

[25] Q. Wang, Y. Wang, L. Chen, A green composite hydrogel based on cellulose and
clay as efficient absorbent of colored organic effluent Carbohydrate polymers, 210
(2019) 314-321. https://doi.org/10.1016/j.carbpol.2019.01.080

[26] Y. Pan, H. Xie, H. Liu, P. Cai, H. Xiao, Novel cellulose/montmorillonite mesoporous composite beads for dye removal in single and binary systems, Bioresour. Technol. 286 (2019) 121366. https://doi.org/10.1016/j.biortech.2019.121366

[27] P. Sirajudheen, P. Karthikeyan, M. C. Basheer, S. Meenakshi, Adsorptive removal of anionic azo dyes from effluent water using Zr(IV) encapsulated carboxymethyl cellulose-montmorillonite composite, J. Environ. Chem. Ecotoxicol. 2 (2020) 73-82. https://doi.org/10.1016/j.enceco.2020.04.002

[28] M. Ul-Islam, T. Khan, J. K. Park, Nanoreinforced bacterial cellulose-montmorillonite composites for biomedical applications, Carbohydrate Polymers 89 (2012) 1189- 1197. https://doi.org/10.1016/j.carbpol.2012.03.093

[29] W. Sajjad, T. Khan, M. Ul-Islam, R. Khan, Z. Hussain, A. Khalid, F. Wahid, Development of modified montmorillonite-bacterial cellulose nanocomposites as a novel substitute for burn skin and tissue regeneration, Carbohydr. Polym. 206 (2019) 548-556. https://doi.org/10.1016/j.carbpol.2018.11.023

[30] J. F. Siqueira, H. P. Lopes, Mechanisms of antimicrobial activity of calcium hydroxide: a critical review, International Endodontic Journal 32 (1999) 361-369. https://doi.org/10.1046/j.1365-2591.1999.00275.x

[31] G. Borkow, N. Okon-Levy, J. Gabbay, Copper oxide impregnated wound dressing: biocidal and safety studies, Wounds. 22 (2010) 301-310.

[32] J. Cabezas-Pizarro, M. Redondo-Solano, C Umaña-Gamboa, M. L. Arias-Echandi, Antimicrobial activity of different sodium and potassium salts of carboxylic acid against some common foodborne pathogens and spoilage-associated bacteria, Revista Argentina de Microbiologia, 50 (2018) 56-61. https://doi.org/10.1016/j.ram.2016.11.011

[33] T. Maneerung, S. Tokura, R. Rujiravanit, Impregnation of silver nanoparticles into bacterial cellulose for antimicrobial wound dressing. Carbohydrate polymers, 72 (2008) 43-51. https://doi.org/10.1016/j.carbpol.2007.07.025

[34] D. Demircan, S. Ilk, B. Zhang, Cellulose-organic montmorillonite nanocomposites as biomacromolecular quorum-sensing inhibitor, Biomacromolecules 18 (2017) 3439-3446. https://doi.org/10.1021/acs.biomac.7b01116

[35] B. LaSarre, M. J. Federle, Exploiting quorum sensing to confuse bacterial pathogens, Microbiol. Mol. Biol. Rev. 77 (2013) 73−111. https://doi.org/10.1128/MMBR.00046-12

[36] F. Nazzaro, F. Fratianni, R. Coppola, Quorum sensing and phytochemicals, Int. J. Mol. Sci. 14 (2013) 12607−12619. https://doi.org/10.3390/ijms140612607

[37] J. Olivero-Verbel, A. Barreto-Maya, A. Bertel-Sevilla, E. E. Stashenko, Composition, anti-quorum sensing and antimicrobial activity of essential oils from Lippia alba, Braz. J. Microbiol. 45 (2014) 759−767. https://doi.org/10.1590/S1517-83822014000300001

[38] E. Cavaleiro, A. S. Duarte, A. C. Esteves, A. Correia, M. J. Whitcombe, V. Elena, A. Sergey, I. Chianella, Novel linear polymers able to inhibit bacterial quorum sensing. Macromol. Biosci. 15 (2015) 647−656. https://doi.org/10.1002/mabi.201400447

[39] N. Amara, B. P. Krom, G. F. Kaufmann, M. M. Meijler, Macromolecular inhibition of quorum sensing: enzymes, antibodies, and beyond, Chem. Rev. 111 (2011) 195−208. https://doi.org/10.1021/cr100101c

[40] S. Ilk, N. Sağlam, M. Ozgen, F. Korkusuz, Chitosan nanoparticles enhances the anti-quorum sensing activity of kaempferol, Int. J. Biol. Macromol. 94 (2017) 653−662. https://doi.org/10.1016/j.ijbiomac.2016.10.068

[41] M. Horue, M. L. Cacicedo, M. A. Fernandez, B. R. Kladniew, R. M. T. Sánchez, G. R. Castro, Antimicrobial activities of bacterial cellulose - Silver montmorillonite nanocomposites for wound healing, Mater. Sci. Eng. C 116 (2020) 111152-111178. https://doi.org/10.1016/j.msec.2020.111152

[42] A. M. Fernández Solarte, J. Villarroel-Rocha, C. Fernández Morantes, M. L. Montes, K. Sapag, G. Curutchet, R. M. Torres Sánchez, Insight into surface and structural changes of montmorillonite and organomontmorillonites loaded with Ag, C. R. Chimie. 22 (2019) 142-153. https://doi.org/10.1016/j.crci.2018.09.006

[43] Y. Tokiwa, B. P. Calabia, C. U. Ugwu, S. Aiba, Biodegradability of Plastics, Int. J. Mol. Sci. 10 (2009) 3722-3742. https://doi.org/10.3390/ijms10093722

[44] Q. Wang, J. Cai, L. Zhang, M. Xu, H. Cheng, C. C. Han, S. Kuga, J. Xiao, R. Xiao, A bioplastic with high strength constructed from a cellulose hydrogel by changing the aggregated structure, J. Mater. Chem. A 1(2013) 6678-6686. https://doi.org/10.1039/c3ta11130j

[45] Q. Wang, J. Guo, D. Xu, J. Cai, Y. Qiu, J. Ren, L. Zhang, Facile construction of cellulose/montmorillonite nanocomposite biobased plastics with flame retardant and gas barrier properties, Cellulose 22 (2015) 3799-3810. https://doi.org/10.1007/s10570-015-0758-0

Adv. App. of Micro and Nano Clay – Biopolymer-based Composites   Materials Research Forum LLC
Materials Research Foundations **125** (2022) 302-322   https://doi.org/10.21741/9781644901915-13

[46] K. Wu, Z. Wang, H. Liang Microencapsulation of ammonium polyphosphate: preparation, characterization, and its flame retardance in polypropylene, Polym. Compos. 29 (2008) 854-860. https://doi.org/10.1002/pc.20459

[47] L. Wang, M. Sánchez-Soto, J. Fan, Z. P. Xia, Y. Liu, Boron/nitrogen flame retardant additives cross-linked cellulose nanofibril/montmorillonite aerogels toward super-low flammability and improved mechanical properties, Polym. Adv. Technol. 30 (2019) 1807-1817. https://doi.org/10.1002/pat.4613

Keyword Index

About the Editors

Dr. Amir Al-Ahmed is working as a Research Scientist-II (Associate Professor) in the Interdisciplinary Research Center for Renewable Energy and Power Systems (IRC-REPS), at King Fahd University of Petroleum & Minerals (KFUPM), Saudi Arabia. He graduated in chemistry from the Department of Chemistry, Aligarh Muslim University (AMU), India. Then completed his M.Phil. (2001) and Ph.D. (2004) in Applied Chemistry from the Department of Applied Chemistry, AMU, India, followed by three consecutive postdoctoral fellowships in South Africa and Saudi Arabia. During this period, he worked on various multidisciplinary projects, in particular, conducting composites, electrochemical sensors, nano-materials, polymeric membranes, electro-catalysis and solar cells. At present, his research activity is fundamentally focused on the 3rd generation solar cell devices, such as, low band gap semiconductors, quantum dots, perovskites, and tandem cells. At the same time, he is also working on energy storage technologies, such as, heat storage, evaluation of electricity storage devices and dust repellent coating for PVs. He has worked on different NSTIP, KACST and Saudi Aramco funded projects in the capacity of a principle and co-investigator. Dr. Amir has eight US patents, over 60 journal articles, invited book chapters and conferences publications. He has edited ten books with Trans Tech Publication, Springer, Elsevier, Materials Research Forum LLC, and several other books are in progress. He is also the Editor-in-Chief of an international journal "Nano Hybrids and Composites" along with Professor Y. H. Kim.

Dr. Inamuddin is working as Assistant Professor at the Department of Applied Chemistry, Aligarh Muslim University, Aligarh, India. He obtained Master of Science degree in Organic Chemistry from Chaudhary Charan Singh (CCS) University, Meerut, India, in 2002. He received his Master of Philosophy and Doctor of Philosophy degrees in Applied Chemistry from Aligarh Muslim University (AMU), India, in 2004 and 2007, respectively. He has extensive research experience in multidisciplinary fields of Analytical Chemistry, Materials Chemistry, and Electrochemistry and, more specifically, Renewable Energy and Environment. He has worked on different research projects as project fellow and senior research fellow funded by University Grants Commission (UGC), Government of India, and Council of Scientific and Industrial Research (CSIR), Government of India. He has received Fast Track Young Scientist Award from the Department of Science and Technology, India, to work in the area of bending actuators and artificial muscles. He has also received the Sir Syed Young Researcher of the Year Award 2020 from Aligarh Muslim University. He has completed four major research projects sanctioned by University Grant Commission, Department of Science and Technology, Council of Scientific and Industrial Research, and Council of Science and Technology, India. He has published 200 research articles in international journals of repute and nineteen book chapters in knowledge-based book editions published by renowned international publishers. He has published 150 edited books with Springer (U.K.), Elsevier, Nova Science Publishers, Inc. (U.S.A.), CRC Press Taylor & Francis Asia Pacific, Trans Tech Publications Ltd. (Switzerland), IntechOpen Limited (U.K.), Wiley-Scrivener, (U.S.A.) and Materials Research Forum LLC (U.S.A). He is a member of various journals' editorial boards. He is also serving as Associate Editor for journals (Environmental Chemistry Letter, Applied Water Science and Euro-Mediterranean Journal for Environmental Integration, Springer-Nature), Frontiers Section Editor (Current Analytical Chemistry, Bentham Science Publishers), Editorial Board Member (Scientific Reports-Nature), Editor (Eurasian Journal of Analytical Chemistry), and Review Editor (Frontiers in Chemistry, Frontiers, U.K.). He is also guest-editing various special thematic special issues to the journals of Elsevier, Bentham Science Publishers, and John Wiley & Sons, Inc. He has attended as well as chaired sessions in various international and national conferences. He has worked as a Postdoctoral Fellow, leading a research team at the Creative Research Initiative Center for Bio-Artificial Muscle, Hanyang University, South Korea, in the field of renewable energy, especially biofuel cells. He has also worked as a Postdoctoral Fellow at the Center of Research Excellence in Renewable Energy, King Fahd University of Petroleum and Minerals, Saudi Arabia, in the field of polymer electrolyte membrane fuel cells and computational fluid dynamics of polymer electrolyte membrane fuel cells. He is a life member of the Journal of the Indian Chemical Society. His research interest includes ion exchange materials, a sensor for heavy metal ions, biofuel cells, supercapacitors and bending actuators.